Transport and Tourism

1

Transport and Tourism

Stephen J. Page

 Longman

Addison Wesley Longman Limited
Edinburgh Gate
Harlow
Essex CM20 2JE
England
and Associated Companies throughout the world

Published in the United States of America
by Addison Wesley Longman Inc., New York

Visit Addison Wesley Longman on the World Wide Web at:
http://www.awl-he.com

First published 1999

ISBN 0 582 32025 9

British Library Cataloguing-in-Publication Data
A catalogue record for this book is available from the British Library

Library of Congress Cataloging-in-Publication Data
A catalog record for this book is available from the Library of Congress

Typeset by 35 in 11/12pt Adobe Garamond
Produced by Addison Wesley Longman Singapore (Pte) Ltd.,
Printed in Singapore

Contents

Preface

The development of tourism studies as a legitimate area of academic study in the 1980s and 1990s has been reflected in the growing literature appearing in specialist journals, the expanding range of textbooks and more specialist research monographs and edited books. As with any subject, its intellectual development is not simply measured in terms of the volume of material published. The effect such literature has on the way in which students and educators are encouraged to think about the subject and pursue new avenues of research is one way of gauging the impact of such studies. The recent methodological and philosophical debates encapsulated in D.G. Pearce and R. Butler (1992) *Tourism Research: Critiques and Challenges* highlights many of the problems associated with tourism research in the 1990s, particularly the failure to link multidisciplinary research together to develop a more holistic understanding of tourism, tourists and their impact. This book is by no means a response to their call for a more explicit methodological concern in tourism research: it is an attempt to highlight how disciplinary research on tourism and transport can be integrated to provide a clearer understanding of the interface and relationship between tourism and transport. This is a major re-evaluation of an earlier text by the author, *Transport for Tourism*, published in 1994 as the initial book in the Routledge (now International Thomson Business Publishing) *Tourism and Hospitality Management Series*.

One consistent problem from which *Transport for Tourism* suffered was a lack of space to deal with issues associated with the transport and tourism interface. Despite the demand for the text, the lack of space meant that key themes such as infrastructure development, transport planning and a greater emphasis on management issues (e.g. marketing) were omitted, due to limits imposed by the publisher. In contrast, *Transport and Tourism* seeks to develop a more rounded student text which has a wider appeal to undergraduate and postgraduate students. In broad terms the initial framework developed in *Transport and Tourism* has been retained and expanded, and new chapters have been added. Data has also been updated and expanded as it is over five years since the original concept for the book was developed. In that period, tourist transport has undergone a number of changes, most notably in the global aviation industry and this book places a stronger emphasis on such issues. The literature base has also expanded rapidly with the publication of new academic journals such as *Transportation Research* (versions A, B, C, D, E and the new version F on the psychology of transport) and the *Journal of Air Transport Management, Transport Policy,* and the *Journal of Transport Geography.* There has also been a continued growth in transport-related articles published in the professional journals and reports produced by Travel and Tourism Intelligence (formerly the Economist Intelligence Unit). In common with other areas of research, the range and nature of data sources and literature has expanded dramatically, often contributing to an information overload. In that context, books such as this offer an important synthesis for students and researchers because a wide range of sources are cited and referenced in an area dispersed across social science, management and other cognate disciplines (e.g. economics). The rise of new technology such as the Internet is acknowledged throughout the book so students and researchers can access up-to-date data sources.

Existing studies aimed at BTEC National and Higher National courses in the UK and general texts published in North America contain common elements concerned with tourism and transport and many of these books deal with the topic admirably in an empirical sense. Yet the relationship between tourism and transport is rarely discussed in any degree of depth in the popular tourism textbooks. For example, the chapter in Cooper et al (1993; 1998) exemplifies the problem with a cursory treatment given to a topic that warrants a more substantive assessment. Such books do not consider what is meant by a tourist transport system and how the needs of the traveller are incorporated into management systems beyond a simplistic level. Recognised undergraduate tourism textbooks generally pay very little attention to the tourism–transport interface. For this reason, an introductory text may help to stimulate some thought and discussion on the role of transport in tourism and vice versa. This book is also part of a new series – *Themes in Tourism,* published by Addison Wesley Longman. The main aim of the series is to introduce and systematically discuss a range of concepts and ideas related to a tourism theme, which are integral to the field of tourism studies within a spatial analytical context. This is developed through the use of supporting case studies while the fundamental principles of tourism are emphasised throughout. In this instance, the book is also designed to be a starting point for those interested in research on transport studies and tourism, because the content and bibliography may serve as an introductory review for more advanced degree level students. Although the bibliography is by no means exhaustive, many of the most significant studies are reviewed.

Academics write books for many different reasons. In this case, the continued paucity of texts and articles which consider transport and tourism in the academic literature has meant that the existing literature still remains either very rudimentary or highly specialised and technical. In some cases, research on transport and tourism is no more than a few paragraphs or chapters found in many of the established tourism textbooks. This has meant that the topic has not been recognised nor has it gained the significance it deserves. After my involvement in Air New Zealand's Consensus Forecasting Exercise in October 1996, it was very evident that transport services were a powerful tool guiding the development of inbound and outbound tourism. In the case of Asia-Pacific, the growing significance of outbound markets and the need for reliable, accurate market intelligence and an understanding of the transport sector were vital. All too often, it is only when academic researchers engage in consultancy that they begin to understand the commercial realities and implications of the subject area they study.

This book is not intended to be a rewrite of the main themes discussed in existing textbooks. Instead it looks at what is sometimes construed as a mundane and specialist area of study dominated by a number of tourism-related publications. This book aims to prompt the reader to consider some of the relationships which exist in providing transport services and facilities for tourists. This assessment of transport and tourism does not claim to present a comprehensive review of the topic. It focuses on some of the key issues which transport providers, decision makers, managers and tourists face in the use, operation and management of tourist transport. If the book raises the profile of transport issues in tourism and stimulates debate amongst its readers, reviewers and critics then it will have succeeded in establishing a consensus of opinion on tourist transport as a legitimate area of study. Published reviews of *Transport for Tourism* certainly stimulated a debate and one hopes that this new text will also encourage researchers to debate the issue further. Inevitably, some people will question some of the ideas raised here, but this can only assist in fostering more discussion in an area frequently overlooked in the tourism literature. The comments and feedback of former reviewers of *Transport and Tourism* were very helpful and have been acknowledged and addressed in this enlarged version.

Inevitably, certain people have helped to shape the thoughts and ideas in this book. My thanks go to Chris Young, Christ Church College, Canterbury, who offered many insights on *Transport for Tourism* while I was still resident in the UK. I am also indebted to Lynne Tunna who typed (and retyped) this manuscript with precision and speed, as well as Pam Andrews. Since my move to New Zealand in 1994, a range of other colleagues have helped shape some of the ideas dealt with in this book. In particular, Michael Hall has been a good sounding board and source of advice as well as a great encouragement. The postgraduate students in my Advanced Travel Management course have also offered a wide range of insights and views on the transport–tourism interface. The advice and assistance from Asia-Pacific's major airline – Singapore Airlines has been invaluable in updating and expanding the case study. The support and help from Craig Astridge, Marketing Manager (New Zealand) and his assistant, Sarah Lawrence, have been most welcome. In particular, the very generous sponsorship received from Singapore Airlines is gratefully acknowledged because it enabled me to return to Europe to update the literature and data sources which are not readily accessible in New Zealand. Massey University also provided a sabbatical period to help me to complete the research and writing of this text. A number of other people also deserve a special mention and in no particular order of importance they are Richard Gibb and Jon Shaw of the University of Plymouth, Graham and Frances Busby, Rigas

Doganis of Cranfield University, Roy Wood, Lesley Pender and Tom Baum of the Scottish Hotel School, Les Lumsdon of Staffordshire University, Geoff and Margaret Robinson, Thea Sinclair of the University of Kent at Canterbury, Jayne Hoose of Christ Church College, Canterbury, Professor Rolf Cremer of Massey University, Kate Thompson of Air New Zealand, Brian Hoyle of the University of Southampton, Marion Bennett, David Airey, David Gilbert and Dick Butler of the University of Surrey, Bruce Prideaux of the University of Queensland, Mark Orams of Massey University, Dr Hugh Sommerville at British Airways, Professor Derek Hall, the Scottish Agricultural College, Ayr, Scotland, and Jon Green at Red Funnel Ferries, Southampton. Other organisations such as London Stansted Airport and the Civil Aviation Authority have also provided a range of useful information. The academic libraries at the University of Sussex, Surrey University, University of Westminster, Massey University, University of Otago, Christ Church College, Plymouth University and Strathclyde University have yielded important research material cited in this text. I am grateful to Dr J. Monin for supporting my sabbatical and to Nanette for her assistance with my research methods classes so I could complete the book. Lastly, I would like to acknowledge the support of Matthew Smith at Addison Wesley Longman who took the original idea for the *Themes in Tourism* series and turned it into reality.

Acknowledgements

The author would like to thank the following for permission to use photographs in the text: The Helicopter Line to use Plate 1.1; Great North Eastern Railways for permission to use Plates 3.1 and 3.2; Singapore Airlines for the cover sheet and Plates 5.1 and 5.2; The British Airports Authority for Plates 7.1, 7.2 and 7.3; The Channel Tunnel Group for Plates 8.2 and 8.3.

I am also indebted to the following for permission to use the tables indicated: Carole Page for Table 2.6; HMSO for Tables 4.5 and 4.6; Travel and Tourism Intelligence for Tables 4.8, 4.9, 4.10, 4.15, 4.16 and 6.6; Blackwell Publishers and the Association of American Geographers for Table 6.7; the *Journal of Travel Research* for Table 6.8 and Appendix 1; John Wiley & Sons for Table 8.1; British Airways for Tables 8.2, 8.3, 8.4, 8.5, 8.6, 8.7, 8.8, 8.9, 8.10; The Town and Country Planning Association for Table 8.12; David Fulton Publishers for Table 6.4, Singapore Airlines for tables in Chapter 5, and Bruce Prideaux for permission to cite the material contained in the case study on Queensland Rail.

I am grateful to the following for permission to reproduce copyright figures: Addison Wesley Longman for figure 1.3 from R. Tolley and B. Turton *Transport Systems, Policy and Planning: A Geographical Approach* (Harlow: Longman, 1995); John Wiley & Sons for figure 1.5 from B. Lamb and S. Davison in L. Harrison and W. Husbands (eds) *Practising Responsible Tourism: International Case Studies in Tourism Planning, Policy and Development* (Chichester: John Wiley & Sons, 1996); Pitman for figures 7.1 and 7.3 from H. Ashford, H. Stanton and C. Moore *Airport Operations* (London: Pitman, 1991); Addison Wesley Longman for figure 9.1 from K. Irons *Managing Service Companies: Strategies for Success* (London: EIU and Addison Wesley, 1994).

If any unknowing use has been made of copyright material, owners should contact the author via the publishers as every effort has been made to trace owners and to obtain permission.

Chapter 1

Introduction

Introduction

Transport is acknowledged as one of the most significant factors to have contributed to the international development of tourism. According to Gayle and Goodrich (1993), in 1991 the international tourism industry employed 112 million people worldwide and generated over $2.5 trillion at 1989 prices. In 1996, 593 million tourists travelled abroad (World Tourism Organisation 1997), which generated a significant demand for tourist transport. In global terms, the expansion of international tourism continues to generate an insatiable demand for overseas travel. Europe remains the most visited of all regions of the world, with half of all global tourist receipts and almost two-thirds of international arrivals in 1996. In 1996, almost 352 million arrivals and US$215.7 billion in receipts were received. Eastern and Central Europe were among the fastest-growing areas to benefit from Western European tourism flows. In contrast, the East Asia-Pacific region remains the area experiencing the highest growth rates, with total arrivals increasing by 9.3 per cent in 1996 to over 87 million with receipts of US$1 billion. These two examples illustrate the scale of international tourism demand, of which a key component is the mode of transport chosen by these travellers. Despite the controversy over the extent to which tourism can be defined as both an industry and a service activity, it is widely recognised that tourism combines a broad range of economic activities and services designed to meet the needs of tourists.

Transport provides the essential link between tourism origin and destination areas and facilitates the movement of holidaymakers, business travellers, people visiting friends and relatives and those undertaking educational and health tourism. Transport is also a key element of the 'tourist experience' (Pearce 1982) and some commentators (e.g. Middleton 1988; Tourism Society 1990) view it as an integral part of the tourism industry.

Transport can also form the focal point for tourist activity in the case of cruising and holidays which contain a significant component of travel (e.g. coach holidays and scenic rail journeys). Here the mode of transport forms a context and controlled environment for tourists' movement between destinations and attractions, often through the medium of a 'tour'. The integral relationship which exists between transport and tourism is demonstrated by Lamb and Davidson (1996 : 264) since

1

transportation is one of the three fundamental components of tourism. The other two are the tourism product (or supply) and the tourism market (or demand). Without transportation, most forms of tourism could not exist. In some cases, the transportation experience is the tourism experience (e.g. cruises, scenic and heritage rail trips, and motorcoach, automobiles and bicycle tours).

Thus, the mode of transport tourists choose can often form an integral part of their journeys and experience, a feature often neglected in the existing research on tourism. However, the interface of transport and tourism does raise the much wider conceptual problem of what is and what is not tourism transport (D.R. Hall 1997). While it is readily acknowledged that there are specialised, dedicated forms of tourism transport (i.e. tourist coaches, charter flights and cruise liners) there are also other forms of transport which are used by both hosts and tourists to varying extents. For example, urban buses, metro systems and scheduled flights to tourism regions are used simultaneously by tourists and local residents and in some cases this can cause competition. Where tourist use of transport modes does occur, competition with other users has wide-ranging economic, environmental, social and political implications for destination areas. The basic difficulty is that few public sector organisations have the resources or inclination to address the issue of tourist–non-tourist use of different forms of transport. Therefore, even though tourist use of transport can be conceptualised at one level, making the distinction is not necessarily feasible in practice. This remains a constant problem that pervades the tourist–transport interface. However, this should not in itself preclude the academic analysis of transport and tourism. So how has transport been viewed in existing textbooks on tourism?

Tourism studies and tourist transport

The majority of influential tourism textbooks are a product of the 1980s and early 1990s, despite some notable exceptions (e.g. McIntosh 1973; Burkart and Medlik 1974, 1975). The rapid expansion in the number of tourism textbooks published is one indication of the emergence of the subject as a serious area of study at vocational, degree and postgraduate level throughout the world. As many national governments recognise the contribution tourism can make to GDP and national economic development, the expansion of their tourism industries has led to a consideration of the immediate and long-term human resource and training requirements. New courses have developed to fill a niche in the educational marketplace and these have generated a demand for course materials to meet the international expansion of tourism education (Goodenough and Page 1993). The range of available textbooks for tourism studies has generally been written from a North American perspective (e.g. Lundburg 1980; Mathieson and Wall 1982; Mill and Morrison 1985; Murphy 1985; Gunn 1988; McIntosh and Goeldner 1990); a European perspective (e.g. Foster 1985; Lavery 1989; Laws 1991; Ryan 1991; Witt et al

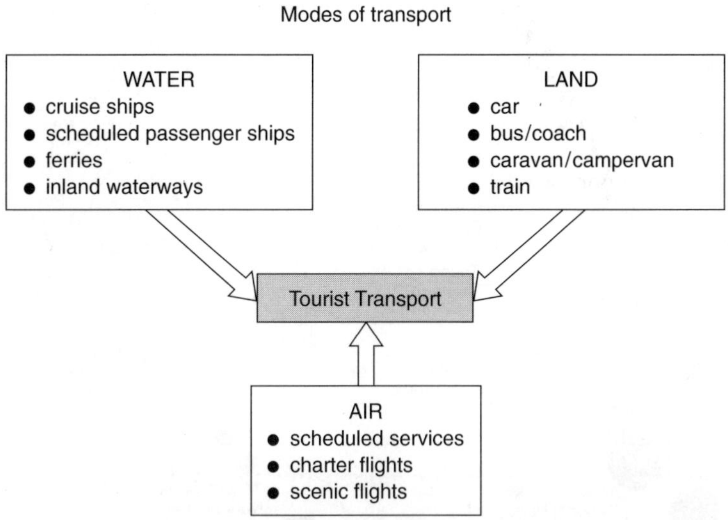

Modes of transport

Figure 1.1   Classification of tourist transport (redrawn from Collier 1994)

1991) or an Australasian perspective (e.g. Pearce 1987, 1992; Collier 1989; Leiper 1990; Bull 1991; Hall 1991; Perkins and Cushman 1993), with few widely available student texts written from an Asian or less developed world perspective.

An examination of these textbooks indicates that travel and transport is a topic frequently cited in relation to its role as a facilitator of the expansion of tourism, as new technology (e.g. the railway and jet engine) and novel forms of marketing and product developments (e.g. package holidays) have contributed to the development of tourism as a mass consumer product. Collier (1994) provides an interesting insight into tourist transport, arguing that three needs need to be fulfilled:

- transporting the tourist from the generating to the host area
- transport between host destinations
- transport within host destinations.

Collier (1994) also classifies tourist transport on several bases (e.g. public or private sector transport; water/land/air transport; domestic and international transport and mode of transport). This classification is expanded in Figure 1.1 which outlines the main modes of transport available, and typifies the approach used in most tourism texts although the issue of scenic flights is commonly omitted (see Figure 1.2 and Plate 1.1). Hall (1991 : 22) highlights the signific-ance of transport since '. . . the evolution of tourism in Australia is inseparable from the development of new forms of transport' and '. . . a clear relationship exists between transport development and tourism growth' (Hall 1991 : 80). The development and expansion of tourist destinations are, in part, based on

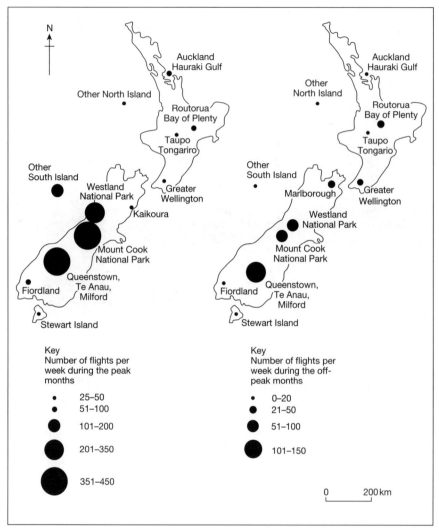

Figure 1.2   The regional distribution of scenic flights in New Zealand in 1993: peak and off-peak season (Source: Ministry of Tourism)

the need for adequate access to resort areas, their attractions and resources. Hence the relationship between transport and tourism is usually conceptualised in most tourism textbooks in terms of accessibility.

Hobson and Uysal (1992 : 209), however, argue that major steps in the development of tourism have been linked with advancements in transport.

> The system . . . creates the structural linkage between origins and destinations . . . [but] . . . the traditional focus on modes of travel often overlooks the underlying reason for the growth of transport communication; that is, the infrastructure that supports and sustains continued growth in the ability of people to travel.

Plate 1.1 Tourist scenic flight, Franz Josef glacier, New Zealand (reproduced courtesy of the Helicopter Line, Auckland)

Tolley and Turton (1995) point to the shrinking of distance by modern forms of international transport (Figure 1.3), and four major phases can be discerned in the evolution of transport technology:

- the transition from horse and windpower
- the introduction of the steam engine
- the development of the combustion engine
- the use of the jet engine.

While such innovations in technology have meant that global shrinkage has occurred (McHale 1969) with reduced journey times, cost reductions and improved capacity, Wackermann (1997 : 35) asserts that the transformations that have taken place as a result of this economic opening-up (and globalisation) of the recreational sector, supported by high-performance means of transport and communication, have made societies less dependent on natural resources and the limitations of distance or of time. Hobson and Uysal (1992) maintain that it is infrastructure which is crucial. They argue that supporting infrastructure has not been able to keep pace with tourism development and therefore congestion may be the biggest constraint facing planners in the millennium.

A number of other textbooks (e.g. Holloway 1989; Mill 1992) have sought to develop this relationship one stage further, by discussing the historical development of tourist travel and accessibility, and the principles governing tourism's expansion within the context of different forms of tourist transport (e.g. air, road, rail and sea travel). Yet tourism studies do not have a monopoly

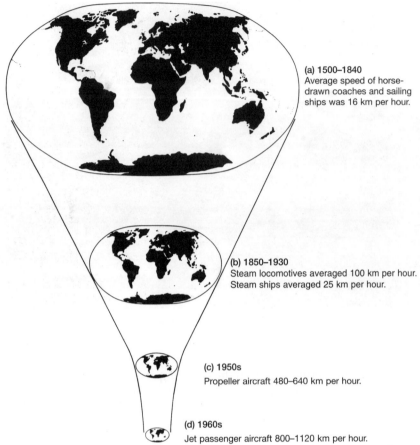

Figure 1.3 The contribution of transport technology to reductions in travel time (redrawn from Tolley and Turton 1995, based on McHale 1969)

on the analysis of transport for tourists. Textbooks on transport studies indirectly discuss the movement of tourists. Many transport studies texts are written from a disciplinary perspective such as economics (e.g. Stubbs et al 1980; Glaister 1981; Bell, Blackedge and Bowen 1983; Banister and Button 1991) while other texts focus on the operational, organisational and management issues associated with different forms of transport (e.g. Button 1982; Button and Gillingwater 1991; Faulks 1990). However, the 'tourist' is rarely mentioned in these books as the term 'passenger' is usually substituted. In fact, a recent text on transport geography (Tolley and Turton 1995) is a case in point, with the tourist also receiving only limited attention in Hoyle and Knowles (1992, 1998). The difficulty here is that the term 'passenger' fails to distinguish between the reasons for tourist movement, implying an impersonal contractual relationship where operators move people between areas on transport systems which are only concerned with the throughput of passengers. In reality, a different situation exists, with transport operators in the 1990s equally concerned with

Plate 1.2 Tourist sightseeing can also use local transport infrastructure such as the Wellington cable car (Source: S.J. Page)

many of the issues facing the tourism industry – particularly customer care and the tourist's experience while travelling. Due to the choice of transport available and the competitive environment for tourist travel in free market economies, transport operators recognise the importance of ensuring that the travel experience is both pleasurable and fulfils consumers' expectations. In state-planned economies, both the demand and supply for tourist transport is regulated by the state and a different political and ideological agenda affects the availability of tourist transport compared with free market economies.

Halsall (1992) identifies the overlap between transport, tourism and recreation, arguing that in reality it is often difficult to distinguish between tourist and non-tourist use of different forms of transport, with the exceptions being dedicated forms of tourist transport such as charter flights and cruises. Former national rail operators such as British Railways (hereafter BR) did not use the term 'tourist', preferring to distinguish between 'business and leisure' travellers when identifying their potential passenger market. Thus the tourist is not explicitly recognised as such, but is regarded as a passenger. In contrast, some tourism researchers recognise the tourist trip as an important feature to examine in its own right (Pearce 1987; Smith 1989), although it receives only scant attention due to the simplistic notions of the tourism–transport relationship. Consequently, the relationship between tourism and transport is rarely discussed in the context of the 'tourist experience'.

Both transport and tourism studies fail to provide an explicit and holistic framework in which to assess the transportation of tourists. For this reason, it is possible to build on the complementarity of these two areas of study to identify the concept of the 'tourist transport system' which highlights the integral role of transport in the 'tourist experience'. This also has the potential to accommodate different approaches to the analysis of tourist travel and transport. What is a 'tourist transport system'?

The tourist transport system: a framework for analysis

To understand the complexity and relationships which coexist between tourism and transport, one needs to build a framework which can synthesise the different factors and processes affecting the organisation, operation and management of activities associated with tourist travel. The objective of such a framework is to provide a means of understanding how tourists interact with transport, the processes and factors involved and their effect on the travel component of the overall 'tourist experience'. Any such framework for analysing tourist transport needs to incorporate the tourist's use of transport services from the pre-travel booking stage through to the completion of the journey and to recognise the significance of the service component. It also needs to incorporate the different modes of transport used by tourists (e.g. air travel by scheduled or charter service, sea travel using ferries or cruise ships and land-based transport including the car, train, coach, motorcaravan, motorbike and bicycle).

One methodology used by researchers to understand the nature of the tourism phenomenon is a systems approach (Laws 1991). The main purpose of such an approach is to rationalise and simplify the real-world complexity of tourism into a number of constructs and components which highlight the interrelated nature of tourism. Since tourism studies is multidisciplinary (Gilbert 1990), a systems approach can accommodate a variety of different perspectives because it does not assume a predetermined view of tourism. Instead, it enables one to understand the broader issues and factors which affect tourism, together with the interrelationships between different components in the system. According to Leiper (1990), a system can be defined as a set of elements or parts that are connected to each other by at least one distinguishing principle. In this case, tourism is the distinguishing principle which connects the different components in the system around a common theme. Laws (1991 : 7) developed this idea a stage further by providing a systems model of the tourism industry in which the key components were the inputs, outputs and external factors conditioning the system (e.g. the external business environment, consumer preferences, political factors and economic issues). As external factors are important influences upon tourism systems, the system can be termed 'open' which means that it can easily be influenced by factors aside from the main 'inputs'. The links within the system can be examined in terms of 'flows' between components and these flows may highlight the existence of certain types of relationships between different components (Figure 1.4). For example:

- What effect does an increase in the cost of travel have on the demand for travel?
- How does this have repercussions for other components in the system?
- Will it reduce the number of tourists travelling?

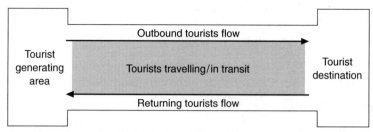

Figure 1.4   A tourism system (redrawn from Page 1994)

A systems approach has the advantage of allowing the researcher to consider the effect of such changes to the tourism system to assess the likely impact on other components.

Leiper (1990) identified the following elements of a tourism system: a tourist; a traveller-generating region; tourism destination regions; transit routes for tourists travelling between generating and destination areas, and the travel and tourism industry (e.g. accommodation, transport, the firms and organisations supplying services and products to tourists). In this analysis, transport forms an integral part of the tourism system, connecting the tourist-generating and destination regions, which is represented in terms of the volume of travel. The significance of transport in the tourism system is also apparent in the model developed by Laws (1991), where a series of smaller sub-systems were also identified (e.g. the transport system) which can be analysed as a discrete activity in its own right while also forming an integral part of the wider 'tourism system'. Thus, a 'tourist transport' system is a framework which embodies the entire tourist experience of travelling on a particular form of transport. The analytical value of such an approach is that it enables one to understand the overall process of tourist travel from both the supplier's and purchaser's perspectives while identifying the organisations which influence and regulate tourist transport. This is aptly summarised by Lamb and Davidson (1996 : 264–65) who highlight the principal relationships that exist between the tourist, transport and their overall experience where

> the purchaser of the tourism product (*the tourist*) must experience the trip to access the product, the quality of the transportation experience becomes an important aspect of the tourist experience and, therefore a key criterion that enters into destination choice. Poor service, scheduling problems, and/or long delays associated with a transportation service, for example, can seriously affect a traveller's perceptions and levels of enjoyment with respect to a trip. Tourists require safe, comfortable, affordable, and efficient intermodal transportation networks that enable precious vacation periods to be enjoyed to their maximum potential.

This highlights the importance of:

- the tourist
- the integral relationship between the transport and overall tourist experience
- the effect of transport problems on traveller perception
- the tourists' requirement for safe, reliable and efficient modes of transport.

To illustrate the relationship between transport and tourism, the following example of Ontario in Canada develops and reinforces many of the issues which are pursued throughout the book.

CASE STUDY   The relationship between transport and tourism: the case of Ontario, Canada

According to Lamb and Davidson (1996) there are a number of interfaces between transport and tourism. In their pioneering study of Ontario, Canada, they provide a range of examples which highlight the multifaceted nature of tourism, transport and the relationship with government (which is also discussed more fully in Chapter 3). Figure 1.5 takes a modal approach to transport (e.g. rail-, air-, bus/coach- and marine-based) and shows some of the functional roles transport infrastructure provides (e.g. linear corridors for tourism uses such as redundant rail routes). In the case of France, the state government investment in the TGV is a clear recognition of the role of intermodal transport for tourism, a feature reiterated by the OECD (1992). Figure 1.5 is also important in emphasising how the tourism industry can harness transport linkages to develop transport-related products (e.g. scenic touring routes by car or bicycle). In the case of Canada, the Canadian province of Ontario illustrates the relationship between transport and tourism.

Ontario is arguably Canada's most prominent tourism region, since in 1991 it accounted for 36 per cent of the national tourism revenue. Nevertheless, in the early 1990s, it was losing market share at a time when the dominant American car-based holidays experienced a decline due to airline deregulation which aided the development of a more competitive domestic tourism business. Tourism Canada (now the Canadian Tourism Commission) saw poor airlinks, an ageing highway infrastructure, airport congestion, declining intercity rail services and the inability to tap into a growing cruise ship industry as key factors affecting the competitiveness of the Ontario tourism industry. This highlighted the key role of transport in the tourism industry.

In 1993, the Ontario Ministry of Tourism, in consultation with industry stakeholders, identified the following priorities for action to improve the situation including:

● Pearson International Airport in Toronto, which received 45 per cent of arrivals to Canada, needed to be upgraded
● Ottawa, Niagara Falls, Toronto and Windsor and Sault Ste Marie were viewed as gateways for American tourists and needed to be developed as the starting point for regional tours
● Ontario's highways, air and waterways needed improving.

A study developed by Ontario's Ministry of Transportation (Ontario Ministry of Transportation 1993) specifically focused on the transport–tourism linkages in Ontario and is relatively rare in highlighting the important integral relationships and synergies that exist. For example, Lamb and Davidson (1996 : 268) argue that in terms of air transport, the tourism industry requires:

● good geographical coverage and frequent services from the main source areas
● competitive fare structures
● attractive, user-friendly airport infrastructure (i.e. customs, ticketing, baggage handling and airport-related services).

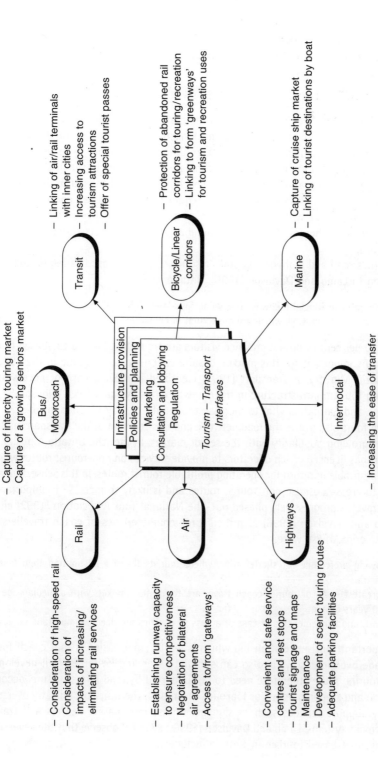

Figure 1.5 Linkages between tourism and transport (redrawn from Lamb and Davidson 1996)

Transit
- Linking of air/rail terminals with inner cities
- Increasing access to tourism attractions
- Offer of special tourist passes

Bicycle/Linear corridors
- Protection of abandoned rail corridors for touring/recreation
- Linking to form 'greenways' for tourism and recreation uses

Marine
- Capture of cruise ship market
- Linking of tourist destinations by boat

Bus/Motorcoach
- Capture of intercity touring market
- Capture of a growing seniors market

Rail
- Consideration of high-speed rail
- Consideration of impacts of increasing/eliminating rail services

Air
- Establishing runway capacity to ensure competitiveness
- Negotiation of bilateral air agreements
- Access to/from 'gateways'

Highways
- Convenient and safe service centres and rest stops
- Tourist signage and maps
- Maintenance
- Development of scenic touring routes
- Adequate parking facilities

Intermodal
- Increasing the ease of transfer
- Coordinating schedules and fares

Tourism – Transport Interfaces
- Infrastructure provision
- Policies and planning
- Marketing
- Consultation and lobbying
- Regulation

Because much of the existing volume of rail traffic is intercity in character and focused in the more densely populated southern part of the province, rail transport was seen as playing a lesser role in Ontario due to the discontinuation of services and the concentration of business in high volume traffic corridors. This is complemented by scenic rail tours such as the *Agawa Canyon* route and *Polar Bear Express.* In 1989 and again in 1992, feasibility studies were undertaken to consider the viability of a Windsor–Quebec City high speed rail corridor. Not only would the American market have access to it at Windsor and Kingston, generating visitor trips, but it would permit time savings, allowing Southern Ontario and Quebec to be visited. Other growth opportunities the project might stimulate were related to the intermodal airlinks at Pearson International Airport, the coach industry and the cruise ship industry on the Great Lakes. Although this would require a subsidised rail corridor to keep rail fares competitive with air, it was seen as a viable option to revitalise Ontario's tourism industry, thereby developing a number of new products through transport innovations.

Ontario's road and highway networks also have an important bearing on tourism. According to Lamb and Davidson (1996), Ontario possesses:

- 25,000 km of provincial highways (including four-lane highways used by tourists)
- Many two-lane highways which are more rural in character.

In fact, 80 per cent of out-of province visitors are from the USA and car-based touring remains important for this market, despite experiencing a decline in the 1980s. The Ontario Ministry of Transport (1993) identified improving ease of access and upgrading tourism infrastructure in the gateway cities (Niagara Falls, Windsor, Sault Ste Marie and Sarnia) as vital to increase the length of stay for such visitors.

Coach tours also utilise the road network, and this segment of the transport market has seen growing popularity with the senior market. In fact the Ontario Ministry of Transport has licensed coach operators to provide services on non-tourist routes which are less profitable in return for operating profitable tourist routes. In this sense, tourism is used to cross-subsidise non-tourist routes and is likely to increase in importance as rail travel is progressively phased out. The National Tour Association (1992) cited in Lamb and Davidson (1996) identified three characteristics of coach travellers in North America. These were:

- a growing preference for shorter trips, where tourists fly to a destination then take a coach trip
- the greatest users of tourist coach travel are the senior market, with a median age of 66–70 years of age
- comfort, and convenience and ease of access to coaches are deemed important.

One important tourism relationship which is worth emphasising is that coach tours encourage overnight stays and so can stimulate spending en route. Well-developed scenic touring routes therefore need to be supported by road infrastructure, accommodation and stopping points as lookouts. In Ontario, the transport ministry oversees highway support services such as signage, tourist information centres and traffic management systems. Lamb and Davidson (1996 : 271–72) observe that the key issues affecting road-based tourism in Ontario include:

- *the quality of infrastructure*, since Ontario's highways are continuing to deteriorate
- *road congestion*, especially at locations such as Lake Ontario, which may deter visitors from exploring such attractions and scenic routes
- *signage* is seen as the highest priority for action by Ontario's tourism ministry and private sector tourism stakeholders to encourage the spontaneous visit and to facilitate the unplanned visit.

Other initiatives related to tourist transport which Lamb and Davidson (1996) review include:

- improving intermodal connections between Pearson International Airport and Central Toronto
- developing stronger linkages between Niagara Falls and Toronto rail, ferry and other transport modes
- the development of new linear land and water corridors which integrate various forms of transport to explore the scenic regions of Ontario.

While Lamb and Davidson's (1996) invaluable study identifies the links between transport and tourism, it also highlights the need for regions or states to consider the concept of seamless transport systems for tourists. This means that the individual transport networks which exist for each mode of tourist transport need to be planned and integrated into a holistic framework. This will ensure that the tourists' experience of transport is a continuous one which is not characterised by major gaps in provision and a lack of integration. For instance, airports need to be linked to their central tourist districts so that visitors transfer from one mode of transport to another with relative ease after long flights when tiredness and disorientation can mar their experience. Recognising such linkages in terms of tourism can also yield invaluable business opportunities (e.g. the growth in airport shuttle companies) where ancillary transport operators can leverage business from the provision of convenient door-to-door services for travellers. So how does this book aim to integrate tourist travel more fully into the study of tourism?

Structure of the book

This book is not intended to be a comprehensive review of transport for tourism in view of the objectives of the new *Themes in Tourism* series. It is designed as a framework in which the reader can gain a clearer understanding of the tourist transport system and some of the ways in which one can analyse its provision, operation and the factors which influencing it. One objective of this book is to overcome the existing perception of tourist transport as a passive element in the tourist's experience which has to be endured to reach a destination area (cruising and touring excepted). The actual process of travelling is an integral part of the tourist's experience even though it is perceived as less important than the activities and pursuits of tourists in the destination. The book offers a number of perspectives of tourist transport which the reader may find a useful starting point for further research on transport and tourism.

One underlying theme emphasised throughout the book is that transport for tourism constitutes a 'service' which is increasingly judged by consumers and providers in relation to the quality, standards and level of satisfaction it engenders. For this reason, both a systems approach and an emphasis on the multidisciplinarity of tourist transport help to transcend the rather fragmented view of this aspect of tourism studies.

In Chapter 2, the multidisciplinarity of tourist transport is examined, drawing upon the concepts and approaches used in economics, geography, marketing and management. Each area of study provides a useful insight into the specialised nature of research on tourism and transport studies which is rarely discussed in terms of the way each complements our understanding of the tourist transport system. As Leiper and Simmons (1993) show, researchers from different disciplines consider various aspects of tourism depending on their background and focus and this inevitably means that they consider specific inputs, outputs and external factors which affect the tourist transport system. For this reason, it is useful to examine some of the common approaches and concepts used by different disciplines in analysing tourist transport. Chapter 3 considers the role of transport policy and planning, its effect on operational and consumer issues and the progress towards a common transport policy in the EU. It also examines how government policy can affect the development of tourist transport systems. A case study of tourist travel by rail in the UK is used to illustrate how national transport policy objectives are implemented in an era of privatisation and the significance for tourist experiences of rail service provision. This is followed in Chapter 4 by an analysis of the demand aspects of tourist travel and the data sources available to tourism researchers. Different sources are examined in relation to domestic tourism in China, transport-specific sources on air travel and cruising, and outbound travel from Japan. The chapter also reviews methods of forecasting future tourist travel. Chapter 5 looks at the supply of tourist transport, focusing on the supply chain and how companies with transport interests seek to exercise control over the distribution and quality of tourist travel services. In Chapter 6 the focus is on the management of tourism supply issues and the response of the US domestic airline market to deregulation to illustrate how concentration in the aviation market has affected the supply of services. In Chapter 7 the management theme is developed a stage further with a focus on transport infrastructure – the role of the airport as a terminal facility where airlines and travellers interact as part of the tourist experience. In Chapter 8 the human and environmental impact of tourist travel and the operation of different modes of tourist transport are discussed using a case study of British Airways and the chapter ends with a discussion of the potential for developing sustainable tourist transport and the role of cycling. Chapter 9 concludes the book, examining some of the main issues which are likely to affect tourist transport in the late 1990s and in the next millenium, particularly the role of globalisation and privatisation in transport systems. Other issues such as service quality are also reviewed as part of the move towards Total Quality Management in tourist transport systems.

Questions

1. To what extent is tourist transport considered as an area of study in transport and tourism studies?
2. What are the main linkages between transport and tourism?
3. What is the value of a systems approach to the analysis of tourist transport?
4. What are the main features of a system?

Further reading

McIntosh, R.W. (1973) *Tourism Principles, Practices and Philosophies*, Columbus Ohio: Grid Inc.
This was the first tourism textbook to be published.
Lamb, B. and Davidson, S. (1996) 'Tourism and transportation in Ontario, Canada', in L. Harrison and W. Husbands (eds) *Practising Responsible Tourism: International Case Studies in Tourism Planning, Policy and Development*, Chichester: Wiley.

The concept of a tourism system is dealt with in:

Laws, E. (1991) *Tourism Marketing: Service Quality and Management Perspectives*, Cheltenham: Stanley Thornes.
Leiper, N. (1990) *Tourism Systems: An Interdisciplinary Perspective*, Palmerston North: Massey University.
Wheatcroft, S. (1978) 'Transport, tourism and the service industry', *Chartered Institute of Transport Journal*, 38 (7): 197–206.

Chapter 2

Analysing and managing tourist transport: multidisciplinary perspectives

Introduction

Tourism, like transport studies, is a multidisciplinary field of study which has borrowed and refined concepts and theories from other subjects as it establishes itself as a legitimate area of academic study. This poses a number of problems for researchers when exploring the relationship between transport and tourism in the context of tourist transport systems. For example, what approaches and methods of study should one use to analyse tourist transport systems? In most cases, research is based on those social sciences disciplines with an interest in tourism and/or transport studies. This has an important bearing on the analysis of tourist transport systems because the types of question a researcher asks, and the focus of their work, are often determined by their disciplinary background. In other words, each subject area has its own range of concepts and way of viewing the world which builds upon the knowledge and research in that area.

While many subject areas have made distinctive contributions to the study of tourism, no one discipline is all-embracing enough to understand the complexity of the tourist transport system. Social science subjects such as social psychology, sociology, and business and management studies have an interest in tourism and transport studies but there is a relative paucity of published research which analyses the tourist transport system.

However, for the purpose of this book, there are a number of subject areas identified for consideration since they have made a direct contribution to the analysis of tourist transport systems. These are:

- economics
- geography
- marketing
- management

Although other cognate areas such as logistics, planning, environmental science and behavioural sciences (e.g. psychology) do have a bearing, these are discussed throughout the book where their contribution is evident. However, the discussion in this chapter emphasises the main subject areas whose contribution is documented and concludes with a focus on management which tends to provide a practical setting in which the contribution of each subject area is harnessed for a practical purpose – the management of tourist transport systems and different components within the system.

It must be stressed that this chapter does not attempt to provide a comprehensive review of the literature and main areas of research on transport and tourism. A wide range of books have already been published in economics, geography and marketing which provide an insight into tourism and transport although none have developed a particular focus on transport for tourism. The approach adopted here is to outline the main principles each subject uses and to illustrate the distinctive contribution it makes to the analysis of tourist transport. This is then developed in more detail with a supporting case study. Since this is only an introductory text, readers are directed to specialised studies for a more detailed insight into particular issues.

The economist and tourist transport

The economist's approach to the analysis of tourist transport is based on two distinct areas of research: transport economics (e.g. Starkie 1976; Beesley 1989) and tourism economics (e.g. Bull 1991; Sinclair 1991; Sinclair and Stabler 1997) and each area of study uses similar concepts to understand how the tourist transport system functions. For this reason, it is useful to consider what issues are examined by economists as a basis for a more detailed discussion of the concepts they use.

What is economics?

Like many social science subjects, there is little agreement on how to define an area of study such as economics. However, according to Craven (1990 : 3) 'economics is concerned with the economy or economic system . . . [and] . . . the problem of allocating resources is a central theme of economics, because most resources are scarce'. Therefore Craven (1990 : 4) argues that economics is the study of methods of allocating scarce resources and distributing the product of those resources, and the study of the consequences of these methods of allocation and distribution.

What is meant by scarcity and resources? The term 'scarcity' is used to illustrate the fact that most resources in society are finite and decisions have to be made on the best way to use and sustain these resources. Economists define resources in terms of:

- natural resources (e.g. the land)
- labour (e.g. human resources and entrepreneurship)
- capital (e.g. man-made aids to assist in producing goods)

and collectively these resources constitute *the factors of production* which are used to produce commodities. These commodities can be divided into:

- goods (e.g. tangible products such as an aircraft)
- services (e.g. intangible items such as in-flight service – see Laws and Ryan 1992)

and the total output of all commodities in a country over a period of time, normally a year, is known as the *national product*. The creation of products and services is termed *production* and the use of these goods and services is called *consumption*. Since, in any society, the production of goods and services can only satisfy a small fraction of consumers' needs, choices have to be made in the allocation of resources to determine which goods and services to produce (Lipsey 1989). The way in which goods and services are divided among people has been examined by economists in terms of the distribution of income and the degree of equality and efficiency in their distribution. Many of these issues are dealt with under the heading of 'microeconomics' which Craven succinctly defines as:

> . . . the study of individual decisions and the interactions of these decisions . . . [including] . . . consumers' decisions on what to buy, firms' decisions on what to produce and the interactions of these decisions, which determine whether people can buy what they would like, whether firms can sell all that they produce and the profits firms make by providing and selling. (Craven 1990 : 4)

Microeconomics is therefore concerned with certain issues, namely:

● the firm
● the consumer
● production and selling
● the demand for goods
● the supply of goods

Economists also examine a broader range of economic issues in terms of *macroeconomics* which is concerned with

> the entire economy and interactions within it, including the population, income, total unemployment, the average rate of price increases (the inflation rate), the extent of companies' capacities to produce goods and the total amount of money in use in the country. (Craven 1990 : 5)

Therefore, macroeconomics is mainly concerned with:

● how the national economy operates
● employment and unemployment
● inflation
● national production and consumption
● the money supply in a country

Within micro- and macroeconomics, both transport and tourism economists examine different aspects of the tourist transport system which is based on the analysis of the concepts of demand and supply.

Demand

Within economics, the concern with the allocation of resources to satisfy individuals' desire to travel means that transport economists examine the *demand* for different modes of travel and the competition between such modes in relation to price, speed, convenience and reliability. Economists attempt to understand what affects people's travel behaviour and the significance of transport as something which is rarely consumed for its own sake: it is usually demanded as a means of consuming some other goods or service (i.e. commuting to work or the travel component of a holiday). According to Mill (1992 : 83–4), the demand for tourist transport is also characterised by:

- its almost instantaneous and unpredictable nature, which requires operators to build overcapacity in the supply to avoid dissatisfied travellers;
- the variability in demand, ranging from *derived demand* (where tourist transport is a facilitating mechanism to achieve another objective, such as business travel) to *primary demand* which is the pursuit of travel for vacation purposes;
- non-priced items (e.g. service quality, reliability and punctuality).

Transport economists have developed mathematical models to analyse the trip-making behaviour of travellers (Ortuzar and Willumsen 1990), the factors influencing demand and why variations occur in the trip-making behaviour of consumers due to relationships between socioeconomic factors (e.g. age, income, profession and family status) and the effect of macroeconomic conditions (e.g. the state of the economy). In contrast tourism economists have examined the demand for travel and tourist products, recognising the significance of demand as a driving force in the economy. This stimulates entrepreneurial activity to produce the goods and services to satisfy the demand (Bull 1991). More specifically, tourism economists examine the *effective demand* for goods or services which is the aggregate or overall demand over a period of time. Since income has an important effect on tourism demand, economists measure the impact using a term known as the *elasticity of demand*. As Bull (1991 : 37) has shown, it is measured using a ratio calculated thus:

$$\text{Elasticity of demand} = \frac{\text{percentage change in tourism demand}}{\text{percentage change in disposable income}}$$

in relation to two equal time periods. The significance of this concept is that the demand for goods to fulfil basic needs (e.g. food, water and shelter) is relatively unchanging or *inelastic* while the demand for luxury items, such as holiday and pleasure travel, is variable or *elastic*, being subject to fluctuations in demand due to factors such as income or price. Thus, elasticity is used to express the extent to which tourists are sensitive to changes in price and service. For example, primary demand is usually more elastic than derived demand. In tourist transportation, researchers recognise the importance of price, which is acknowledged as a more complex issue than income, due to the varying impact of exchange rates, the relative prices of destinations and the high level of competition between destinations for tourists. Furthermore, the different elements which comprise the tourism product (e.g. transport, accommodation and attractions) are complementary and it is difficult to separate out one individual item as exerting a dominant effect on price since all are interrelated in terms of what is purchased and consumed.

To assess the impact of price on the demand for tourism, economists examine the *price-elasticity of demand*, where an inverse relationship exists between demand and price (Bull 1991). For example, it is generally accepted that the greater the price, the less demand there will be for a tourism product due to the limited amount of the population's disposable income which is available to purchase the product. Price-elasticity is calculated thus:

$$\text{Price-elasticity} = \frac{\text{percentage change in quantity of tourism product demanded}}{\text{percentage change in tourism product price}}$$

Other contributory factors which influence the demand for tourism include the impact of tourist taxation, the amount of holiday entitlement available to potential tourists as well as the effects of weather, climate and cultural preferences for holidaymaking which are expressed in terms of seasonality. These factors also need to be viewed in the context of the economics of each specific mode of tourist transport (e.g. rail, coach, air and sea travel) to understand how demand is incorporated into the operator's supply of the service. In view of the variation between different modes of transport, readers should consult more detailed studies (Bell et al 1983; Button 1982; Glaister 1981 and Stubbs et al 1980).

Supply

Economists are also interested in the *supply* of a commodity (e.g. tourist transport), which is often seen as a function of its price and the price of alternative goods. Price is often influenced by the cost of the factors of production, but in the case of tourist transport, it is difficult to identify the real cost of travel. For example, state subsidies for rail travel in Europe are used to support the supply of services in the absence of a major demand for social reasons (Whitelegg 1987). As a result of subsidies, the price charged may not always reflect the true cost, particularly where tourist transport providers use cross-subsidies in their operations. Cross-subsidisation implies that profits from more lucrative routes are used to support uneconomic and unviable services to maintain a route network, thereby increasing the choice of destinations. According to Bull (1991 : 78), the supply of tourist transport can be characterised by:

- major capital requirements for passenger carriage (e.g. the cost of aircraft, passenger trains and ferries)
- government regulations and restrictions to monitor the supply which is determined by state policy
- competitive reaction from other businesses involved in tourist transport
- a high level of expertise required to operate and manage tourist transport enterprises.

Bull (1991) suggests that the principle questions which economists are interested in from the supply side are:

- what to produce
- how to produce it
- when, where and how to produce it

From the transport operator's perspective, the main objective in supply terms is to maximise profitability from the available capacity, which is usually expressed in terms of the *load factor*. Transport companies can maximise passenger revenue by minimising costs and pricing their product or service at a competitive rate. Certain travel markets are very price-sensitive, which means that consumers may be easily persuaded to switch to another operator or mode of transport if the price rises beyond a critical level (the demand for youth travel on express coach services is a good example of a price-sensitive market). Despite price-sensitivity, airlines and other modes of transport distinguish between scheduled routes which operate a regular timetabled service and charge higher fares, and charter services operated on behalf of tour operators to carry holidaymakers who have purchased a transport-only component or package holiday from the tour operator. The price differential for scheduled and charter passengers is reflected in the passenger load factor which the scheduled airline needs to reach to achieve a profit (see the case study of Singapore International Airlines in Chapter 5). Scheduled routes charge a higher tariff but operate on a lower load factor compared to charter flights where a lower unit cost is charged but a higher load factor (often 90 per cent) is needed to yield a profit. Seasonality in the demand for tourist transport services may affect the load factor, and peak usage at popular times means that transport operators use premium pricing to manage the supply and maximise profit to offset losses in times of limited demand. To illustrate the importance of pricing, the following case study examines the issue in relation to air travel.

CASE STUDY   The pricing of air travel

According to Hanlon (1996) the way in which airlines price their products is a complex process, with a multitude of fares available according to when one travels; whether travel is off-peak or peak; the class of travel; the length of stay at a destination and whether it includes a Saturday night in Europe. In addition, when and where the ticket is purchased influences the price, together with a host of other factors. Even though there is extensive variation in fare levels across routes, fare levels generally taper with distance so that the fare per km is lower on a long-haul route than a short-haul route. But global anomalies exist. For example, fare levels in Europe have been higher per route km than in North America, while fare levels between Europe and Asia-Pacific have been lower than between Europe and Africa. In the USA, Milman and Pope (1997) highlight the inherent problem associated with the demand for air travel. It is described as stochastic, meaning that the number of reservations and actual trips may vary, implying that airlines need to price their product in a price-sensitive manner to ensure that sales and seat prices maximise revenue for each departure. The various price strategies used include:

- *single-pricing*, where a uniform price is charged for passengers on the same flight (e.g. People Express Airlines)
- *multiple pricing*, which involves segmenting the market and offering empty seats at a lower price

- *yield management*, involving management planning, and computer programming to optimise revenue by selecting the optimum mix of travellers
- *leg-based pricing* where seat allocation is used to maximise revenue on each sector of a flight
- *origin–destination pricing models*, where complex linear programming techniques are used to compare whether a multi-leg passenger trip will generate more revenue than a single-leg trip, and to allocate seats accordingly
- *expected marginal seat revenue*, which is explained in detail by Vasigh and Hamzee (1996).

Because airline fares can vary and offer so many possible fare classes for one route, it raises the issue of price discrimination.

Hanlon (1996) examines the role of price discrimination, where the producer (the airline) charges different prices for various units of the same commodity where there are similar unit costs in the cost of supply. In this instance, peak and off-peak costs are not a form of discriminatory pricing because the price differential reflects the extra costs of meeting peak demand. Discriminatory pricing exists where prices differ more than costs, especially when price–cost margins vary and some customers are paying a higher differential. In many cases of price discrimination, the major factor is the difference in demand elasticity. The principal explanation advanced in the economics literature is related to the inverse elasticity rule: the firm charges more where the demand is low and less where it is high (Hanlon 1996). Firms need to be able to prevent customers who have paid low prices from reselling to those who would be charged high prices.

A yield management system linked to the airline's computer reservation system employs complex algorithms combined with historical booking data to forecast the scenario which enhances revenue from the seat allocation. By restricting the availability of lower-priced fares, airlines are able to safeguard against revenue dilution in an environment where many of the costs of airline operation are relatively fixed for each departure (see Figure 2.1). The innovative simulation model discussed by Belobaba and Wilson (1997) includes the role of passenger choice behaviour and a yield management system was employed to evaluate the impact on market shares, traffic and revenues of each competing airline in a hypothetical market situation. Simulation has to be undertaken given the commercial sensitivity of airline data and competition. A model was therefore used which included four fare options for a typical US domestic airline market (full coach fare through to a 14-day advance purchase non-refundable excursion fare priced at 40 per cent of the full fare). Historical booking data was used to forecast demand on future flights and to optimise fare class booking limits.

The simulation exercise highlighted a 'first mover' advantage, where the airline which initiates yield management (YM) gains the competitive advantage. This results in a better fare mix of traffic and, especially, protects seats for late-booking, high-fare passengers. More interesting was the finding that the airline without YM can in fact be hurt in terms of revenue by not having a YM system. The carrier with a YM system is able to dump unwanted low-fare passengers onto the flights of the carrier without YM, filling up that carrier's capacity and leaving it without adequate last-minute seat availability for the highest-fare passengers (Belobaba and Wilson 1997 : 9).

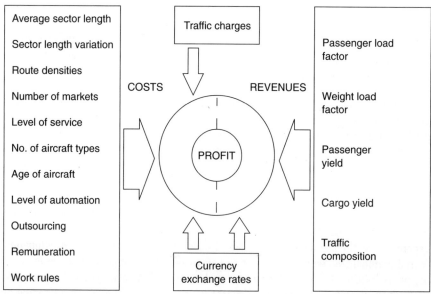

Figure 2.1   The cost factors in airline operations (redrawn from Seristö and Vepsäläinen 1997)

Airlines are able to prevent people reselling airline tickets by making them non-transferable. Airline research shows that business travellers have relatively price-inelastic demand and are not able to book far in advance. In contrast, the price-elastic traveller can book a long time in advance and be flexible, and their concern is with the benefit of travelling at a lower price. Such price discrimination provides airlines with opportunities to segment the market with price discrimination and reason for travel as major factors in the segmentation. Indeed Belobaba and Wilson (1997 : 3) confirm that

> the practices of differential pricing and yield management, although confusing to most air-line passengers, have, over the past decade, been embraced by most airlines. The incremental revenues generated by these practices contribute to the coverage of the predominately fixed operating costs of a scheduled flight departure. By offering a range of fare options at different price levels on the same flight, airlines attempt to segment the total demand for air travel according to the different sensitivities to price and the need for travel flexibility of business and leisure travellers. To ensure that low-fare leisure passengers do not con-sume all of the seats on high-demand flights, airlines employ yield management (YM) systems.

These yield management systems forecast the demand for flights and calculate the number of seats which should be available for sale at each fare class to maximise flight revenues. A number of studies have estimated the benefits to individual airlines (e.g. Belobaba 1987; Weatherford and Bodily 1992), where revenue gains of 2–5 per cent occur from implementing a (YM) system. Yet the study also indicated that if another competitor implemented a YM system, both saw improvements in revenue, indicating the ability of carriers to catch up. This suggests that YM is not a zero-sum

game and that YM may force passengers to pay fares nearer to what they are prepared to pay rather than lower fares resulting from competitive pressures between
the carriers. From the consumers' perspective, discriminatory pricing which is enhanced
by YM is undesirable if the price is raised for some, but desirable if the price declines.
However, there is a general trend in pricing airline tickets, where those most affected
by discriminatory pricing (e.g. business travellers) bear higher costs. In contrast, those
high costs of travel help to subsidise the costs borne by discount travellers. Of course,
problems occur when the market share of the more lucrative business travel fares
decline for a particular airline, reducing the cross-subsidisation that the high revenue
passengers contribute to the overall profitability of its operation.

While business travellers often complain of the discriminatory pricing used to
justify higher standards of service, comfort and flexibility in reservations, much
of the cost is often borne by businesses as an expense rather than by individual
travellers. Whether such pricing enhances or hinders competition depends upon
the perspective one takes. For example, discriminatory pricing can allow airlines to
experiment with fare levels. Yet it can also be viewed as a tool which companies
with a dominant market position can use to weaken the competition while increasing prices in other markets. Thus, where such competition occurs, travellers who can
be flexible in terms of travel arrangements for their journey, enjoy a high degree of
price elasticity. As a result airlines set fares at a costs plus a low profit margin for
travellers with a high degree of price elasticity.

Although this is desirable for the consumer, it assumes that competitors cannot
sustain business at low profit margins if their scale of operation is not able to carry
such marginal pricing. In fact some notable examples, such as the Laker Sky Train
which was engaged in a price-war on the lucrative North Atlantic routes in the
1970s, complained of one other form of pricing – predatory pricing.

Predatory pricing is where a dominant business pursues a policy of eliminating,
deterring or restricting competition. This usually involves selling at below cost to
destroy the competition. While Hanlon (1996) debates the rationality of such a policy,
since the losses for the dominant business are much greater than for the competition, the dominant business will expect to recoup the losses once the competition
is removed. In some cases, an outright takeover or merger is more rational, but predatory pricing may ease the way to a takeover depending on the personalities in the
organisations involved. In the USA, a paper by Areeda and Turner (1975) developed
a surrogate measure of predatory pricing: it exists where the price is set below a
business's short-run marginal cost. A marginal cost is generally seen as the 'addition
to total cost resulting from the last unit of output. It refers to those elements of cost
that can be avoided or escaped if the last unit is not produced' (Hanlon 1996 : 172).
Yet in air transport, the marginal cost may be low and seats are a perishable commodity which cannot be stored if supply is greater than demand. For airlines, the
marginal cost is low for filling an unsold seat when the flight will depart even if
the aircraft has seats left unsold. Even where low fares are sold to attract travellers
to fill seats which would otherwise remain empty, these are still priced above the
marginal cost. (See Hanlon 1996 : 172–86 for more discussion of how to assess predatory pricing and the mechanisms to assess marginal costs.) Non-price forms
of predatory pricing can also be used by airlines seeking to stifle competition by

increasing supply through increased frequencies so that competitors find it difficult to break into a market where supply exceeds demand. Without regulation, Computer Reservation Systems (CRSs) and Frequent Flyer Programmes (FFPs) may also disadvantage smaller airlines unable to access CRSs and offer FFPs at the same reward rate on routes facing competition. Since FFPs have been used to target customer loyalty among high-yielding passengers (e.g. first and business class), the gains may be very high, for although they may only comprise 20 per cent of passenger volumes they may contribute 50 per cent of revenue.

Discussion of the pricing of air travel illustrates that principles of economics underpin the economic behaviour of airlines. In addition, the free market economy can result in a wide range of competitive situations. In practice, there is often regulation by national and supranational organisations responsible for air travel such as the Civil Aviation Authority in London and the European Commission in Brussels. These bodies seek to adjudicate between what is acceptable and unacceptable competition and what constitutes predatory behaviour. As Hanlon (1996 : 183) rightly concludes,

> The crux of the matter is to find the dividing line between genuine revenue-enhancing yield-management techniques and practices aimed directly at undermining the economics of a competitor's operation . . . The Civil Aviation Authority approaches the problem by tracking fare developments and trends over time on a selected number of routes.

Although such monitoring can assist in the process of establishing predatory pricing, the volume of seats sold at each fare level also needs to be considered. Where predatory behaviour occurs on international routes, this falls outside the jurisdiction of national regulatory bodies and investigations can take a long time to institute. Despite the suspected existence of predatory behaviour, it is often difficult to distinguish from aggressive competitive behaviour.

Economists also have an interest in macroeconomic issues associated with the supply of tourist transport services and the structure of the market system in which companies operate. For example, economists have examined the effect of company strategy in the tourism and transport sectors in response to market competition which may affect the management, operation and provision of services to consumers. These different market conditions may range from near-perfect competition to a situation where the three following conditions may occur (see Bull 1991 : 60–5 for a more detailed discussion):

- *oligopoly*, where the control of the supply of a service is by a small number of suppliers
- *monopoly*, where exclusive control of services is by a single supplier
- *duopoly*, where two companies control the supply of services

For the consumer, such activities may have a significant effect on the choice, price and degree of competition which occurs. In some cases, a monopoly situation or a variant may lead to anti-competitive practices and may not necessarily be in the public's interest, although arguments contrary to this view have also been expressed in relation to the break-up of British Rail's (BR) provision of rail services in the UK (see Chapter 3). Transport and tourism economists have also retained an interest in the macroeconomic effects of tourist transport on national economies. Tourist use of transport is a major contributor to the balance of payments when examining the economic benefits of tourism. Such considerations have an important bearing on public or private sector investment decisions when examining the costs and benefits of building new tourist transport infrastructure (e.g. the construction of a new airport). Issues related to the employment-generating potential of new tourist transport infrastructure and the effect on income generation for local economies feature prominently in these investment decisions. Economists also use complex research techniques such as multiplier analysis (see Archer 1989) to evaluate the secondary or indirect economic benefits of additional tourist expenditure for local areas. There is also a growing awareness among economists of the environmental costs of tourist transport (Banister and Button 1992; Perl et al 1997).

Geography and tourist transport

Within geography, the study of tourist transport has largely been undertaken by transport geographers (see Knowles 1993 for an excellent review of recent studies in transport geography) and tourism geographers (Pearce 1990, 1995a). The main concern of geographers when considering tourist transport can be related to three key concepts which characterise the study of geography:

- *space*: area, usually the earth's surface
- *location*: the position of something within space
- *place*: an identifiable position on the earth's surface

Geographers are therefore interested in the spatial expression of tourist transport as a vital link between tourist-generating and tourist-receiving areas. In particular, geographers are concerned with the patterns of human activity associated with tourist travel and how different processes lead to the formation of geographical patterns of tourist travel at different scales, ranging from the world to the national (e.g. country), regional (e.g. county) and local levels (e.g. an individual place). Previous geographical research on transport has looked at its role in different regions, its impact on economic development in terms of accessibility, the effect on the environment and the role of policy making (see Farrington 1985). In many of the popular transport geography textbooks (e.g. Hay 1973; Lowe and Moryadas 1975; Adams 1981; White and Senior 1983; Barke 1986; Hoyle and Knowles 1992, 1998; Tolley and Turton 1995) there are a number of fundamental concepts which geographers use to analyse the spatial components of transport. Geographers have typically analysed travel as a response to satisfy a human desire for movement and the spatial outcome of such journeys. They have also considered the spatial variables in the transport system (e.g. location and places) and how these affect the costs and production of other social and economic activities. For the geographer, transport facilitates the process of movement which has economic and budgetary costs, while behavioural factors (e.g. perception and preferences for particular forms of transport) determine the journey in terms of the available infrastructure and routes. In analysing the transport system, geographers have considered:

- the linkages and flows within a transport system
- the location and places connected by these linkages (usually referred to as 'centres' and 'nodes')
- the system of catchments and relationships between places within the network
  Source: Modified from Taafe and Ganthier (1973).

In some respects, transport tends to be viewed as a passive element by geographers in a tourism context rather than as an integral part of tourism and recreational activity. In fact comparatively little progress has been made by researchers in this area of study since the influential studies undertaken in the 1960s and 1970s by geographers (e.g. Wall, 1971; Patmore, 1983). While useful syntheses such as Halsall (1992) provide bibliographies of past studies, few of the studies reviewed by Halsall (1992) were written by geographers in the 1980s and 1990s explicitly to develop an understanding of the way in which transport facilitates and in some cases conditions the type of recreational and tourism activity which occurs. Much of the research activity by geographers in the 1990s published in the *Journal of Transport Geography* has been pre-occupied with air travel (Chou 1993; Shaw 1993; Feiler and Goodovitch 1994; O'Connor 1995). The continued neglect of transport in the analysis of tourism has also been compounded by the perception of other geographers who did not concede the importance of tourism as serious areas for research until the late 1980s and 1990s (Hall and Page 1999). Even then, it is often relegated to a minor research area despite the rapid growth in tourism and recreational courses in most countries.

Recent texts such as Tolley and Turton (1995) do at least highlight the need to consider tourism in the context of transport geography. Yet even Tolley and Turton do not devote much space to the topic and few other studies have sought to place transport for recreation and tourism on the research agenda beyond the notable occasional studies by the Transport Study Group of the Institute of British Geographers (Tolley and Turton 1987; Halsall 1982). Where previous studies exist (e.g. Halsall 1992), the emphasis has been on modal forms of recreational and tourist travel (e.g. land-based transport and the use of trails, water-borne transport and heritage transport for nostalgic travel – see Page 1993a, 1994a). This does not adequately develop the contribution that geographers can make to the management, planning and development of transport for recreational and tourist travel.

CASE STUDY   The geography of tourist transport in Ireland: air travel

Recent research by Horner (1991) on airport and air-route development in Ireland illustrates how the geographer can analyse tourist transport provision. Ireland is

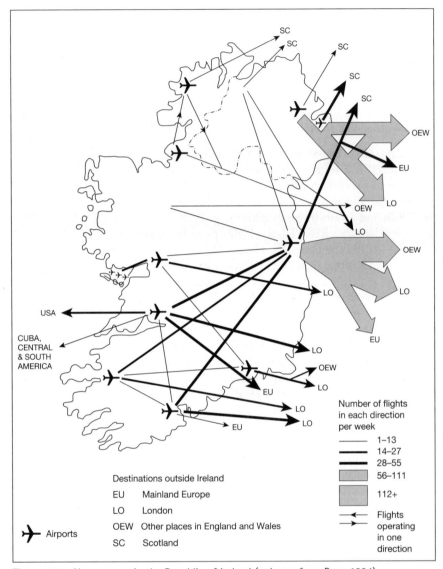

Figure 2.2   Air transport in the Republic of Ireland (redrawn from Page 1994)

often perceived as peripheral to many of Europe's main international tourism markets, and air traffic is the dominant form of transport used by inbound and outbound tourists (see Page 1993b for a more detailed discussion of peripherality and tourist travel in Ireland). As Figure 2.2 shows, the main air routes and the amount of traffic carried on the routes is represented as a flow line. Each *flow* is between an origin (e.g. Ireland) and a series of destinations. The origin points or *nodes* are the airports at which visitors arrive and from which they depart. For Ireland, the main flows are between:

- Dublin/London and mainland Europe
- Belfast/Belfast City and London
- Ireland and other places in England, Wales and Scotland.

These airports are the main entry and exit points for tourists using air transport and are called *gateways* (see Pearce 1987). From Figure 2.2 it is also possible to examine how the nodes are related to catchment areas the airports serve, the type and frequency of services offered and the degree of spatial organisation. Figure 2.3 shows that between 1975 and 1990 there have been changes in the degree of accessibility provided by the growth in smaller regional airports in Ireland. In 1990, most areas within Ireland were within a one-hour drive of an airport, which gives rise to a distinct pattern of airport provision. The geographer analyses this pattern in terms of the spatial hierarchy which exists in it by looking at the number and type of scheduled flights operated. Horner (1991) provided a ranking based on these two features to identify a hierarchy of airports which was a function of their airstrip size. The hierarchy comprised:

I	Major airports capable of handling all sizes of aircraft, including 747 Jumbo jets (Belfast International, Dublin and Shannon)
IIa	Airports able to accommodate medium-sized jets operating scheduled services (e.g. Cork, Belfast City and Knock)
IIb	Similar airports without scheduled services (Baldonnel)
III	Airports with scheduled services capable of accommodating small jets (Carrickfin, Co. Galway, Farranfare, Co. Kerry, Waterford, Co. Sligo)
IV	Other airstrips with shorter, hard-surface landing facilities
V	Other landing areas, mainly grass

The hierarchy of airports provides scheduled services, and the linkages and flows between the airports indicate which routes passengers use between the major gateways and smaller regional airports. This leads to a pattern of 'hubs' (gateways) and 'spokes' (smaller airports) and a distinctive spatial network. Geographers have developed methods of assessing how different nodes are connected within the network (e.g. the degree of connectivity) and how accessible different nodes are within the network (see Lowe and Moryadas 1975 for a more technical discussion). In the case of Ireland, the dominant feature is the limited amount of north–south cross-border traffic in the network. The significance of Great Britain/mainland Europe as origins and destinations for much of the traffic is based on a small number of hubs in Ireland. Aside from Dublin and Belfast, Shannon is an important node, since it is 'a mandatory stopping place for scheduled transatlantic flights originating or terminating in the Republic of Ireland' (Horner 1991 : 43).

Transport geographers are also interested in the air transport network in Ireland in terms of how it evolved, the spatial processes which shaped the development of the network and how geographical models and theories might be used to examine the flows and interactions in the network. Clearly, the role of government policy in shaping and regulating the network would also be of significance and this point is developed further in Chapter 3. How have these concepts been used by tourism geographers researching tourist use of transport?

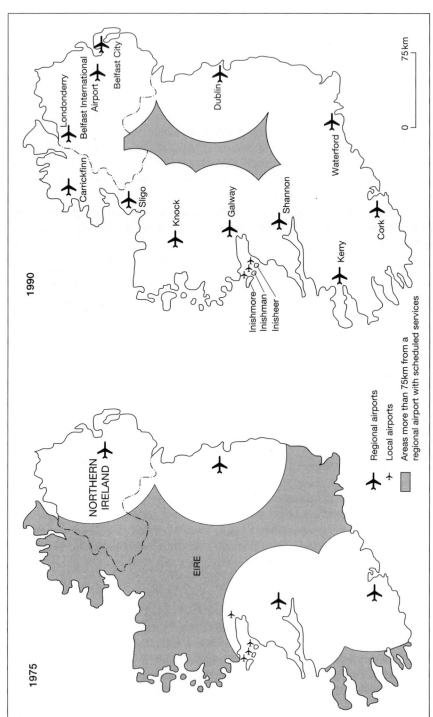

Figure 2.3 Changes in access to air travel in the Republic of Ireland, 1975–90 (redrawn from Page 1994)

Plate 2.1 In remote rural areas, such as North Hokianga, New Zealand, access for car-travellers is essential to facilitate tourism and resident access as the 12-car ferry at Rawene shows, which is operated by the Far North District Council (Source: Dr C. Johnson)

Tourism geography and tourist transport

According to Pearce (1979), geographical research on tourism initially focused on:

- the spatial analysis of the supply and demand for tourism
- the geography of tourist resorts
- tourist movements and flows
- the analysis of the impact of tourism
- the development of models of tourist space to understand the evolution and expression of tourism in specific locations

Since the 1970s, tourism research has examined the patterns and processes associated with the development of international and domestic tourism. One particular skill which the geographer has contributed to the study of tourism is the ability to synthesise (i.e. sift, search and make sense of) the approaches and analysis of tourism phenomena undertaken by other disciplines. During the 1980s geographical research on tourism moved from a traditional concern with regional case studies and descriptions of tourism in particular locations, to a more systematic and analytical approach (Hall and Page 1999). This is reflected in the recognition of how a spatial perspective can contribute to the analysis of tourism. How have geographers viewed tourist transport?

According to Mansfeld (1992 : 58) 'the use of various modes of transport in getting to the destination has several spatial consequences' associated with the distance travelled, amount of time involved in travelling and the mode of transport used. In many cases, geographers have not distinguished, or found it easy to distinguish between tourist and recreational travel. Pearce (1990 : 28) acknowledges that advances in transport technology have altered the patterns of tourist flows and made tourist travel more flexible and diffuse. Prior to the expansion of car ownership and mass air travel, the patterns of tourist travel were linear. It was constrained and confined to transport corridors (e.g. river bank railway lines) or the destinations served by sea. As the following discussion of the car shows, it has been a major catalyst in making patterns of travel more diffuse.

The car and tourist travel

In the postwar period, the growth of car ownership has not only made the impact of recreational and tourist travel more flexible, it has caused overuse at accessible sites. This ease of access, fuelled by a growth in road building and the upgrading of minor roads in many developed countries has been a self-reinforcing process leading to overuse and a greater dominance in passive recreational activities. Probably the most influential study of car usage among recreationalists was Wall's (1972) study of Kingston-upon-Hull in 1969. Wall identified the two principal types of studies used to analyse recreational activity – namely site studies (of particular facilities or areas) and national studies such as the widely cited Pilot National Recreation Survey (British Travel Association and University of Keele 1967, 1969). Wall supplemented data from the national survey with a regional sample of 500 Hull car-owners in 1969. While the results are now very dated, his study highlighted the importance of seasonality and timing of pleasure trips by car and the dominance of the car as a mode of transport for urban dwellers. It also highlighted the role of the journey by car as a form of recreation in itself as well as the importance of the car as more than just a means of transport. In fact the study has a degree of similarity with other studies that followed, such as Coppock and Duffield's (1975) survey of recreation and the countryside. This focused on patterns and processes of recreational activity in Scotland and mapped and described patterns of recreational travel, especially those of caravanners. At the same time, the study noted the tendency for many recreationalists not to venture far from their car at the destination, a point reiterated by Glyptis's (1981) innovative and seminal study of recreationalists using participant observation in Hull. Wall (1971) also found that the majority of pleasure trips were day trips less than 100 km away from Hull, being spatially concentrated in a limited number of resorts along the Yorkshire coast and southerly part of the region. Other research (Wager 1967) highlighted the versatility of the car and its use to venture into the reaches of the North York Moors National Park. Other researchers' findings (e.g. Burton 1966) explained the attraction of the car to recreationalists in that it allowed them to enjoy the countryside and observe its visual characteristics rather than have physical contact with it.

More recent research by Eaton and Holding (1996) identified the growing scale of such visits to the countryside by recreationalists in cars. In 1991, 103 million visits were made to National Parks in the UK (Countryside Commission 1992), the most popular being the Lake District and Peak District parks. In terms of car usage, car traffic was estimated to grow by 267 per cent by the year 2025 from the levels current in 1992 (Countryside Commission 1992). In fact, the greatest impact of rising car usage has been seen in the decline in public transport usage for recreational trips. Yet many National Parks seem unlikely to be able to cope with the levels of usage predicted for the year 2025, given their urban catchments and the relative accessibility by motorway and A roads in the UK. Eaton and Holding (1996) review the absence of effective policies to meet the practical problems of congestion facing many sites in the countryside in Britain. While schemes such as the Sherpa Bus routes in Snowdonia National Park have been introduced (Mulligan 1979) they have not been incorporated into any wider policy objectives for transport. This combines with a failure to design public transport to suit the needs and perception of users to achieve reductions and solutions to congestion in National Parks. Thus it is clear that the car poses a major problem, not only for urban areas and their use by commuters and recreationalists, but also in terms of the sheer growth in volume in areas not designed for large numbers of car-users. This problem is especially acute when spatially concentrated at 'honeypots' (locations which attract large numbers in a confined area) in National Parks.

The UK Tourism Society's response to the Government Task Force on Tourism and the Environment (English Tourist Board/Employment Department 1991) highlighted the impact of the car by commenting that

> no analysis of the relationship between tourism and the environment can ignore transportation. Tourism is inconceivable without it. Throughout Europe some 40 per cent of leisure time away from home is spent travelling, and the vast majority of this is by car . . . Approaching 30 per cent of the UK's energy requirements go on transportation . . . [and] . . . the impact of traffic congestion, noise and air pollution . . . [will] . . . diminish the quality of the experience for visitors.

Various management solutions exist, such as encouraging urban areas to develop 'intervening opportunities' in the urban fringe through networks of Country Parks in the UK (see Harrison 1991) which are capable of absorbing large numbers of visitors seeking a countryside experience. Such parks are able to fulfil the need for local recreational demand and to reduce the impact of transport on other sites. However, car ownership in the UK developed rather later than in North America, and for this reason it is pertinent to focus on one region in North America where it has had a major influence on recreational and tourist travel – the Great Lakes area.

Chubb's (1989) study of the Great Lakes region of North America also highlighted the role of the car in cross-border travel. The region, which contains the world's largest complex of fresh water, also contains a diversity of recreational and tourism resources, including lakes, forests, large park areas, cottages and resort complexes (Figure 2.4). Recreational cottages line many of

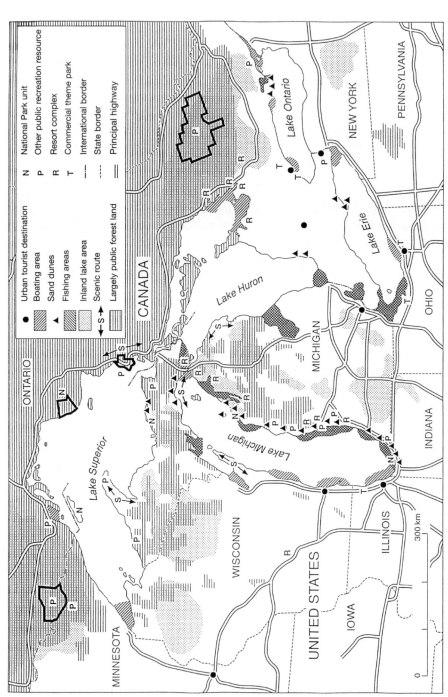

Figure 2.4 The Great Lakes region in North America (redrawn from Chubb 1989)

Legend:
- • Urban tourist destination
- ▨ Boating area
- ▲ Sand dunes
- ▥ Fishing areas
- Inland lake area
- S→ Scenic route
- Largely public forest land

- N National Park unit
- P Other public recreation resource
- R Resort complex
- T Commercial theme park
- — — International border
- ----- State border
- ═══ Principal highway

Table 2.1   Recreational and tourist flows in the Great Lakes region

Type of activity	Destinations of Ontario residents	Destinations of American residents	Motivations	
			Primary	Secondary
Short-term shopping or entertainment	Niagara Falls New York Buffalo Detroit	Windsor Niagara Falls Ontario Toronto	Novelty Proximity Prices	Exchange rate
Longer-term shopping or entertainment	Minneapolis Chicago Cleveland	Toronto	Goods and services available Novelty	Exchange rate Sales Tax rebate
Visiting friends and relatives	Adjacent areas in USA, especially E. Michigan	Adjacent areas in Ontario, especially S.W. portion	Early history Settlement patterns	Proximity
Power or sailboat cruising	Mackinac City Detroit area Cleveland	Toronto Owen Sound Parry Sound	Proximity to sheltered waters	Marina facilities Shore facilities
Travel to own cottages		Lake Erie and Huron shorelines Lake areas across S. Ontario	Proximity Time of acquisition	Ethnic origin
Long-term fishing excursion	Saginaw Bay Western Michigan Salmon areas	Georgian Bay interior lakes in S.E. and Northern Ontario	Chance of success Fish species Size	Environmental facilities
Automobile touring	Various routes around one or more of the four Great Lakes		Well known Natural Commercial	Facilities

Source: Modified from Chubb (1989).

the lakes as second home development (Coppock, 1977) and other seasonal residences are distributed throughout the region, to which owners drive for weekend and vacation use from May to September. A range of resort areas also exist. The region also experiences a high usage of recreational vehicles such as caravans and camper vans. Table 2.1 outlines a range of tourism flows which complement the recreational travel within the region. For example, the majority of cottage owners who travel into Ontario are from the United States. The cross-border flows of car traffic produce a range of spatial forms which are related to the configuration of the Great Lakes and location of hotels, motels and campsites. As Chubb (1989 : 300) illustrated, a number of straight-line and circuit routes exist which cross the international border between Canada and the USA. The straight-line routes include the highway north from Minneapolis to Thunder Bay; Interstate 75 north through Michigan, along Lake Superior; the Niagara Falls–Toronto–Georgian Bay route. The circuit routes which incorporate a touring component and cross the border include the Lake Ontario route (800 km) which is the shortest; the Lake Erie route (1,000 km) which includes Niagara Falls; and the Lake Superior and Lake Huron routes, each of which is approximately 1,600 km in length. Therefore, it is clear that while the car has an overwhelming influence on the choice of recreation and tourism, it is also used for passive recreation, especially day trips. Likewise, in North America, distance is not so much of an inhibiting factor for recreational and tourist travel as the example of the Great Lakes implies. Distances are greater and the car is a fundamental catalyst in the region's development of a recreational and tourism infrastructure to serve the needs of visitors and recreationalists from within Ontario and from the USA.

Both air transport and car travel have provided new opportunities for more flexible patterns of travel though, as Sealy (1992) suggests, air travel and the expansion of international tourism are largely a nodal transport system dependent upon the airports (the nodes) and the flights (the flows) serving them. In the case of the Mediterranean, Pearce (1987) indicates that the expansion of charter airlines has provided a closer link between the tourism markets and potential destinations, and the increase in the geographical range of charter aircraft (i.e. increased flying time) and reduced costs of air travel have led to an expansion in the scale and distribution of tourism in the Mediterranean.

One further concept which tourism geographers have examined is the patterns of tourist transport – namely the tour. A tour is a tourist-oriented form of travel and Pearce (1987) examines the patterns and flows of tourist traffic in terms of preferred routes of travel. In the case of New Zealand, Forer and Pearce (1984) examine the tour itineraries of coach operators to provide information on the patterns of travel and circuits of coach tours. They found that on the North Island of New Zealand, a series of linear tours existed between the major gateways – Auckland and Wellington, with tourists visiting popular resorts and attractions en route. On the South Island, a more complex series of looped tours existed. Many of these often originated and ended at Christchurch (the second largest gateway) as tourists explored the diverse range of landscapes and scenic locations associated with National Parks such as Mount Cook, Westland and Fiordland. Pearce and Elliot (1983) developed a statistical technique – the Trip Index – to examine the extent to which places visited by tourists were major destinations or just a stopover. The Trip Index was calculated thus:

$$\text{Trip Index} = \frac{\text{Nights spent at the destination}}{\text{Total number of nights spent on the Trip}} \times 100$$

A Trip Index of 100 means that the entire trip was spent at one destination and a value of zero would mean that no overnight stay was made on the entire journey.

Other features which the tourism geographer has examined in terms of tourist transport include the use of the private car and public transport (Halsall 1992). Pearce (1990) also notes the importance of transit services from airports to city centres, as well as tourist use of transport to tour sites in major cities such as the London Underground system. In fact, the Docklands Light Railway which connects Central London to London Docklands has also become a popular form of tourist transport in its own right (Page 1989b, c; Page 1995a). Thus, geographers have undertaken research on tourist transport systems at different spatial scales in 'terms of mode, routes and types of operation (e.g. scheduled/non-scheduled)' (Pearce 1990 : 29) and the spatial patterns, processes and networks which facilitate tourist travel, thereby making destinations more accessible. To illustrate how geographers can produce a synthesis of different disciplinary perspectives (e.g. environmental science) and the application to tourist transport, a case study follows on recreational boating. The case study highlights how the coexistence of recreational and tourist use of a particular form of tourist transport – boating – can generate a range of problems which require solutions that can be addressed from a management perspective (which is discussed later in the chapter).

CASE STUDY   Tourist and recreational boating on the Norfolk Broads

The Norfolk Broads (hereafter the Broads) is a wetland region in East Anglia (Figure 2.5) which was created in the medieval period through a series of flooded peat diggings. The region comprises a wetland area focused on a number of rivers such as the Bure, the Yare and the Waveney and their tributaries in the eastern part of Norfolk and northern part of Suffolk. Although the region, with 200 km of tidal rivers and 3,640 ha of water space, is not large compared to the previous example of the Great Lakes region in North America, it is an area of natural beauty used for intensive recreational boating. It was identified as a potential candidate for National Park status but was not included under the original designations. In 1989 the area was accorded virtual National Park status when the Broads Authority was established by the 1988 Norfolk and Suffolk Broads Act. This granted the Broads Authority the same autonomy as a National Park in terms of finance, policy and administration and thus it receives a 75 per cent grant from central government (Glyptis 1991). Recreational boating dates back to the 1870s when the early wherries (local sailing vessels which carried cargo) began to carry passengers for pleasure. In the 1880s, John Loyne pioneered the hire boat industry and in 1908 the present H. Blake and Co. was established to rent purpose-built vessels to visitors who travelled to the area by rail. Thus, the use of transport for recreational day use and for much longer holiday use led to the development of a particular form of recreation and tourism which was water-based and transport-dependent. In 1995, boat companies owned 1,481 motor cruisers and launches which were hired to approximately 200,000 visitors a year, although the number of boats hired has fluctuated during the 1980s and 1990s. A number of environmental concerns have developed over the last 30 years related to their use, the two most serious being:

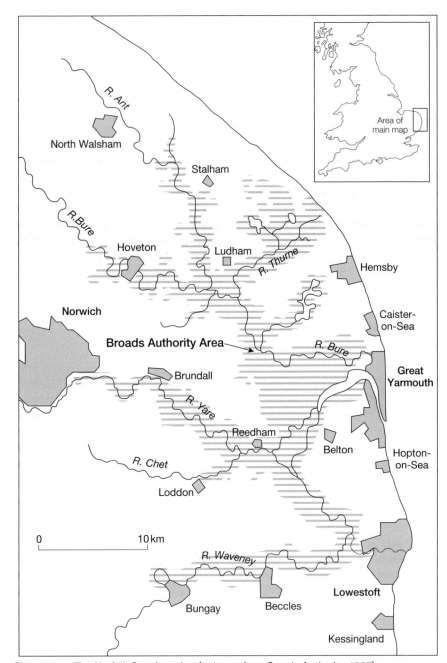

Figure 2.5   The Norfolk Broads region (redrawn from Broads Authority 1997)

- erosion of river banks and the destruction of the protective reed fingers, which have implications for all river bank users and farmers
- enrichment of the nutrient content of the water with nitrates and phosphates. This has contributed to the process known as eutrophication, which leads to an excessive growth in a limited number of plant species and the extinction of the diverse range of flora and fauna.

The Norfolk Broads is a recreational resource under pressure. Given the region's proximity to the urban centres in South East England, the Midlands and towns in East Anglia, access to this series of interconnected linear parks has undoubtedly contributed to these pressures. In essence, there are two competing arguments for and against the continued growth and development of recreational boating in the Broads. On the one hand, there is the physical environmental debate which emphasises the damage motor boats cause to the local ecosystem through their wash turbidity, as their propellers cause a suction effect on the river bed. This also impacts upon reed beds and causes abrasion on the river banks as well as polluting the water through petrol and oil emissions. However, boats alone are not the main cause of environmental damage as agricultural fertilisers and domestic sewage are also a major contributor to over-enrichment and species extinction problems. Anglers may also be equally to blame for damage to river banks. Yet the problem with the hire boat industry is that it is spatially concentrated (see Table 2.2) and heavy usage occurs at weekends in the northern parts of the river system (e.g. Thurne Mouth and the middle reaches of the river Bure at Horning and Wroxham/Hoveton and to a lesser degree the lower reaches of the river Ant). This places intense pressure upon attractions within that region, especially during the very short summer season. For example, a Broads Authority boat movement census in 1994 highlighted that on one peak Sunday in August, there were 6,296 craft movements recorded at 14 census points. In contrast, the upper reaches of many rivers are protected by their relative inaccessibility and in a few cases, by low bridges.

In contrast to the environmental arguments, there are powerful economic arguments for continuing to promote the development of recreational boating. The hire boat industry is estimated to contribute £25 million to the local economy and generates approximately £9,000 for the economy for each boat rented (Broads Authority 1997). In 1991, a survey of boatyards by the Broads Authority found that in the 105 yards responding, boating employed 1,622 people: 884 full-time, 148 part-time and 590 on a seasonal basis. Furthermore, recreational and tourist spending indirectly contributes to 5,000–5,500 jobs in the hospitality sector, local tourist attractions as well as in the local marine industry. Even though boating has been in recession within the area in the 1990s, boatyard diversification to develop facilities for the needs of visitors (e.g. the construction of accommodation) has also posed planning problems for the Broads Authority since such diversification changes the traditional image of the area.

Given these two competing arguments over recreational boating, it is pertinent to ask who is responsible for decision making and planning for boating in this region. Much of the decision making is fragmented across a number of public and private sector bodies, each with particular powers, responsibilities and visions relating to boating. These include:

- National government working through the Countryside Commission has debated the issue of the most appropriate administrative structure for the region, which led to the Broads Authority being established. Yet other government departments such as the Ministry for Agriculture, Food and Fisheries has a very different view of the region's function from the Countryside Commission, which is not recreational or conservationist in focus.
- The Broads Authority can regulate planning proposals for new boatyards and extensions to existing ones. It can also develop new waterside facilities such as public moorings in accordance with its overall management plan (Broads Authority 1997).

Table 2.2   Hire craft operators on the Broads

Location	Number of operators	Motor cruises	Day launches	Sailing cruisers or day craft
River Waveney				
Burgh St Peter	1	–	2	–
Burgh Castle	1	–	1	–
St Olaves	2	9	2	–
Somerleyton	1	8	–	15
Beccles	4	42	17	–
Oulton Broad	5	23	16	–
Total	14	82	38	15
River Chet				
Loddon	5	46	3	–
River Yare				
Reedham	2	24	–	–
Brundall	12	190	2	–
Thorpe	5	113	7	–
Total	19	327	9	0
River Bure				
Acle	4	86	10	–
S. Walsham	1	9	2	–
Upton	1	–	–	16
Horning	7	128	32	19
Wroxham/Hoveton	14	209	124	13
Belaugh	1	74	–	–
Total	28	506	168	48
River Thurne				
Ludham	2	11	–	17
Potter Heigham	3	124	69	1
Martham	2	19	7	29
Hickling	1	9	6	2
Total	8	163	82	49
River Ant				
Ludham Bridge	1	–	7	1
Sutton	1	12	–	–
Stalham	4	357	6	2
Wayford	2	6	9	–
Total	8	375	22	3

Source: Broads Authority (1997) based on the Broads Authority *Register of Boats 1995*.

● The Great Yarmouth Port and Haven Commissioners are responsible for all navigation on the Broads and rivers. They license boats and have imposed speed restrictions, enforced by river inspectors. They can also regulate river traffic charges.
● The East Anglian Tourist Board and its parent body, the English Tourist Board, can promote the region to visitors together with local districts. Each body has a degree of choice over the

image it chooses to portray and the markets it targets, and could even advocate alternative tourist activities in the region.

- The hire boat companies and their marketing agencies have undertaken publicity campaigns
to try and influence the behaviour of visitors and to encourage a greater consciousness of
the natural environment. Boat companies could restrict the number of vessels hired at
particular places, and at certain times if they felt it was in their long-term interest.

- A variety of other bodies have responsibilities, such as the regional water authority (Anglian
Water), while the National Rivers Authority has powers over water quality and land drain-
age. There are also a large number of pressure groups which have a strong lobby function
such as the National Farmers Union, the Nature Conservancy Council and the local Broads
Society. Although no one group has direct influence over the hire boat industry, they can
exert pressure on decision makers in other bodies to influence planning outcomes.

This case illustrates that it is not so much the volume of boats but rather the
intensity of use in time and space as well as the behaviour of users which conflicts with
the natural environment. For this reason, it is worthwhile focusing on the possible
solutions to the recreational problems which this form of transport poses. There is no
all-embracing solution which will deal with the conflicts. Coexistence with the fragile
ecosystem is therefore a major challenge for recreation management. Environmental
deterioration is not a problem that can be solved quickly, but the Broads Authority
has a management plan for the area which is viewed as a long-term vision for the
Broads. It is a workable strategy and takes into account the fact that although a
general decline in recreational boating holidays occurred in the late 1980s and early
1990s, demand is likely to develop again in the new millenium. Furthermore, the
1981 Wildlife and Countryside Act has also highlighted the problem of reconciling
agricultural and commercial practices with environmental conservation. Since any
form of management for the Broads needs to be accompanied by objectives which
inform the strategic direction, sustainability is a prime concern for the region (Broads
Authority 1997). The Broads Authority has developed a number of policies to address
the issue of coexistence of the recreational boating industry with the authority's
underlying objective, which is to foster sustainable management and development.
Each of these policies not only requires coordination of the various agencies involved
with the hire boat industry, but also an ongoing dialogue with all parties affected by
the Broads management plan. By fostering an ongoing partnership between public
and private sector groups, the management plan will remain a workable strategy for
the future and ensure the boating industry is an integral but managed element of
tourism and recreational activities available in the region. The case study also illustrates
that there is a growing concern over management of the impacts of recreational
boating and that sustainable policies and practices are developed in relation to recreational and tourism transport systems.

Marketing and tourist transport

When one considers the contribution of economics and geography to the
analysis of tourist transport, it is evident that a wide range of research papers
and books exist. In the case of marketing, the number of studies focused on

tourist transport is limited because there is a tendency for marketing to be more visible and results-oriented than based on the academic analysis of good practice. Promotional material and advertising campaigns constitute a major investment in time, money and creative thinking, with companies reluctant to highlight good practice which might undermine their future business potential. According to Horner and Swarbrooke (1996) the diversity of transport modes used in tourism (i.e. air, rail, water, road and off-road transport) makes it difficult to generalise about transport and marketing. This is further complicated by the private, public and voluntary sector organisations involved in transport for tourism which all have different marketing objectives. In addition, a conceptual problem exists: sometimes transport may be sold as a self-contained product, sometimes as part of a large composite product (e.g. an inclusive tour). In order to understand how marketing is integral to the analysis of tourist transport, it is therefore useful to examine the principles used in marketing and the activities undertaken by marketers.

What is marketing?

According to Kotler and Armstrong (1991), marketing is a process whereby individuals and groups obtain the type of products or goods they value. These goods are created and exchanged through a social and managerial process which requires a detailed understanding of consumers and their wants and desires so that the product or service is effectively delivered to the client or purchaser. Within tourism studies, there has been a growing interest in marketing (e.g. Middleton 1988; Jefferson and Lickorish 1991; Laws 1991; Lumsdon 1997a; Horner and Swarbrooke 1996; Seaton and Bennett 1996) compared with transport studies, which has tended to employ marketers when required to deal with such issues. More recently, new marketing-for-tourism texts have begun to incorporate material on marketing transport services, although this usually tends to focus on airlines (Seaton and Bennett 1996; Horner and Swarbrooke 1996). In transport studies, marketing has assumed less importance than operational and organisational issues, but there is a growing awareness that 'transport operators are seeking to augment their basic product with add-on services that generate more income but also satisfy more of the consumers' needs' (Horner and Swarbrooke 1996 : 319). Gilbert (1989) considers the growth and establishment of marketing within tourism and the critical role of a consumer orientation among transport providers. For example, British Airways explained its financial turnaround from a loss of £544 million in 1981/82 to a profit of £272 million in 1983/84 in terms of a greater marketing orientation based on recognising customer needs and attempting to satisfying them (Gilbert 1989). In this respect, marketing has a fundamental role to play in analysing tourist transport. Within marketing, three key areas exist:

- strategic planning
- marketing research
- the marketing mix

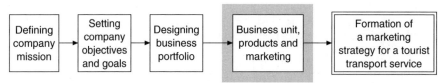

Figure 2.6 Strategic planning for tourist transport (based on Kotler and Armstrong 1991)

Strategic planning

Within any business or company, there is a need to provide some degree of order or structure to its activities and to think ahead. This is essential if companies are to be able to respond to the competitive business environment in which organisations operate. For this reason, a formal planning process is necessary which is known as *strategic planning*. According to Kotler and Armstrong (1991 : 29), strategic planning can be defined as 'the process of developing and maintaining a strategic fit between the organisation's goals and capabilities and its changing marketing opportunities'.

Businesses need to be aware of their position in the wider business environment and how they will respond to competition and new business opportunities within an organised framework. To illustrate how strategic planning operates and its significance to tourist transport, it is useful to focus on the structured approach devised by Kotler and Armstrong (1991). As Figure 2.6 shows, the first stage is the definition of an organisation's purpose, which requires a company to consider:

- What business is it in?
- Who are its customers?
- What services do its customers require?

Following the definition of purpose, a company may incorporate these principles into a *mission statement* (see David 1989). This provides a focus for the company activities, as can be seen in the case of BR's *Passenger's Charter*, which identified the following aims as part of its mission for customer service:

- A safe, punctual and reliable train service
- Clean stations and clean trains
- Friendly and efficient service
- Clear and up-to-date information
- A fair and satisfactory response if things go wrong

Although a mission statement may be used for public relations purposes, as in the case of BR, it is necessary to set objectives and goals. For example, the BR *Passenger's Charter* required the organisation to consider:

- What were its business objectives?
- Was it seeking to improve its market share of tourist travel by rail?
- Was the overriding business concern to improve the financial turnover and profitability?

- If so, what marketing objectives would need to be set to achieve these goals?
- Was the overall marketing objective to improve the public image of BR and to emphasise customer care and service standards?

Within this context, BR had to prepare a marketing strategy which acknowledged its business and marketing objectives and identified the resource requirements to achieve internal targets (e.g. achieving a greater market share of tourist travel) as well as the implications for research, sales and promotion and how this translates into an overall benefit for BR in a given timescale, such as a one-year or five-year period.

Obviously, the *Passenger's Charter* was only one aspect of strategic planning for BR's passenger services in 1992, but it does emphasise how important marketing and strategic planning are in launching such a new business initiative. The next stage following the setting of objectives and goals is termed the *business portfolio*. Here the company analyses its own products or services in terms of its own business expertise and how competitors' products and services may affect them. This is frequently undertaken as a SWOT analysis, which considers:

- the **S**trengths
- the **W**eaknesses
- the **O**pportunities
- the **T**hreats

of its products and services in the business environment.

For those operators who may wish to develop a strategy which incorporates an element of organisational growth and expansion, a number of options exist. As Horner and Swarbrooke (1996 : 325) show, these can be divided into:

- *Marketing consortia*, where a group of operators cooperate to create and develop a product such as the Euro Domino joint railway ticket, permitting rail travel on different European railways.
- *Strategic alliances*, where different businesses agree to cooperate in various ways. This has varied by sector in the tourism industry, but may involve marketing agreements or technical cooperation. However, in the airline industry, Gallacher and Odell (1994) observed that 280 alliances existed among 136 airlines, 60 per cent of which were formed after 1992. Hanlon (1996 : 201) outlined the nature of such alliances among airline companies, which included joint sales and marketing; joint purchasing and insurance; joint passenger and cargo flights; code sharing; block spacing; links between frequent flyer programmes; management contracts; and joint ventures in catering, ground handling and aircraft maintenance. The strategic value of such alliances is reconsidered in Chapter 6, as they can result in 'horizontal alliances . . . [involving] . . . firms selling the same product . . . vertical alliances . . . [being] . . . those with suppliers, distributors or buyers. And external alliances . . . with potential entrants or with the producers . . . in other industries' (Hanlon 1996 : 204). The main concern for governments, as Chapter 3 will show, is that the transport industry may be overconcentrated among a limited number of companies or consortia that could eventually lead to market domination and higher prices. For example, Youssef and Hansen's (1994) study of the Swissair and SAS alliance

Table 2.3 British Airways franchise airline agreements, 1997

Airline	Number of destinations served	Airline base	Date franchise began
British Mediterranean	6	Heathrow Airport	February 1997
British Regional Airlines	37	Isle of Man	January 1995
Brymon Airways[1]	13	Plymouth and Bristol Airports	August 1993
City Flyer Express	14	Gatwick Airport	July 1993
Cornair	12	South Africa	October 1996
GB Airways	17	Gatwick Airport	February 1995
Loganair	16	Glasgow Airport	July 1994
Maersk Air	9	Birmingham Airport	August 1993
Sun Air	9	Billund Airport	August 1996

[1] Brymon Airways was acquired by British Airways in August 1993.
Source: Modified from K. O'Toole (1998) 'Widening the Franchise', *Flight International* 25–31 March, 1998, p. 35.

indicated that on the main hub routes – Copenhagen, Stockholm and Oslo to Geneva – competition was severely reduced. Consequently, the alliance had yielded higher profits for the airlines on the hub-to-hub routes.

- *Acquisition*, which is the purchase of equity in other operations such as the Scandinavian company Stena's acquisition of European ferry operations, especially in the UK.
- *Joint ventures*, where operators seek to create new carriers.
- *Franchising*, where major operators use their market presence and brand image to extend their influence further by licensing franchisees to operate routes using their corporate logo and codes (e.g. the British Airways agreement with Maersk, Air UK, Manx Airlines and Loganair); see Table 2.3.
- *Ancillary activities*, the development of which adds value to the operator or organisation's core business. For example, many European ferry operators now offer inclusive tour operations (e.g. P&O European Ferries).

Marketing research

This process is one which is often seen as synonymous with market research (Brunt 1996) but as the following definition by Seibert (1973) implies, in reality it is a much broader concept as 'marketing research is an organised process associated with the gathering, processing, analysis, storage and dissemination of information to facilitate and improve decision-making'. It incorporates various forms of research undertaken by organisations to understand their customers, markets and business efficiency. The actual research methods used to investigate different aspects of a company's business ultimately determine the type

Table 2.4   Categories of marketing research

Research category	Used in	Typical marketing use
1. Market analysis and forecasting	Marketing planning	Measurement and projections of market volumes, shares and revenue by relevant categories of market segments and product types
2. Consumer research	Segmentation and positioning	(a) Quantitative measurement of consumer profiles, awareness attitudes and purchasing behaviour including consumer audits (b) Qualitative assessments of consumer needs, perceptions and aspirations
3. Products and price studies	Product formulation, presentation and pricing	Measurement and consumer testing of amended and new product formulations, and price sensitivity studies
4. Promotions and sales research	Efficiency of communications	Measurement of consumer reaction to alternative concepts and media usage; response to various forms of sales promotion, and sales force effectiveness
5. Distribution research	Efficiency of distribution network	Distributor awareness of products, stocking and display of brochures, and effectiveness of merchandising, including retail audits and occupancy studies
6. Evaluation and performance monitoring studies	Overall control of marketing results	Measurement of customer satisfaction overall, and by product elements, including measurement through marketing tests and experiments

Source: Middleton (1988 : 109).

of research undertaken. The main types of research can be summarised into six categories (see Table 2.4). A number of good introductions to marketing research are available and more recent books on tourism research are recommended as preliminary reading on this topic (e.g. Witt and Moutinho 1989; Veal 1992). Marketing research allows the company to keep in touch with its customers to monitor needs and tastes which are constantly changing in time and space. However, the actual implementation of marketing for tourist transport ultimately depends on the 'marketing mix'.

The marketing mix

The marketing mix is 'the mixture of controllable marketing variables that the firm [or company] uses to pursue the sought level of sales in the target market' (Kotler cited in Holloway and Plant 1988 : 48). This means that for a given tourist transport organisation such as an airline, there are four main marketing variables which it needs to harness to achieve the goals identified in the marketing strategy formulated through the strategic planning process. These variables are:

- *Product formulation* – the ability of a company to adapt to the needs of its customers in terms of the services it provides. These are constantly being adapted to changes in consumer markets.
- *Price* – the economic concept used to adjust the supply of a service to meet the demand, taking account of sales targets and turnover.
- *Promotion* – the manner in which a company seeks to improve customers' knowledge of the services it sells so that those people who are made aware may be turned into actual purchasers. To achieve promotional aims, advertising, public relations, sales and brochure production functions are undertaken. Not surprisingly, promotion often consumes the largest proportion of marketing budgets. For transport operators, the timetable is widely used as a communication tool, while brochures and information leaflets are produced to publicise products. Some of the other promotional tools used to increase sales include promotional fares, frequent flyer programmes (Beaver 1996) and 'piggy-back promotions . . . where purchasing one type of product gives consumers an opportunity to enjoy a special deal in relation to another product' (Horner and Swarbrooke 1996 : 322) such as the Sainsbury–British Airways promotion in the early 1990s.
- *Place* – the location at which prospective customers may be induced to purchase a service – the point of sale (e.g. a travel agent).

As marketing variables, production, price, promotion and place are normally called the four P's. These are incorporated into the marketing process in relation to the known competition and the impact of market conditions. Thus, the marketing process involves the continuous evaluation of how a business operates internally and externally and it can be summarised as 'the management process which identifies, anticipates and supplies customers' [see the section of this chapter on management studies] requirements efficiently and profitably' (UK Institute of Marketing, cited in Cannon 1989). To illustrate the importance of marketing, particularly promotion, the following case study of Queensland Rail in Australia emphasises how a product in decline can be revitalised and repackaged for the tourist.

CASE STUDY   Marketing tourist transport – Queensland Rail

In the state of Queensland in Australia, the rail services were state-owned in 1997. The railway company Queensland Rail has sought to inject a new lease of life into its loss-making, long-distance passenger services by repackaging its services to meet a niche demand among tourists. Research by Dann (1994) explained the resurgence of interest in rail travel as a function of nostalgia among tourists, a feature also emphasised in the context of heritage tourism by Swarbrooke (1994). Furthermore, Prideaux (1997) cites the research by Taylor (1983) which identified the speed, comfort, amenities and sociability of train travel as appealing for tourists.

Although long-distance passenger services commenced in 1885 in Queensland, with luxury services being added in the 1930s (e.g. the *Sunshine Express*), followed by air-conditioned carriages in the 1950s (e.g. the *Sunlander* operating between Brisbane and Cairns in 1953), the postwar period saw a decline in rail travel. Road transport, notably the rise of the private car, and the expansion of air travel led to a decline in services. The extensive network of passenger services contracted as routes were reduced, lines closed and services cancelled. Although rail sought to introduce innovations to stem the flow of traffic (e.g. car-carrying wagons), its decline mirrored the future of rail in many other countries up to the 1980s (Kosters 1992). This led to a downward spiral of declining passenger numbers and declining services. In addition, Prideaux (1997) pointed to:

- The slow nature of train travel (e.g. in 1997 travel by air between Brisbane and Cairns took $2\frac{1}{2}$ hours compared to 32 hours by the *Sunlander*).
- Services are infrequent, with the *Sunlander* operating six times a week in 1961 and three times a week in 1997.
- Passenger carriages were designed for longevity and recent fashions and tastes are not reflected in 1950s rolling stock.
- The train is inflexible compared to the car, with its convenience and wide availability.
- Concessionary fares for retirement pensioners have not yielded a great deal of revenue and generated intense competition in the peak season, also impacting upon the image of the railway.
- Lack of investment in new rolling stock and new technology, until the introduction of the *Spirit of Capricorn* in 1989 and upgraded track in the 1990s. The introduction of a tilt-technology on the *Spirit of Capricorn* in 1997 (similar to the ill-fated British Rail Advanced Passenger Train in the 1970s and 1980s) will reduce the journey time for Brisbane to Rockhampton from 9 hours 25 minutes to 7 hours.
- An absence of promotion due to public sector principles that did not emphasise commercial practices until the late 1990s.
- A perceived lack of reliability in keeping to the published schedules.
- Rail travel was expensive and not price-competitive with air travel.
- Competition from long-distance coaches, following the deregulation of the sector in 1985. Price and departure flexibility among coach operators combined with mass media advertising and their attractive multi-sector price structures for tourist travel, severely dented rail's market share.

While Queensland Rail sought to address some of these issues in the 1990s, the loss of traditional clientele meant the search for new markets after the state government's corporatisation of Queensland Rail in 1995. This followed the reorganisation of long-distance rail services in the late 1980s into a new group – Traveltrain. Tourism was seen as an area for expansion, with two services – Sunshine Rail Experience and Kuranda Scenic Railway, which departs from Cairns and traverses scenic gorges and travels through virgin rainforest and the jungle village of Kuranda. By 1994, 500,000 passengers a year were traveling on the Kuranda Scenic Railway, making it Australia's sixteenth-largest tourist attraction that same year, which exceeded visits to the Great Barrier Reef.

Following the success of the *Queenslander* tourist train in 1986, the *Spirit of the Outback* and the *Spirit of the Tropics* were introduced in 1993 to help offset losses in other long-distance services. In 1995, the *Savannahlander* was brought in and subsequently won tourism awards. The extent of services in 1996 is outlined in Table 2.5 which is to be complemented by the introduction of the Great South Pacific Express in 1998, a joint venture with the Venice–Simplon Orient Express, estimated to cost A\$34 million (Prideaux 1997). But how has Queensland Rail been able to turn an ailing railway company into a successful tourist venture?

One of the tools used by Queensland Rail was the development of promotional themes to sell individual services, where a number of market segments were carefully nurtured. According to Prideaux (1997), these included:

- Long-distance rail services (e.g. the *Sunlander*) using traditional routes developed in the nineteenth century with free travel for pensioner concessionaires, which contributed to revenue losses, have been upgraded and new traffic sought since pensioner concessions could be curtailed in a climate of public sector retrenchment. In 1995, the *Sunlander* underwent a A\$2.8 million refit with themed carriages and public entertainment areas. This was accompanied by a reimaging of these services to convey a restful travel experience full of discovery and attractions.
- Luxury on wheels, such as the *Queenslander* tourist train introduced in 1986 which will be complemented by the Orient Express train in 1998.
- Nostalgia, where the farmer *Midlander* service from Rockhampton to Winton was relaunched as the *Spirit of the Outback* in 1993. An elimination of pensioner concessions on this service combined with a strong appeal to nostalgia characterised the marketing of this service.
- The youth-budget market, seeking to tap Queensland's growing backpacker market which was estimated at 250,000 travellers in 1993. In 1994, the *Spirit of the Tropics* with its onboard disco was introduced on the Brisbane to Cairns route.
- Trains as tourist attractions as epitomised by the Kuranda Scenic Railway, with the *Gulflander* and *Savannahlander* also marketed in the same way. These utilise a range of heritage themes (e.g. mineral mining) and a strong emphasis on nostalgia.
- Rail-touring, using daytime travel along the Queensland coast between Brisbane and Cairns and overnight stays in motels along the lines of the 1962 Sunshine Rail Experience.

Following corporatisation, Queensland Rail only received state subsidies to operate services the state deemed were important from a public welfare perspective (see Chapter 3 on government policy). The move towards profitability was underwritten by a 10-year A\$4 billion capital works programme to assist with the corporatisation (O'Rourke 1996). Long-distance passenger services remain unviable, with a large network and high costs of operation. The *Sunlander* contributes 34 per cent of Traveltrain's gross revenue, followed by the Kuranda Scenic Rail service with 19 per cent. However, tourism ventures have not noticeably reduced operational losses on long-distance rail services due to what Prideaux (1997) considers to be Queensland Rail's inability to market its services. Until recently travel sales could only be made through railway offices and not through travel agents. However, recent alliances with tour wholesalers such as American Express and Jetset Tours are a major step forward.

Table 2.5 Queensland Rail Traveltrains 1996

Name	Date of introduction	Length of service (km)	Time of journey	Cost (A$) 1996		Facilities	Services per week	Types of service
				Economy	Sleeper			
The Queenslander (Brisbane/Cairns)	1986	1,681	31 hours	N/A	489	Restaurant, lounge car, roomettes	1	Tourist
Sunlander (Brisbane/Cairns)	1953	1,681	32 hours	159	243	Dining car, club car	3	Tourist/ scheduled passengers
Kuranda Scenic Railway (Cairns/Kuranda)	1891	35	1.3 hours	23	N/A	Commentary	26	Tourist
Spirit of the Tropics (Brisbane/Cairns)	1994	1,681	32 hours	135	129	Disco, dining car	2	Tourist/youth
Inlander (Townsville/Cairns)	1954	977	19 hours	125	192	Dining car	2	Passenger
Westlander (Brisbane/Charleville)	1954	777	16 hours	107	172	Club car, dining car	2	Passenger
Spirit of the Outback (Brisbane/Longreach)	1993	1,325	24 hours	158	226	Club car, dining car	2	Tourist
Savannahlander (Mt Surprise/Forsayth)	1994	121	5 hours	35	N/A	Commentary	2	Tourist
Gulflander (Normanton/Croydon)	1891	152	4 hours	35	N/A	Commentary	2	Tourist
Spirit of Capricorn Brisbane/Rockhampton	1989	639	9.5 hours	67	N/A	Light refreshments	9	Commuter
Sunshine Rail Experience	1962	1,681	6 days	1,199	N/A	Club car	19 per year	Rail tour

Source: Prideaux (1997).

Prideaux (1997) also points to the desire of Queensland Rail to reposition itself in the tourism sector, although its pricing structure continues to present a problem. For example, in 1995/96 only marginally more than 20 per cent of passengers on long-distance services paid the full fare (excluding the Kuranda service), with 50 per cent being pensioners paying a nominal charge or travelling free. Services often fail to reach 60 per cent cost recovery on operations, with some generating less than 25 per cent. In contrast, in New Zealand, organisations like Tranz Rail have developed successful tourist trains and perhaps such organisations may offer lessons for Queensland Rail with its privatised operation.

While Queensland Rail is still evolving its marketing of its services, opportunities to expand this role with packaging rail tours like Tranz Rail in New Zealand offer a new direction. The sale of a rail experience, moving Queensland Rail from a promotional to a more sales-oriented role will require new marketing strategies to enhance the marketing effect. Continued service improvements on-board trains combined with a higher degree of cost-recovery need to be a financial target for Queensland Rail. This can be achieved by increasing the profitability of its existing and forecast markets (e.g. tourists) and by developing a more sophisticated marketing philosophy, perhaps with a stronger emphasis on service quality.

As the case study of Queensland Rail has indicated, marketing has a certain degree of synergy with economics as an understanding of economic concepts and the way in which the marketplace works is fundamental to marketing. But the application of marketing principles to the tourist transport system requires one to recognise that one is dealing with a service (see Kotler and Armstrong 1991) and the tourist experience which is embodied in the concept of the service encounter.

Tourist transport as a service

Gilbert (1989) noted the growing importance of service quality and consumer satisfaction in tourism in the late 1980s. In the context of tourist transport, what is meant by a service? Defining the term 'service' is difficult due to the intangibility, perishability and inseparability of services. Kotler and Armstrong (1991 : 620) define these three terms as follows:

- *Service intangibility* – a service is something which cannot be seen, tasted, felt, heard or smelt before it is purchased.
- *Service perishability* – a service cannot be stored for sale or use at a later date.
- *Service inseparability* – a service is usually produced and consumed at the same time and cannot be separated from providers.

Van Dierdonck (1992) argues that the intangible nature of a service is determined by the fact that, unlike manufactured goods, a service is provided and consumed at the same time and same place, making it difficult to define and communicate its form to customers. Even so, it is possible to identify six core elements in a service if it is defined as a product, where each element affects customer perception of the service:

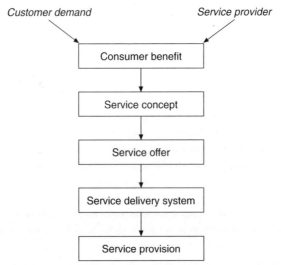

Figure 2.7   The service process (after Cowell 1986)

- the image of the service
- the image of personnel with whom customers interact
- image differences within the same sector as the service provider (e.g. how a service compares with those offered by its competitors)
- the customer group targeted
- the influence of the physical environment in which the service is delivered (e.g. the building)
- the working atmosphere in which the service is formulated, designed and delivered *Source*: Modified from Flipo, cited in Van Dierdonck (1992).

An alternative view of a service is that it constitutes a process rather than an end product, which actually disappears once it has been made. In this respect, a service can be conceptualised as a process which responds to the diverse needs of consumers. Since consumers are not homogeneous it is difficult to standardise a product to meet every need. The process of providing a service which tailors something to meet precise and varied needs is integral to the concept of responsive service provision.

Researchers such as Poon (1989) have argued that the challenge for tourism-related enterprises, such as transport providers, is to respond to the growing sophistication of tourists so that the entire tourism experience meets the expectations of the consumer. In this context, three key issues which emerge are service quality, customer care and the service encounter (Gilbert 1989). From the consumer's perspective, the use of critical incidents (Bitner et al 1990) in the service process (e.g. where the service delivery breaks down) has been used to analyse the consumer's service encounter and how they view it under adverse conditions. If we accept that there is a consensus among marketing researchers that service provision needs to be seen as an ongoing process, how does this process operate? Cowell (1986) examined this process as a four-stage system (see Figure 2.7) with the provider trying to offer a service in response to actual or perceived customer demand. The process is based on the following concepts: consumer benefit, the service concept, the service offer and the service delivery system.

The consumer benefit

At the outset, the supplier of a service tries to understand what the consumer wants and how they may benefit from the service. At this stage, a detailed understanding of consumer behaviour is required which recognises the relative importance of the factors influencing the purchase decision (Qaiters and Bergiel 1989). These include social, economic, cultural, business and family influences and how they condition and affect the attitudes, motives, needs and perception of consumers. In the case of tourist transport services, a significant amount of research on the social psychology of tourists has examined what holidays tourists choose, the mode of transport selected and the factors affecting their decision making as consumers (Javalgi et al 1992; Mansfeld 1992). Following the consumer benefit stage, the service provider translates the assessment of consumer demands into a service concept.

The service concept

At this point, the supplier examines the means of producing a service and how it will be distributed to consumers. Marketing research at both the consumer benefit and service concept stage is essential to assist in identifying the specific market segment to target and the nature of the consumer/producer relationship (e.g. is the service to be sold direct to the public or via a different distribution channel such as a travel agent?). Lastly, the producer identifies and develops the image which is to be associated with the service. Having established what the service will comprise in concept form, it is developed further into the service offer.

The service offer

At this point, the service concept is given more shape and developed within precise terms set by managerial decisions which specify:

- the elements – the ingredients
- the form – how it will be offered to consumers
- the levels of service – what the consumer will expect to receive in terms of the quality and quantity of the service.

The composition of the service elements is discussed in detail by Gronroos (1980). The form of the service concept also needs to be considered in terms of how the corporate image will be communicated to the public. Furthermore the service levels, the technical aspects of service quality and how it is rendered also need to be assessed as part of the service offer. Despite the significance of the service offer, there is little evidence available to suggest that consumers judge service quality in a definitive way. For example, Lovelock (1992a, b) acknowledges that services such as tourist transport, which have a high degree of customer contact, need to be recognised when entering the last stage – service delivery.

The service delivery system

This is the system which is developed to deliver the service to the customer and will comprise both the people responsible for different aspects of the service experience and the physical evidence such as the transport and environment in which it is delivered (Bitner 1992). The tourist's experience of these components is embodied in the *service encounter* (Laws 1991). It is in the service delivery system that barriers may occur in the provision of a satisfactory encounter (see Thornberry and Hennessey 1992) and a great deal of marketing research has been directed to identifying deficiencies, critical incidents and ways of over-coming dissatisfaction in this area (Bitner et al 1990). The pursuit of excellence in service delivery (Peters and Waterman 1982; Berry and Parasuraman 1991) has meant companies monitoring what the consumer wants and then providing it. In this context marketing assumes a critical role both in terms of research and communication with customers. By providing quality in service provision, it may help to develop customer loyalty in the patronage of tourist transport. As the competitive market for tourist transport intensifies, the demand for service delivery systems which are customer-centred is likely to be an important factor in affecting tourist use of transport services. The consumer is a key player in the service process, being an active participant and important judge of quality (Zeithmal and Berry 1985). It is therefore essential to consider how transport providers can integrate many of the perspectives discussed in the chapter so far, to meet customers' needs and to run a tourist transport business successfully. Attention now turns to the contribution which management can make.

Management studies and tourist transport

Having examined the role of other disciplines and their contribution to the analysis of transport, it is useful to consider one area which draws the others together and harnesses their skills. That area is management. For this reason, it is pertinent to examine what management is, its rationale, the principles which guide managers and the factors which can impact upon the need to manage. This is followed by a case study of airline management. In a purely abstract academic context, management is concerned with the ability of indi-viduals to conduct, control, take charge of, or manipulate the world to achieve a desired outcome. In a practical business setting, management occurs in the context of a formal environment – the organisation (Handy 1989). Within organisations (small businesses through to multinational enterprises), people are among the elements which are managed. As a result, Inkson and Kolb (1995 : 6) define management as 'getting things done in organisations through other people'. In a business context, organisations exist as a complex interaction of people, goals and money to create and distribute the goods and services which people and other businesses consume or require.

Organisations are characterised by their ability to work towards a set of com-mon objectives (e.g. the sale of holidays to tourists for a profit). To achieve their objectives, organisations are often organised into specialised groupings to achieve particular functions (e.g. sales, human resources management, accounts and finance) as departments. In addition, a hierarchy usually exists where the organisation is horizontally divided into different levels of authority and status, and a manager often occupies a position in a particular department or division at a specific point in the hierarchy. Within organisations, managers are grouped by level in the organisation from the:

- *Chief Executive Officer* (CEO) or *General Manager* at the top who exercises responsibility over the entire organisation and is accountable to a Board of Directors or other representatives for the ultimate performance of the organisation.
- *Top managers* are one level down from the CEO and their role is usually confined to a specific function, such as marketing or sales. They may act as part of an executive team who work with other top managers and the CEO to provide advice on the relationship between different parts of the organisation and contribute to corporate goals.
- *Middle managers* fill a niche in the middle of the hierarchy with a more specialised role than the top managers. Typically they may head sections or divisions and be responsible for performance in their area. In recent years, corporate restructuring has removed a large number of middle managers to cut costs and placed more responsibility on top managers or the level below – first line managers.
- *First line managers* are the lowest level of manager in an organisation, but arguably perform one of the most critical roles – the supervision of other staff who have non-managerial roles and who affect the day-to-day running of the organisation.

Managers can also be classified according to the function they perform (i.e. the activity for which they are responsible). As a result, three types can be discerned:

- *Functional managers* manage specialised functions such as accounting, research, sales and personnel. These functions may also be split up even further where the organisation is large and there is scope to specialise even further.
- *Business unit, divisional, or area managers* exercise management responsibilities at a general level lower down in an organisation. Their responsibilities may cover a group of products or diverse geographical areas and combine a range of management tasks, requiring the coordination of various functions.
- *Project managers* manage specific projects which are typically short-term undertakings, and may require a team of staff to complete them. This requires the coordination of a range of different functions within a set time frame.

The goals of managers within organisations are usually seen as profit-driven, but as the following list suggests, they are more diverse:

- Profitability, which can be achieved through higher output, better service, attracting new customers and by cost minimisation.
- In the public sector, other goals (e.g. coordination, liaison, raising public awareness and undertaking activities for the wider public good) dominate the agenda in organisations. Yet in many government departments in developed countries, private sector, profit-driven motives and greater accountability for the spending of public funds now feature high on the agenda.
- Efficiency, to reduce expenditure and inputs to a minimum to achieve more cost-effective outputs.
- Effectiveness, achieving the desired outcome; this is not necessarily a profit-driven motive.

In practical terms, however, the main tasks of managers are based on the management process, which aims to achieve these goals. Whilst management theorists differ in the emphasis they place on different aspects of the management process, there are four commonly agreed sets of tasks. McLennan et al (1987) describe these as:

- *Planning,* so that goals are set out and the means of achieving the goals are recognised.
- *Organising,* whereby the work functions are broken down into a series of tasks and linked to some form of structure. These tasks then have to be assigned to individuals.
- *Leading,* which is the method of motivating and influencing staff so that they perform their tasks effectively. This is essential if organisational goals are to be achieved.
- *Controlling,* which is the method by which information is gathered about what has to be done.

Managing requires a comparison of the information gathered with the organisational goals and, if necessary, taking action to correct any deviations from the overall goals.

Fundamental to the management process is the need for managers to make decisions. This is an ongoing process. In terms of the levels of management, CEOs make major decisions which can affect everyone in the organisation, whereas junior managers often have to make many routine and mundane decisions on a daily basis. Yet in each case, decisions made have consequences for the organisation. To make decisions, managers often have to balance the ability to use technical skills within their own particular area with the need to relate to people and to use 'human skills' to interact and manage people within the organisation, and clients, suppliers and other people external to the organisation. Managers also need these skills to communicate effectively to motivate and lead others. They also need cognitive and conceptual skills. Cognitive skills are those which enable managers to formulate solutions to problems. In contrast, conceptual skills are those which allow them to take a broader view, often characterised as 'being able to see the wood for the trees', whereby the manager can understand the organisation's activities, the inter-relationships and goals and can develop an appropriate strategic response (Inkson and Kolb 1995).

In recent years, there has been a growing recognition that to perform a managerial task successfully, a range of competencies are needed. A competency, according to Inkson and Kolb (1995 : 32) is 'an underlying trait of an individual – for example a motive pattern, a skill, a characteristic behaviour, a value, or a set of knowledge – which enables that person to perform successfully in his or her job'. The main motivation for organisational interest in competency is the desire to improve management through education and training. As Table 2.6 shows, competencies can be divided into three groups:

- understanding what needs to be done
- getting the job done
- taking people with you

While the concern with competencies questions the traditional planning, organising, leading and control model as a description of the management

Table 2.6   The McBar management competencies model

Competency	Components of the model
Understanding what needs to be done	• Critical reasoning • Strategic visioning • Business know-how
Getting the job done	• Achievement drive • Proactivity • Confidence • Control • Flexibility • Concern for effectiveness • Direction
Taking people with you	• Motivation • Interpersonal skills • Concern for impact • Persuasion • Influence

Source: C. Page et al (1994 : 26).

process, an overriding emphasis on the skills managers need to perform tasks also has inherent problems. Research has shown that the view that such skills are generic and can be generalised to all situations is incorrect. In fact, human skills are very much related to personality and individual style and conceptual skills are based on the natural abilities of individuals. However, a certain degree of training and development as well as everyday learning may assist managers to improve their effectiveness. What is critical is the manager's ability to be adaptable and flexible to change, particularly in fast-moving areas such as tourism.

Change is a feature of modern-day management and any manager needs to be aware of, and able to respond to changes in the organisational environment. For example, general changes in society, such as the decision of a new ruling political party to deregulate the economy, have a bearing on the operation of organisations. More specific factors can also influence the organisational environment including:

• *socio-cultural factors*, which include the behaviour, beliefs and norms of the population in a specific area.
• *demographic factors*, which are related to the changing composition of the population (e.g. birth rates, mortality rates and the growing burden of dependency where the increasing number of ageing population will have to be supported by a declining number of economically active people in the workforce).
• *economic factors*, which include the type of capitalism at work in a given country and the degree of state control of business. The economic system is also important since it may have a bearing on the level of prosperity and factors which influence the system's ability to produce, distribute and consume wealth.

- *political and legal factors* that are the framework in which organisations must work (e.g. laws and practices).
- *technological factors*, where advances in technology can be used to create products more efficiently. The use of information technology and its application to business practices is a case in point.
- *competitive factors*, which illustrate that businesses operate in markets and other producers may seek to offer superior services or products at a lower price. Businesses also compete for finance, sources of labour and other resources.
- *international factors*, where businesses operate in a global environment and factors which obtain in other countries may impact on the local business environment. For example, in December 1997 the unexpected financial crisis in South Korea led Air New Zealand to cease its four-times-a-week service to Seoul and from 1 February 1998, Qantas also withdrew services.
- *change and uncertainty* are unpredictable in free market economies, and managers have to ensure that organisations can adapt to ensure continued survival and prosperity, as exemplified by the rapid response of Air New Zealand and Qantas to the financial crisis in South Korea to offset potential losses. Change continually challenges all organisations and change in any one factor within an organisation can impact upon how it functions. Various techniques can be used to help to overcome internal resistance to change within organisations. As Kotter and Schlesinger (1979) observe, these include:

  - education and communication
  - participation and involvement
  - facilitation and support
  - negotiation and agreement
  - manipulation and co-optation
  - explicit and implicit coercion

Change may be vital for organisations to adapt and grow in new environments and the introduction of information technology is one example where initial resistance within businesses had to be overcome. Increasingly, managers are not only having to undertake the role of managing, but also the dynamic role of 'change agent'. Managers have to understand how systems and organisations work and function to create desirable outcomes. It is the ability to learn to manage in new situations where there are no guidelines or models to follow, which Handy (1989) views as the way people grow, especially in a managerial role.

CASE STUDY   Airline management in the 1990s

The airline industry developed as a commercial enterprise during the 1930s as technological advances in aviation enabled companies to develop regular passenger services, cross-subsidised by the provision of freight and air postal services. In the postwar period, modern-day air transport emerged as an international business, providing services and products for a diverse group of users including scheduled and non-scheduled (charter) transport for air travellers and cargo transport for businesses. The airline industry is truly a global business and by the year 2002 commercial world passenger air transport is expected to generate 3,431.4 billion revenue passenger kilometres (RPKs). The growth in global passenger travel by air has risen consistently, with the Boeing Commercial Airline Group (1994) anticipating an average annual rate of growth in air travel of 5.2 per cent through to the year 2013, led by growth in the Asia-Pacific region. This growth is 'equivalent to adding the capacity of one of the world's largest airlines, with a fleet of more than 600 airplanes, to the system, each and every year' (Boeing Commercial Airline Group 1994 : 4). Such a rate of change is not new for the airline industry, since it has a history of dramatic change over a short period of time dating back to the introduction of commercial aviation services in the 1920s and 1930s.

Airline management as a concept

Airline management is a concept which all commercial airline companies have taken seriously due to the rate and pace of change in the competitive global business environment in which they operate. It is also a volatile business where unpredictable external factors (for example, the Gulf War) can dramatically affect their ability to attract customers and may threaten their profitability – 'the bottom line'. Consequently, airline management may loosely be defined as the process whereby individual (and groups of) airlines seek to organise, direct and harness their resources, personnel and their business activities to meet the needs of their organisation and customers in an effective and efficient manner. In the case of the airline industry, the ability to respond to change and to adapt is a key challenge for management. Profitability of the airline is the principal objective of privately owned airlines. Yet in reality, many airlines began life as state-owned enterprises, due to the capital-intensive nature of this business activity, which requires a steady and predictable long-term stream of revenue to absorb the high capital and operating costs. The airlines of many smaller countries are still subsidised and maintained for political reasons. For example, state ownership of airlines persists in order to ensure the prestige of the national carrier in international relations, to maintain a degree of control over tourist arrivals, and to provide competition with other carriers to prevent a monopoly or duopoly situation from arising. Therefore, airline management may, in some cases, be motivated by political factors that outweigh the forces of the free market system and normal competition.

To understand the scope and process of airline management, a focus upon the issues which all airlines have to deal with on a day-to-day basis (operational issues) and on a longer-term basis (strategic issues), together with the marketing of airlines, illustrates the wide range of issues and problems which require management to ensure the airline is functioning in an efficient and effective manner. In reality, the management process for airline companies is a continuous activity which requires a predetermined structure to ensure that all the business activities are adequately integrated to meet the needs of internal customers and external business needs (i.e. those of the customer or purchaser of services and products).

A typical management structure for one airline – Air New Zealand – is shown in Figure 2.8. It illustrates the scope and range of activities with which airline management is involved in this type of organisation. For example, airline managers not only have to oversee, coordinate and direct the activities of airline operations (domestic and international airline business), they also have to manage diverse

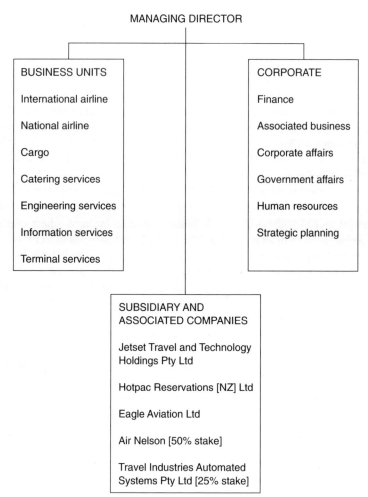

Figure 2.8   The management structure of Air New Zealand

activities which affect the organisation's main business (for example, ground hand-ling, planning, human resource management and reservations). For the organisation to operate in an effective manner, airlines typically allow different elements to oper-ate as separate business units, with a general manager reporting to a management board. This helps to achieve a degree of integration within the organisation while delegating responsibility for specialised functions to individual business units. How-ever, before considering the management issues typically affecting international and domestic airlines, it is important to realise that 'airline managers are not free agents. Their actions are circumscribed by a host of national and international regulations. These are both economic and non-economic in character and may well place severe limitations on airlines' freedom of action' (Doganis 1991 : 24). As a result, regulation is a major factor affecting the business environment.

The global airline business

The recent experiences of the international airline industry would indicate that while collectively it lost US$2.7 billion in 1990, due in part to the Gulf War and the recession, the mid-1990s have seen these fortunes change for many of the top airline companies. Although the Asia-Pacific region continues to offer the greatest growth prospects for airline operators, the 'international aviation marketplace is heterogeneous in nature. For example, the airline industry in Canada, Australia and New Zealand can be considered similar in that each has few dominant airlines, few high volume routes, and relatively large, but thinly populated geographic expanses to serve' (Go 1993 : 180). In contrast, Europe contains three major carrier segments (scheduled, regional and charter carriers) while the international trend is towards a few large transnational airlines dominating and controlling the business. British Airways' goal in the 1990s is to become a global carrier and it has entered into strategic alliances to overcome competition by cooperating with other partners and by the purchase of airlines in other countries (for example, US Air). It has also purchased a stake in other airlines to gain access to the main airline markets in those regions and countries in which it does not operate in order to achieve a global mode of operation. Within Europe, strategic alliances have resulted from the prospect of the establishment of a single continental aviation market. Other airlines have pursued a strategy of vertical and horizontal integration (see Chapter 6 for more detail) to diversify their business interest in aviation so that they now operate subsidiary companies offering airport catering, passenger services, tour wholesaling, aviation maintenance facilities, pilot training, computer reservation systems, taxi services, property development and quality service training in addition to the core function of running an international airline. In this context, many airlines have adapted different management responses to the changing global environment for the airline industry in the 1990s. While technological developments in the 1980s assisted in the expansion and cost-reduction strategies of global airlines, the 1990s are being heralded as the decade of global aviation. The challenge for international and domestic airlines is to remain competitive and to develop the enhanced management and marketing skills acquired in the 1980s as the customer became the main focus of business activity, replacing the earlier focus on operational issues.

Airline marketing and management in the 1990s

One of the principal concerns of airline managers is the matching of the supply of services with the actual and forecast demand. Such a concern underpins all aspects of airline management and planning because without a fundamental understanding of marketing and planning, it is not possible to address issues such as aircraft selection, route planning, scheduling, product planning and pricing, advertising and long-term strategy, since each area requires an analysis of demand. The demand for airline services is also directly related to supply, since qualitative factors such as the quality of airline service (on the ground and in-flight) affect its supply features. Airline planning is therefore part of the management process, with supply adjusted to meet demand. The more competitive and unregulated the market, the greater the degree of planning and adjustment needed to match supply and demand. This is normally undertaken under the auspices of the marketing process. Such issues pose an even greater challenge for airline managers since the growing global competition for air travel has meant the 1990s are proving to be the decade of the air traveller as a consumer: not only have airlines been forced to reduce operating costs to compete for business, but they have been required to focus on the needs of their customers. If mangers ignore the needs of their customers, their companies are unlikely to remain key players in the global airline business of the 1990s.

A common misconception in airline management during the 1970s was that marketing is simply about selling the final product. Airline marketing is now more complex than was recognised in the 1970s, since it is vital to the management process of deciding what to produce and how it should be sold. Thus, 'the role of airline marketing is to bring together the supply of air services, which each airline can largely control, with the demand, which it can influence but not control, and to do this in a way which is both profitable and meets the airline's corporate objectives' (Doganis 1991 : 202). If the essence of airline marketing is to satisfy customer needs, the airline needs to be market-oriented, as opposed to being dominated by an operational bias. For example, British Airways explained its financial turnaround from a loss of £544 million in 1981/82 to a profit of £272 million in 1983/84 in terms of its greater marketing orientation, based on recognising customers' needs and setting about satisfying them. Similar examples exist among other international airlines, such as Scandinavian Air Services (SAS) which has used marketing and a focus on service quality to retain a competitive edge.

The role of marketing in the management of airline services can be summarised as a four-stage process:

1 Identify markets and market segments using research methods and existing data sources and traffic forecasts.
2 Use the market analysis to assess which products to offer. This is known as product planning. At this stage, price becomes a critical factor. Therefore product planning is related to:
   (a) market needs identified from market research
   (b) the current and future product features of competing airlines and the cost of different product features
   (c) assessing what price the customers can be expected to pay for the product
3 Develop a marketing plan to plan and organise the selling of the products. Sales and distribution outlets need to be considered, together with a detailed programme of advertising and promotion.
4 Monitor and review the airline's ability to meet service standards, assessed through sales figures, customer surveys, analysis of complaints and long-term planning to develop new service and product features.

Retaining a competitive edge: new marketing techniques for airline managers
One area which international airlines have consistently targeted through advertising and the pursuit of service quality is business travel and the luxury market – first-class travel. This is because it is the most profitable segment of the market for scheduled airlines, with premium pricing and high profit margins per seat, compared to lower yields in economy class, where profitability is attained by achieving high load factors. In the competitive airline business, the purchasing power of corporate

business accounts, often in excess of US$100 million a year, has become the target for airline managers seeking to nurture this profitable segment. This is reflected in the estimated 150.6 million business trips undertaken in the USA in 1991. Airline managers have adopted aggressive marketing strategies, negotiating contracts and tying companies to one airline company. This is also reflected in the emergence of frequent flyer programmes (FFPs).

Frequent Flyer Programmes
These programmes were originally developed in the USA in May 1981 by American Airlines (AA) and nearly 30 million people in the USA have accumulated points for free travel. However, only 15 per cent of the participants ever redeem their points for flights. FFPs have, in part, been developed to attract business following deregulation. Aided by powerful computer reservation systems (CRSs), AA used its database of passengers to identify frequent flyers, who were enrolled into a 'Very Important Travellers Club'. The reward for loyalty was free air travel on a route which consistently had a high ratio of unsold seats – Hawaii. While AA's Advantage FFP aimed to maximise revenue yield and its load factor, it also assisted in developing Dallas as a logical hub and capturing business from competing airlines. AA's Advantage scheme was closely followed by other companies launching similar schemes (United, TWA and Delta) with AA management recruiting British Airways to its FFP to add an international dimension. One consequence of FFPs was the greater utilisation of the 'hub and spoke' system.

To broaden the marketing appeal of FFPs, non-airline travel services (car rental companies, tour operators and hotel chains), airline-affiliated credit card purchases (mileage credits added according to the volume of spending) and other services now attract FFP points. It remains a fertile area for continued competition among airlines to offer the most attractive FFPs. While FFPs are a basic element of airline marketing strategy in the USA, they have also been introduced by European and other global carriers. In the Asia-Pacific region, Singapore Airlines, Malaysian Airlines and Cathay Pacific have agreed to establish a joint venture FFP to compete with the mega-carriers (that is, British Airways, United and AA). This appears to be the way forward for FFPs as carriers group together to broaden their scope and FFPs may be a basis for reinforcing the strategic alliances many airlines already operate.

Clearly, marketing is a key component of airline business strategies in the 1990s and a number of airlines have decided to withdraw from non-core activities, such as the hotel business, baggage handling, ground-support activities and catering. This has sometimes been a response to spiralling costs, which may be controlled through the use of contracted-in services. For example, in an attempt to reduce costs, the state-owned Irish airline, Aer Lingus, sold its holding in the Copthorne chain of airline hotels to reduce costs and to focus on core airline activities. Airline management has attracted very little attention from management researchers, probably because it remains a complex industry to understand fully unless one has worked closely with an airline organisation. Such organisations are wary of outsiders due to the highly competitive nature of their business and the prospect of 'insider knowledge' being used to the advantage of other airlines. A great deal of the available research data on airlines and their management remains confidential to the individual organisations and their employees, and only commissioned management consultants are usually privileged to consult it.

Despite these constraints, the airline industry is in a constant state of flux, adapting to changing trends. Computer reservation systems, controlled by the global airlines (for example, AA) now offer these organisations an unparalleled control over the passengers' choice of services, allowing them to become brokers of other travel services. This is transforming the airline business by offering a profitable business activity which may outstrip the actual profitability of individual airline operations. This has meant that airline managers need to be versatile, adaptable and more will-ing to view their business activities as part of a global travel and tourism industry. As customers' expectations increase, technological advances in aircraft design (for example, the new Boeing 777) may see the cost of air travel stabilise, if the cost of aviation fuel remains fairly constant. The future shape of the global aviation industry is likely to contain a limited number of global mega-carriers, with smaller airlines integrated into their operations by strategic alliances and other devices (for example, part-ownership by the larger carriers). A more hierarchical system of route provision is also likely to evolve and airlines will be forced to deal with greater volumes of passengers using a 'hub and spoke' system of provision to achieve efficient operations (see Chapter 6 for more detail), with passenger volume tied to aircraft type. The major constraint on this rapidly evolving business activity for airline managers will be the availability of uncongested airspace and airports with sufficient capacity. This is already affecting air travel in Europe and the USA, while the environmental lobby regularly opposes airport expansion near to major urban centres. Airline managers will need to be increasingly focused on ensuring that their operational requirements can be fulfilled and their environmental impacts monitored and minimised to reduce local opposition to air transport. For example, many airports have introduced bans on aircraft which do not meet noise emission levels. These managerial issues have to be addressed and accommodated within the day-to-day operation and longer-term planning by airline managers so that passengers are not adversely affected by delays, congestion and inadequate planning.

Summary

This chapter has shown that different disciplines have developed a range of concepts and distinctive approaches to the analysis of different aspects of the tourist transport system. The case studies have highlighted how these concepts can help us to understand different features of the tourist travel experience and the interaction of the consumer and producer in tourist transport systems. The economists' analysis of tourist transport systems is based on the demand and supply issues associated with the use and provision of different modes of transport and the implications for the future (i.e. forecasting, which is dealt with in Chapter 4). In contrast, the geographer has largely focused on the spa-tial analysis, organisation and distribution of tourist patterns of travel, while the transport geographer has considered the policy, management and planning issues associated with the provision of transport, sometimes in combination with economists (see the contributions in Banister and Button 1992). In marketing, the interest in tourist transport is developing within the existing literature, although the contribution should not be understated because marketers have identified the importance of a more consumer-oriented focus for tourist transport provision. The concern for service quality which has emerged in the marketing literature has started to permeate tourist transport provision, particularly in the airline sector (see Laws 1991 for a fuller discussion) which is explored further in Chapter 3. In the case of management, tourist transport is often viewed as a practical and process-oriented activity which is fundamental to the organisation, with control, leading and planning functions necessary to ensure a vibrant, profitable and functional business organisation.

It is clear from the discussion of the contributions made by economists, geographers, marketers and managers to the analysis of tourist transport systems, that these contributions need to be integrated into a more coherent framework, which is possible if a systems approach is adopted. In reality, the decision-making functions undertaken by transport operators and public sector organisations associated with tourism and transport invariably use a variety of economic, geographical, marketing and management principles in their everyday work to plan and develop tourist transport services. The concern with enhancing the tourists' travel experience, so that it meets with the preconceived notion of service quality and provision, has become a concern not only for operators but also for government policy making. It is the issue of government policy making and its effect on tourist transport systems which is the focus of the next chapter.

Questions

1. What concepts characterise the way economists, geographers, marketers and management researchers approach tourist transport?
2. Prepare a SWOT analysis for a tourist transport service with which you are familiar.
3. How is the use of yield management assisting tourist transport operators to maximise their revenue?
4. What are the main qualities needed to manage a tourist transport business?

Further reading

The following references are a good starting point from which to examine the contribution each discipline makes to the analysis of tourist transport.

The economist

Gray, H. (1982) 'The contribution of economics to tourism', *Annals of Tourism Research* 9: 105–25.
Sinclair, M.T. and Stabler, M. (1997) *The Economics of Tourism*, London: Routledge. This is a comprehensive and very thorough review but a sound grasp of economic concepts and principles is required to gain the maximum benefit from reading it.

The geographer

Pearce, D. (1995) *Tourism Today*, London: Longman, 2nd edition.
   This remains the most authoritative source on the geographer's analysis of tourism, using models and spatial analysis to understand the dynamics of travel and tourism activity.
Mansfeld, Y. (1990) 'Spatial patterns of international tourist flows: towards a theoretical approach', *Progress in Human Geography* 14 (3): 372–90.
   This provides an interesting approach to the analysis of tourism flows.
Page, S.J. (1994) 'European bus and coach travel', *Travel and Tourism Analyst* 1: 19–39.
   This offers a broad analysis of the patterns of land-based coach travel in Europe.

Marketing

Gilbert, D. (1989) 'Tourism marketing – its emergence and establishment', in C. P. Cooper (ed) *Progress in Tourism, Recreation and Hospitality Management*, Volume 1, London: Belhaven, 77–90.
   This provides a good insight into the development of tourism marketing.

The following references have a limited amount of material on marketing tourist transport:

Horner, S. and Swarbrooke, J. (1996) *Marketing Tourism, Hospitality and Leisure in Europe*, London: International Thomson Business Press, Chapter 20.
Seaton, A. and Bennett, M. (1996) *Marketing Tourism Products: Concepts, Issues and Cases*, London: International Thomson Business Press, Chapters 15 and 19.

The Yorkshire Moors National Park Authority has appointed a tourism and transport officer to assist with the administration of the hail and ride Moorbus Scheme. The Moorbus Scheme offers an integrated route network to provide access to the principal leisure and recreational sites and attractions in the National Park. The Internet site is http://www.country.goer.org/nymoors

Management

Foster, D. (1985) *Travel and Tourism Management*, London: Macmillan.

Chapter 3

The role of government policy and tourist transport: regulation versus privatisation

Introduction

The development of tourism in specific countries is a function of the individual governments' predisposition towards this type of economic activity. In the case of outbound tourism, governments may curb the desire for mobility and travel by limiting the opportunities for travel through currency restrictions, such as in the case of South Korea in the 1980s, while still encouraging inbound travel. Similarly, in the USSR under the former communist rule, the opportunities for domestic tourism were controlled by limiting the supply of holiday infrastructure. However, such examples are not the norm because most governments seek to maximise the domestic population's opportunities for mobility and travel by the provision of various modes of transport to facilitate the efficient movement of goods and people at a national level. In fact, the development of transport to facilitate inbound tourism is often motivated by government's desire to increase earnings from tourist receipts.

To achieve the government's objectives for transport to facilitate tourist travel, policies are formulated to guide the organisation, management and development of tourist and non-tourist transport (Starkie 1976; Knowles 1989). This is normally followed by specific planning measures which seek to implement the policies (Banister 1994). These will often require national governments to invest in – or encourage private sector organisations to invest in – new infrastructure which may be dedicated to tourist travel (and cargo) such as airports. Conversely, and more typically, governments may need to develop policies and planning measures which expand and develop national, regional and local infrastructure to accommodate tourist and recreational travel alongside the use for commuting and non-tourist travel. D.R. Hall (1997) rightly highlights the conceptual problem of distinguishing between what is tourist transport and what is not. Government policies often highlight this problem since while specialised, dedicated forms of tourist transport exist (e.g. tourist coaches, package flights and cruise liners), there are other forms of transport shared by hosts and tourists to varying extents. For example, urban buses and metro systems may be used by tourists, but their *raison d'être* is primarily for local transport and this is reflected in generalised transport policies which often do not accommodate use by tourists. In Singapore, the state-subsidised mass rapid transport system is integrated with local bus services to provide the population with an efficient public transport system. Yet like many other metro systems it offers incentives for tourist use through the provision of 'tourist explorer tickets'. Indeed, there are few forms of transport which tourists are unlikely to use but the separation of tourist and non-tourist use is not usually an issue for policy makers. In a theoretical and more abstract context, research by sociologists interested in tourism, such as Urry (1990) and his work on the *tourist gaze* invokes notions of post-tourism, where all transport is potentially for and of tourism as the tourist gaze is supported by the tourist gaze (D.R. Hall 1997).

The transport policies developed by national governments are influenced by their changing attitudes, outlook and political ideology. This often manifests itself in terms of the level of expenditure on capital investment, infrastructure provision and policies to *facilitate* (Department of Employment/English Tourist Board 1991) or *constrain* tourist travel. Griffith (1989) discusses the effect of international airways sanctions as an example of government policies formulated to achieve a political goal against South Africa.

Likewise, when governments fail to acknowledge that policies and action are required to facilitate the development of tourist transport systems, capacity problems and bottlenecks may develop where tourism is imposed on inadequate national infrastructure. Gormsen (1995) documents the situation in China where 'serious capacity problems can be observed, not only for the airline companies, but also for railways, road traffic and river navigation, despite the fact that since the establishment of the PRC [People's Republic of China] all of these systems have been improved to a large extent' (Gormsen 1995 : 82). As Table 3.1 shows, while the railway network, asphalt roads and air capacity have been expanded under government direction (see Hayashi et al 1998 for an up-to-date review), the growth in patronage has outstripped supply as 'undercapacity at all levels of the transportation system remains a problem. Foreign tourists who must pay higher prices than citizens of China . . . receive preferential travel on the public transportation lines' (Gormsen 1995 : 84). In an attempt to improve the efficiency of domestic air travel, the state airline, Air China, was divided into seven regional airlines after 1985. A further five Air China-related airlines and sixteen private airline companies are now in operation. By increasing the price of domestic air travel, airlines have managed to suppress demand to address capacity constraints. In the case of China, it is evident that even when policies are put in place to expand infrastructure, demand can often outstrip supply.

As Banister (1994) observes, forecasting the demand for transport planning is an Achilles' heel which has not improved even with growing technical sophistication in the methods used. Whitelegg (1993) may have highlighted one of the problems that needs to be recognised in a European context: 'leisure, recreation and tourism trips . . . are major growth areas in travel behaviour' (Whitelegg 1993 : 160). Page (1998) also argues that from a spatial perspective, tourist and recreational travel cannot be modelled and explained using conventional models and principles in transport geography. This is because is does not conform to principles of distance optimisation and least-distance

Table 3.1 The development of China's transport system 1949–87

Year	Number of public vehicles			Rail (1,000 km)	Road (1,000 km)	Number of passengers (millions)				
	Trains (1,000)	Buses (1,000)	Aircraft			Rail	Road	Water	Air	Total
1949	3	17	–	30	32	103	18	15	0.27	137
1952	5	19	45	35	55	163	45	36	0.02	245
1957	8	34	105	43	121	312	237	87	0.07	638
1962	10	47	245	55	263	741	307	163	0.17	1,221
1965	10	60	287	58	304	412	436	113	0.27	963
1970	11	75	293	70	411	524	618	157	0.22	1,300
1975	14	173	357	77	511	704	1,013	210	1.39	1,929
1978	15	257	382	85	647	815	1,492	230	2.31	2,539
1980	16	351	393	86	662	921	2,227	264	3.43	3,417
1985	21	795	390	95	750	1,121	4,764	308	7.47	6,200
1987	23	1,115	402	99	982	1,124	5,936	389	13.10	7,464

Source: Modified from Gormsen (1995) based on the National Statistics Bureau, China's Statistics Publishing House, 1989 and Historical Statistics by Provinces, China 1949–90, National Statistics Bureau, China's Statistics Publishing House, 1990.

patterns of travel. Therefore, government policy cannot necessarily assume that tourists will seek to use infrastructure developed for national economic goals. As Chapter 2 noted, the car may be the principal agent driving tourist and recreational trips. The UK's Department of Transport observed,

> tourism and leisure depend heavily on travel and transport. In Britain, travelling by road represents the dominant mode for visitors to get to and from their final destinations or to see places of interest . . . [and] . . . the tourism industry, in common with many other industries, also depends heavily on roads for the transport of the goods and services people need. (Department of Transport 1987 : 1)

In fact with over 21 million cars licensed in the UK, roads are assuming an even greater role in the patterns of tourist activity, despite problems associated with congestion in urban areas.

In the Republic of Ireland, the Department of Transport and Tourism recognises that:

> the development of an enhanced internal transport network fully integrated with appropriate access infrastructure and services is a fundamental requirement for the future development of the tourism industry. This requires the provision of good quality access transport services and facilities together with a satisfactory internal transport infrastructure and services. (Government of Ireland 1990 : 24)

Although the UK's Department of Transport and Government of Ireland highlight the interrelated nature of transport and tourism, in practice the responsibility for transport and tourism is often fragmented. It is frequently dispersed across different government departments, and where little inter-departmental liaison occurs, there is poor coordination of policies affecting tourist transport (see Banister and Hall 1981 for a public policy perspective) if they exist at all.

Public sector involvement in the tourist transport system at national government level is designed to facilitate, control and in some cases regulate or deregulate the activities of private sector transport operators with a view to 'looking after the public's interests and providing goods whose costs cannot readily be attributed to groups or individuals' (Pearce 1990 : 32). (Also see Chapter 5 on airline deregulation.) Since the private sector's primary role is revenue generation and profit maximisation from the tourist transport system, the government's role is to promote and protect the interest of the consumer against unfair business practices, and to ensure that safety standards are maintained to protect the interests of employees in large- and small-scale transport operations. Oum (1997 : 3) argues that in the case of Asian airlines 'government policy makers and bureaucrats regulating the airline industry would agree that their long-run objectives are essentially two-fold: to attend to the needs of consumers (the travelling public and air cargo shippers) and to ensure a strong and viable airline industry within the country'.

This chapter examines the role of government policy in relation to tourist transport provision and how it can impact upon various modes of transport. The concept of transport policy is introduced and its meaning is discussed in relation to political variants of transport policy in free market economies.

Emphasis is placed upon different political ideologies and how these have affected transport policies for tourism with the rising tide of privatisation in many societies. The significance of transnational organisations such as the EU and progress towards transport policies in aviation and rail travel are discussed.

The role of government policy and tourist transport

The term 'policy' is frequently used to denote the direction and objectives an organisation wishes to pursue over a set period of time. According to Turner (1997), the policy process is a function of three interrelated issues:

- the intentions of political and other key actors
- the way in which decisions and non-decisions are made
- the implications of these decisions.

Policy making is a continuous process and Figure 3.1 outlines a simplified model of the policy process which is applicable to the way tourism transport issues are considered by government bodies. Hall and Jenkins (1995) examine the issue of policy making in a tourism context. The journal *Transport Policy* offers many relevant articles on this issue with regard to different aspects of transport, including tourist and recreational use. National policy is normally formulated by government organisations with economic and social factors in mind, without an explicit concern for tourism, even though transport networks are used for tourist and non-tourist travel. The development and shape of transport policy is partly affected by the existing infrastructure which has resulted from major public and private sector investment to achieve general

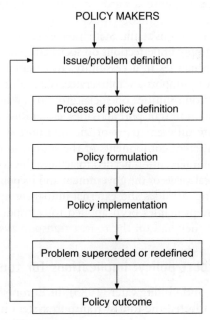

Figure 3.1   The policy making process

and specific transport objectives. It may also be the case that other government policies (e.g. defence) impact upon transport. For example, many innovations in civil aviation have been developed as a spin-off from defence policy, particularly military research and development. In addition, defence infrastructure is sometimes used for civilian purposes, as in the case of the former Eastern Europe (D.R. Hall 1993a). Changes to the underlying infrastructure cannot be executed quickly because of the major capital costs, planning procedures and the time delay in responding to the demand for new infrastructure. For example, the expansion of an airport to accommodate a forecast growth in tourist arrivals may take between five and ten years from its inception to the completion stage. Such developments also have to be set against the increased environmental impact (e.g. noise, pollution, waste and greater numbers of tourists – see Chapter 8) resulting from additional flights (see Somerville 1992). Transport policy is a reactive element of government activity as changes in society and the demand for tourist and non-tourist travel require a certain degree of continuity and change in policy to meet new trends and activity patterns among the population. What political and ideological principles affect transport policy?

National transport policies have been characterised by a range of approaches which span a spectrum from a free market orientation to those based on planned resource allocation (Farrington 1985). The market-oriented view has been pursued on the premise that centralised state control of transport produces an unwieldy and often unresponsive service requiring unnecessarily high subsidies from state taxation. By introducing a greater degree of private sector involvement and competition, it is argued that improved services should result and the need for public subsidies should diminish. In contrast, supporters of the regulated planned response towards state involvement in transport have pointed to classical economic theory which recognises that, in a free market economy, supply imperfections result. State intervention in the market economy is justified to rectify supply imperfections on social efficiency and environmental grounds to avoid inequalities in accessibility. In situations where inadequate levels of demand exist to support a viable service, state subsidies may be required to provide access for communities on social grounds (Whitelegg 1987). Thus government intervention in transport activities, either directly or through other agencies to coordinate different parts of the transport system, is essential to bring order to the different components of the system so that they operate in harmony. The level of intervention needed to achieve this coordinating function depends on the political views of the government and its policy objectives, which are subject to changing attitudes and external influences upon policy formulation. How have these principles been applied to transport policy in the past and what effect have they had on the tourist transport system?

Interpreting transport policy: implications for tourist travel

According to Button and Gillingwater (1983), in Europe and North America transport policy has affected the tourist transport system in a number of ways although it is based on two underlying economic principles:

- allocative efficiency (the use of resources and the price mechanism to achieve efficient access to transport and travel)
- political obligations (the need for the state to protect the public interest in transport provision).

These principles have had an important effect on the provision of transport for tourist travel in terms of the development, expansion and regulation of different modes of transport to facilitate access to tourism resources. Modes of transport which enable mass travel developed at different times in Europe and North America but the effects were similar in terms of making tourist travel more available. Technological innovations and their commercial exploitation (e.g. the motor car), and their diffusion to different social groups during the late nineteenth and twentieth centuries, have been shaped by transport policy to achieve the twin goals of allocative efficiency and political obligations.

In historical terms, the principles of allocative efficiency and political obligations have been interpreted in different ways by governments and Button and Gillingwater (1983) identify four distinct phases in transport policy:

- *The Railway Age*, which, in the UK, led to heavy investment in the provision of infrastructure that made seaside resorts accessible to the working classes after the 1870s. Government promotion of railways in the private sector dominated the period, except during the wartime period (1914–18) when state control emerged to coordinate and manage the railways in the 'national interest' of efficiency.
- *The Age of Protection*, which characterised the 1920s and 1930s, saw the emergence of road transport, particularly the rise of the private car and coach travel in an unplanned manner. This led governments to intervene in the marketplace to avoid massive cost-cutting among private operators as competition intensified due to the growing number of small transport operators. Such intervention was justified on the basis that the effects of major competition might have led to a reduction in the number of operators, following a price war to secure passengers and market share. It was expected that this would be followed by a much reduced route network and poorer level of service. In the USA, this was characterised by the 1935 Motor Carriage Act which protected the Greyhound Bus Operations, providing one major operator with a virtual monopoly on interurban bus travel.
- *The Age of Administrative Planning*, which emerged in the postwar period (i.e. after 1945) and superseded the Age of Protection, saw the private car emerge as a potent force for tourist and recreational travel. One consequence was the growing financial weakness of railways, although urban growth continued to dictate the need for large, efficient urban passenger transport systems. The nationalisation of railway networks and other forms of public transport epitomise this era in transport policy. The 1960s saw a growing burden of state subsidies to support public transport and some attempts to radically restructure the transport network (e.g. the Beeching Report in the UK and subsequent rationalisation of the rail network). In the UK, the 1968 Transport Act sought to reorganise public transport and one consequence for tourist travel was the establishment of the National Bus Company, with the responsibility for express coach travel. In the USA, the 1962 Urban Mass Transportation Act provided Federal Grants for two-thirds of public transport projects. Efficiency in provision was interpreted by government policy in terms of integration and coordination of transport planning which reduced wasteful competition (Button and Gillingwater 1983).

- *The Age of Contestability* has characterised the period since the early 1970s in the USA (and the 1980s in the UK), based on the pursuit of the principle of deregulation to achieve 'allocative efficiency in transport policy' (Farrington 1985). By creating efficient transport operations and reducing public subsidies, the private sector is seen as the main panacea for efficient transport operations (Knowles and Hall 1992). In the UK the sale of the state-owned airline, British Airways, and its emergence as a profitable private sector company is cited by supporters of this political philosophy as one of the main successes of privatisation and deregulation. It is interesting to note that the UK/North American experience in transport policy in the 1970s and 1980s has not been adopted and endorsed in many other European countries where the state remains committed to investment in tourist transport systems.

In the UK context, Banister (1994) identified the essential need to recognise that 'society is now in transition from one based on work and industry to one in which leisure pursuits dominate – the post industrial society' (Banister 1994 : 68). The implication is that policy making needs to recognise the significance of meeting the transport needs of this society. In a detailed review of 'the age of contestability' and transport policy in the UK between 1979 and 1991, Banister highlighted a number of major features including privatisation and a growing move towards meeting the transport needs of the leisure society. He cites the following actions of the Conservative government as instrumental in this, with the underlying premise of reducing public expenditure in transport:

- Transport Act 1980 – deregulation Part 1
- Transport Act 1981
- Transport Act 1982
- Transport Act 1983
- Transport Act 1985 – deregulation Part 2
- Channel Tunnel project launched – February 1986
- Terminal 4 opened at Heathrow Airport in 1986 and Piccadilly Line Underground extension to Heathrow Airport completed to Terminal 4
- British Airways privatised – January 1987
- British Airports Authority privatised – June 1987
- Docklands Light Railway opened – July 1987
- London City Airport opened – October 1987
- British Airways/British Caledonian merger – November 1987
- National Bus Company privatisation – April 1988
- New terminal at Stansted Airport opened – Spring 1991

As Banister (1994 : 69) explains, 'many of the policies have been justified on . . . criteria, such as ideology (for example deregulation), democracy (for example widening share ownership), efficiency (for example, private enterprises are more efficient than public ones), and accountability'. The Treasury was the prime mover in these new policy directions, constraining expenditure on investment and renewal of transport infrastructure by the state. The emphasis in policy was a shift from state investment in public transport to investment in roads to meet a demand for car use. While the government increasingly turned to the private sector for capital and investment in the 1990s, it merely provided expenditure that was required in the 1980s to keep up with demand. Banister (1994 : 70) acknowledges that

In the early 1980s there was a concern over the increasing levels of public support for bus and rail services with subsidy levels for each amounting to about £1,000 million. The market philosophy is to ensure that all services are provided at a level and a price which is determined competitively. Intervention should only take place when the market is seen to fail, for example, for social need, but even then the intervention should be specific . . . and that there should be no cross subsidisation between services.

The policy changes in the 1980s and up to the mid-1990s were characterised by:

- transferring transport from the public sector to the private sector through privatisation or denationalisation
- the introduction of competition into public transport services and greater regulatory reform
- the pursuit of greater levels of private capital in the large transport infrastructure projects such as the Channel Tunnel.

In fact, privatisation is not a process in government policy making which is confined to the UK and USA. For this reason, attention now turns to privatisation in the Asia-Pacific aviation sector.

Privatisation in Asia–Pacific aviation

Oum (1997) indicates that the Asia-Pacific region is one of the fastest-growing regions of the world for scheduled air travel. In the period 1980–90, commercial air traffic expanded by 7.95 per cent per annum compared to 5.8 per cent growth per annum in North America. Forecasts from the International Air Transport Association (IATA) indicate that passenger traffic in Asia-Pacific will rise from 31 per cent of world scheduled traffic in 1990 to 41 per cent in the year 2000 and 51 per cent in 2010. While the currency crisis in Asia in 1997/98 may impact upon these forecasts, it is evident that the region will become the world's most important airline market. Oum (1997 : 1), however, argues that 'even though the major airlines of Asia belong to the world's fastest growing airline market, they have remained relatively small in terms of network size, traffic volume and operating revenue, compared with major carriers in the United States and Europe'. Part of this is related to the restrictive bilateral agreements which also protect the home market. There has been considerable research in the aviation sector which compares the productivity of airlines under government control with their subsequent performance as privately owned enterprises. Most of the studies indicate superior performance under conditions of privatisation (Findlay and Forsyth 1984; Oum 1995) although not in every case (Oum and Yu 1995). Even the countries of Eastern Europe are privatising their national flag carriers as central state planning is removed (D.R. Hall 1993b).

Table 3.2 indicates that prior to 1985 there were isolated cases of privatisation but much of the privatisation occurred post-1985. For example, while the case of Singapore Airlines indicates that the government remains the majority

Table 3.2 Airline privatisation in the Asia-Pacific region 1985–95

Airline	Country	1985 ownership		1995 ownership	
		Private	Public	Private	Public (%)
American Airlines	USA	X		X	
Aeromexico	Mexico		X	X	
Aero Peru	Peru		X		20
Air Canada	Canada		X	X	
Air China	China		X		100
Air Lanka	Sri Lanka		X		100
Air New Zealand	New Zealand		X	X	
Air Nugini	Papua New Guinea		X		100
Air Pacific	Fiji		X		79.6
Air India	India		X		100
All Nippon Airways	Japan	X		X	
Ansett Australia	Australia	X		X	
Avianco	Colombia	X		X	
Biman Bangladesh	Bangladesh		X		100
Canadian Air International	Canada	X		X	
Cathay Pacific Airways	Hong Kong	X		X	
China Airways	Taiwan	X		X	
Continental Airways	USA	X		X	
Delta Airlines	USA	X		X	
Garuda	Indonesia		X		100
HAL (Hawaiian)	USA	X		X	
Indian Airlines	India		X		100
Japan Air Systems	Japan	X		X	
Japan Airlines	Japan		X	X	
Korean Air	South Korea	X		X	
Ladeco	Chile	X		X	
Lan Chile	Chile			X	
Malaysia Airlines	Malaysia		X		30
Merpati	Indonesia		X		30
Mexicana	Mexico		X		35
Northwest Airlines	USA	X		X	
Pakistan International	Pakistan		X		57.4
Philippine Airlines	Philippines		X		33
Qantas	Australia		X		75
Royal Brunei	Brunei		X		100
Saeta Air Ecuador	Ecuador	X		X	
Singapore Airlines	Singapore		X		54
Thai International	Thailand		X		93.7
United Airlines	USA	X		X	
Vietnam Airlines	Vietnam		X		100

Source: Modified from Forsyth (1997).

shareholder, it is largely a private airline. In Oceania, the majority of airlines are privately owned, though many of the smaller airlines of the Pacific Island states remain state-owned to retain a degree of control over tourist arrivals (Forsyth and King 1996) and for strategic reasons related to accessibility and the independence of major carriers. In Pacific South American states, airlines have been privatised as well as those in North America, Canada and Mexico. In Asia, the situation is mixed. While some states have privatised their airlines and others intend to do so, a considerable number remain state-owned. Most of the privatised companies are the large successful airlines. As Forsyth (1997a : 53) observes, 'there is a strong correlation between per capita income levels and private ownership. In the richer countries in the region, such as Japan, Korea, Singapore, Taiwan, Malaysia, along with the colony of Hong Kong, all the airlines are privately owned. By contrast, there are few examples of private ownership of airlines in the poorer countries', while a number of governments have allowed the introduction of private airlines (e.g. Asiana in Korea, Sempati in Indonesia and Dragonair in Hong Kong) to compete on international routes.

Where privatisation has occurred, it can take a number of forms:

- listing on the stock market (JAL and Singapore Airlines)
- sale of the airline to other companies in the same country
- sale of the airline to strategic shareholders

Privatisation also allows overseas investment in airlines through direct investment by shareholders and equity holdings through strategic alliances. In contrast, there has been less interest in the privatisation of airports in Asia, with only Australia taking Britain's lead in privatising all of its airports. The argument advanced to justify privatisation of airlines is the efficiency gains which will result. In the case of airports, it is widely recognised that the greatest proportion of costs are capital as opposed to operating costs. Therefore the potential for efficiency gains is limited (although a more detailed discussion can be found in Chapter 7). However, some of the aspects of airport operation which private firms may perform efficiently include:

- terminal operation
- retail outlet operation
- building and operating runways
- contracted-out services

However, Forsyth (1997a : 62) maintains that 'investment policy, and ownership of the main facilities may best be left to the public sector'. In an Asia-Pacific context, privatisation has certainly opened up the opportunities for the region's airlines to become part of a global aviation industry through foreign investment, alliances and cooperation motivated by commercial motives in the absence of state policies to protect state airlines. As Wheatcroft (1994 : 24) argues, 'in the long term, the privatisation of airlines seems certain to contribute to a reduction in protectionism in international aviation policies'.

Given that many of the airlines in developing countries within Asia-Pacific have remained under state control to assist with the development of the tourism industry, the following case study is introduced of a developing country outside of the region – Zimbabwe. The case study highlights some of the principles developed in the discussion of government policy and tourist transport to illustrate how the state carrier can be used to protect and then develop tourism. The case study also addresses a neglected aspect of tourism and transport within the literature, namely the situation in less developed countries, particularly those in Africa. It also provides an interesting counterbalance to the situation in Asia-Pacific and an opportunity to assess the extent to which the state has adopted different policy responses for tourist transport operations.

CASE STUDY   Air transport and tourism in Zimbabwe

Simon's (1996) innovative study *Transport and Development in the Third World* highlights the relationship of transport to the development process. It also illustrates the comparative neglect of the less developed world by transport researchers and a lack of critical debate on policy directions. This is also the case in relation to tourism and transport. In the case of Zimbabwe, the national government has identified tourism as a central theme in the plans for economic development (Turton and Mutambirwa 1996). In 1990, some 10 years after independence, tourism was the fourth-largest earner of foreign revenue for Zimbabwe. In global terms, however, Africa attracts less than 3.5 per cent of world arrivals and Zimbabwe ranks seventh in terms of earnings from tourism in Africa, with less than 25 per cent of those in Kenya. De Kadt (1979) and Britton (1982) identified the significance of tourist transport in expanding tourism in developing countries. Turton and Mutambirwa (1996 : 453) rightly argue that 'air transport is essential to this process of tourism expansion in Zimbabwe, both in respect of long-haul flights from overseas and the provision of adequate domestic services'. Both international arrivals and domestic travel require appropriate policies to facilitate the development of tourist activity and travel. Visitor arrivals in Zimbabwe have increased since 1980 (Figure 3.2) (Zinyama 1989), although the length of stay has decreased from 9.3 days in 1985 to 5.3 days

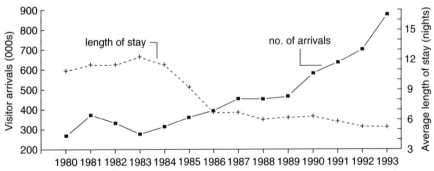

Figure 3.2   Visitor arrivals in Zimbabwe and the average length of stay 1980–93 (redrawn from Turton and Mutambirwa 1996)

Table **3.3** Tourist arrivals in Zimbabwe (Jan–March 1994) by
mode of transport

Method of arrival	Number of visitors
Air transport	
Harare	34,646
Bulawayo	1,958
Rail transport	1,446
Road transport	
Beit Bridge	14,782
Victoria Falls	56,996
Plumtree border post	8,141
Chirunda	20,630
Total	138,599

Source: Based on Turton and Mutambirwa (1996).

in 1993. While a proportion of arrivals crossing the South Africa–Zimbabwe border are
classified as visiting friends and relatives, over 90 per cent of arrivals are classified
as tourists. A marked seasonality in arrivals is evident, with arrivals peaking in July–
September and over Christmas, closely corresponding to peak vacation periods in the
northern hemisphere.

As Table 3.3 shows, a significant number of tourist arrivals who stay one night or
more are from South Africa and Zambia. Turton and Mutambirwa (1996) argue that
if the day-trippers crossing the South Africa and Zambia border are removed from
arrivals (Figure 3.3), 76 per cent of arrivals are from these two adjacent countries.
Between 1985 and 1993 arrivals from these two countries increased by 190 per cent
and 116 per cent respectively. Aside from African arrivals, Table 3.4 shows that
United Kingdom (UK) arrivals are the largest non-African source market. But they
only represent 5.5 per cent of all arrivals followed by those from Europe and Anglo-
America. In terms of transport, over 75 per cent of arrivals enter and leave Zimbabwe
by road, using the Beit Bridge from South Africa, the Plumtree border from Botswana,
the Chirundu and Victoria Falls crossings from Zambia and crossings from Mozambique
(Figure 3.3). Approximately 20 per cent of arrivals use long-haul and regional services,
with Harare International Airport the main gateway. This is complemented by lesser
numbers of arrivals from South Africa who use Victoria Falls and Bulawayo airports
(Figure 3.3). In the period 1987–94, arrivals at Harare increased from 537,348 to
769,777.

Within Zimbabwe, Air Zimbabwe is subject to a high degree of government
control through the Ministry of Transport under the provisions of the Air Zimbabwe
Act (1989). All decisions on the purchase of aircraft, marketing investment and the
structure of air fares are subject to ministerial approval. Prior to the economic
structural adjustment programme in 1991, which weakened the Zimbabwe dollar
and increased air fares, 40 per cent of the foreign revenue earned by Air Zimbabwe
was placed in government reserves. The result was a lack of funds for the airline

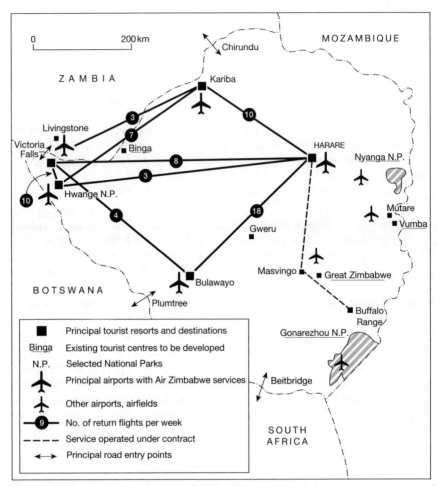

Figure 3.3   The air transport system in Zimbabwe (redrawn from Turton and Mutambirwa 1996)

Table 3.4   Visitor arrivals in Zimbabwe 1985–93

Region	Arrivals (000s)				
	1985	**1987**	**1989**	**1991**	**1993**
Africa	260	262	338	514	735
North & South America	16	15	17	15	25
Europe	48	53	68	64	99
Australia & New Zealand	6	8	13	13	15
Asia	3	3	3	4	6

Source: Based on Turton and Mutambirwa (1996).

to market itself effectively in a global environment. While Turton and Mutambirwa (1996) examine the development of air services in Zimbabwe prior to and after independence, they highlight the early problems the airline faced, including a loss-making tour-operating business (Flame Lily Tours). In particular, the reluctance of government to allow the airline to raise prices in relation to market conditions led to infrequent but major price increases which adversely affected inbound operators using Air Zimbabwe services in 1986 and again in 1992.

Turton and Mutambirwa highlight the limitations imposed by government on investment in tourism due to limited access to foreign exchange funds. The outcome prior to the 1990s was Air Zimbabwe's inability to 'provide an adequate fleet of air-craft for its internal routes and the supply of vehicles to transport visitors between airports and hotels and on game park tours' (Turton and Mutambirwa 1996 : 456). In terms of the management and operating policies of Air Zimbabwe, long-haul services date to the former Air Rhodesia–British Airways service between London and Harare, a joint daily non-stop service. In the 1980s, Air Zimbabwe added other services between Harare and Frankfurt, Lisbon, Sofia and Athens highlighting the link to the European markets. A service to Perth, operated by Qantas using a seat-lease agreement, was also added.

Following the economic structural adjustment programme in 1991, government policy relaxed restrictions on foreign airlines. As a result, Air France launched a Paris–Luanda–Harare service in October 1992. This was followed by KLM's reintro-duction of an Amsterdam–Johannesburg–Harare route. Both the Air France and KLM routes were suspended in the early years of independence to protect Air Zimbabwe. The explanation for the suspension given in the Annual Report of the Air Zimbabwe Corporation (1983) was that Air Zimbabwe's competitors operated larger aircraft on the European routes and so it was at a competitive disadvantage. Both Swissair and Lufthansa have also added flights to Johannesburg which then terminate in Harare. However, as Turton and Mutambirwa (1996 : 458) observe, 'current load factors on these routes suggest that long-haul capacity is in excess of market demand and vigorous fares discounting is taking place'. If transport policy has been characterised by a loosening of state control in aviation, Simon (1996) examines the effect of government policy on Harare's bus services between 1988 and 1993, which have moved from private to mixed ownership with government participation in stage bus operations (Maunder and Mbara 1995) and then moved back to private ownership in 1993. If this experience offers lessons for tourism and aviation, it is that government policy formulation is in a constant state of evolution to meet changing circumstances and ideological shifts.

Air Zimbabwe maintains domestic air services for international tourists with 80 per cent of traffic tourist-oriented on the Harare–Victoria Falls–Hwange National Park–Harare circuit (Figure 3.3). In contrast, on the Harare–Bulawayo route, only 20 per cent of traffic is tourist-oriented. Air Zimbabwe operations on international and domestic routes have been severely constrained by terminal facilities at Harare and smaller airports. A lack of suitable aircraft has also hampered operations. Although a new terminal is planned for Harare, funding the development remains a problem. The Department of Civil Aviation, responsible for aviation services, raised landing fees in 1994 and 1995 by over 1,000 per cent to be in line with European landing fees, although the level of service is not comparable with that in Europe. As a result, these policies may affect the viability of certain foreign airline operations and may thereby affect future arrivals. Victoria Falls airport was upgraded (i.e. a new passenger terminal was added for the Commonwealth Heads of Government Conference in 1991), but it has no night-flying capability. Airports at Kariba and Mutare also have limited runway capability, unable to cater for Boeing 737s. As a result, it is only possible to operate one Air Zimbabwe BAE 146 and some Fokker 50s. The use of appropriate aircraft for the services remains a major operational dilemma for the airline.

After independence, Air Zimbabwe used Boeing 707 aircraft on long-haul flights and in 1983, a Canadian overseas airline consultancy recommended the use of a fleet of Boeing 737s for domestic operations and 747s for long-haul flights. Turton and Mutambirwa (1996) examine the subsequent purchase of a limited number of aircraft and the problems resulting from operating F-50s on a hire-purchase agreement for domestic services. While Air Zimbabwe's revenue is largely generated from domestic services, some fares are below operating costs, which may sometimes lead to state subsidies for tourist services. However, inconsistent policies resulting from state control of the airline and provision of services which are in the public interest mean that Air Zimbabwe cannot effectively meet the competing demands for air services. The inflexible nature of management decision making at Air Zimbabwe was identified by Turton and Mutambirwa (1996). A restructuring of the airline to operate on a commercial basis and a more enterprising approach was seen as vital in the late 1990s.

Competition on domestic routes has been introduced, with Zimbabwe Express leasing BAe 146 aircraft from the government on the Harare–Kariba–Victoria Falls and Harare–Bulawayo routes. While the introduction of competition has aided route capacity, domestic services cannot always cope with the inbound arrivals from Europe on the range of airlines now operating international services. As a result, arrivals often have to stay overnight in Harare, though this has added to the development of urban tourism in the capital. However, proposals for additional direct flights to Victoria Falls from other countries may impact on Harare's gateway status. South African Airways can use Victoria Falls for scheduled and charter flights in conjunction with Air Zimbabwe. Arrivals have increased from 8,370 to 59,000 between 1987 and 1994. There is also pressure from South African Airways to increase capacity on this route. Furthermore, Air Zimbabwe has lost market share by the introduction of long-haul services from rival airlines. This may be compounded by plans in South Africa to promote a South African tourism product of which Zimbabwe will be just one component. This could result in increased charter traffic to regional airports at Victoria Falls and Hwange. It could also mean Air Zimbabwe loses a proportion of its existing domestic tourism market as foreign carriers terminate services in Johannesburg instead of Harare.

The contribution which Air Zimbabwe 'has made to the promotion of tourism has been seriously restricted by the rigidity of state control over finance, aircraft acquisition, pricing and operational policies. . . . The recent liberalisation of air transport within the country is seen as providing the opportunity for Air Zimbabwe to offer improved services between Harare and the major resorts and to develop routes to new tourist areas, all at competitive fares and no longer subject to ministerial approval . . . Air Zimbabwe will have to adjust to . . . the challenge to its domestic service monopoly by other airlines in their bid to link up major Zimbabwe resorts within an international southern African framework' (Turton and Mutambirwa 1996 : 461). From a policy framework, this study highlights the ever-changing nature of tourist transport and the importance of flexibility in government policy to accommodate new business conditions and the difficult balance to be achieved between ideology and effective management to facilitate tourism development through transport.

Since changes in transport policy within individual countries have repercussions on both tourism and on other countries, attention now turns to the role of policy formulation and development among transnational organisations such as the European Union. The single European Act has sought to introduce greater competition and a reduction in state subsidies to achieve greater efficiency in transport provision (see the special issue of the *Journal of Air Transport Management* 3/4 1997 on this theme). Even this is not without problems as the example of the Common Transport Policy indicates.

Towards a common transport policy for the European Union ——

In the case of aviation, the EU policy 'has changed dramatically over the last decade after a move from aviation markets being a series of heavily regulated, discrete bilateral cartels, dominated by nationally owned flag carriers, to a structure which promises by the end of the century to have become a liberalised multinational civil aviation market' (Button 1997 : 170). Accordingly, Button (1996) views this as evolving into an institutional structure similar to that which developed in the USA (see Chapter 6). Aviation was initially excluded from the EU's Common Transport Policy in 1957, with individual countries regulating their own domestic aviation markets. In the international arena, bilateral service agreements were used to regulate air travel. Policies existed which regulated scheduled fares, market entry and service provision.

Changes have occurred in the EU aviation market:

- Individual countries (e.g. the UK) did not free the market entirely; instead the procedure for allocating licences was more liberal. Other countries like France, Italy and Germany have been more reluctant to institute domestic liberalisation.
- Since the mid-1980s, bilateral agreements have been liberalised between member states (e.g. the UK–Netherlands agreement in 1984), which relaxed market entry restrictions. Changes on the UK–Ireland routes indicate that lower fares resulted.
- External changes, such as the USA's 'open skies' policy have assisted EU member states in developing more liberal bilateral agreements.
- In the late 1980s, the EU began to develop a common policy following court rulings which led European politicians to consider liberalisation measures (Stasinapoulos 1992, 1993).

Since 1998, competition rules have been applied to aviation. The EU's three air transport liberalisation packages aimed at removing restrictions on access, price and capacity. The First Package in January 1988 was modest and agreed to a move towards a single market. The Second Package in June 1990 removed 'government to government capacity-sharing arrangements, introduced in stages the notion of double disapproval of fares and prevented governments discriminating against airlines, provided that they conformed with technical and safety standards' (Button 1997 : 172). It also made foreign participation easier, making fifth freedom rights more available (the right of an airline to carry traffic from one point in a foreign country to a point in another foreign country, when this service is linked from an airline's home country to the point in the first foreign country, French 1995 : 45). The final element, the Third Package, launched in January 1993, led to a regulatory framework which emerged in 1997 that is similar to the US domestic market. In other words, since 1997, full cabotage has been permitted and fares are no longer regulated. Foreign ownership among EU carriers can occur although the EU has reserved the right to intervene in the market under certain circumstances: to freeze the market if it becomes unbalanced and to prevent downward spirals on air fares (see Button 1997 for more detail and French 1995 for the effect on regional airlines). Aviation, however, is not the only area for action by the EU as rail travel has also been considered.

Rail transport and rail networks are state-owned in all member states of the EU (despite changes in the UK), and they receive subsidies to assist with the operation of uneconomic services. Yet there are certain governments (e.g. the UK and Sweden) which have sought to introduce greater competition, aspects of privatisation and a more explicit market orientation for rail passenger transport after many years of state control. As Page (1993c) notes, state ownership of rail transport has produced organisations managing national railway networks which are characterised by:

- a close relationship with central government due to state ownership and the provision of operating subsidies
- periodic changes in the political will of national governments to subsidise rail
- constraints on investment planning and a forward-looking approach to development
- a lack of management freedom prior to the 1980s
- an inability to prevent loss of market share in rail travel in the 1970s and early 1980s.

The close relationship between the management of railways and the interests of national governments has meant that policy making has been largely inward-looking and focused on national rather than EU-wide concerns. Gibb and Charlton (1992) have confirmed this view in their analysis of the EU's role as an organisation intended to foster cooperation in transport policy among its member states. They found that in the case of rail travel, the wider considerations of the EU have been undermined by the vested interests of national governments in their own rail networks. As a result, the EU has made little progress towards developing a common rail policy due to two principal objectives of EU transport policy which consistently cause conflict:

- the desire to liberalise rail transport to achieve free trade policies in the move toward the Single European Market[1]
- the need to harmonise the conditions of competition in rail services in pursuit of social intervention policies.

[1] In the Single European Act passed in June 1987, railway transport was not included.

The EU 'Proposal for a Council Directive on the Development of Community Railways' (COM (89) 564 final), which was finalised in July 1991, embodied these principles. It proposed that railway operations and infrastructure should be the responsibility of two separate organisations, so that railway companies can make commercial decisions about long-distance passenger services. This is designed to facilitate an expansion in international rail passenger services where the user entering a national network pays a user fee, thereby reducing the possibility of cross-subsidisation of railway networks by national governments. While the EU aimed to introduce freedom of access to national railway networks across its member states, the changes required by national governments to achieve these objectives imply an end to the highly regulated environment in which state-managed railway enterprises operate. To date, there has been a failure to agree the principles necessary to develop a Common Transport Policy, and rail transport is no exception to this. The prospects for providing access to national railway networks and the introduction of competitive operations are remote, due to the challenge to existing national markets and services. In the absence of an EU Common Transport Policy what should be the main components of an effective state policy for tourist transport?

The Organisation for Economic Cooperation and Development (OECD) has argued that effective state policies for tourist transport should be integrated so that tourism and transport concerns work in harmony rather than in isolation. Transport policy needs to consider tourism as a related activity which is directly influenced by the objectives pursued in national and regional transport policy. In fact Sinclair and Page (1993) examine the role of policy formulation in a transfrontier region (the Euroregion – in North East France and South East England) in relation to transport, tourism and regional development. This is an interesting example where cross-border cooperation and planning for transport may assist in coordinating the variety of bodies involved in implementing tourism and transport planning in two different countries. It also illustrates how a wider policy-making framework can assist in the recognition of the vital link between transport planning and tourism to promote economic development in local areas.

Yet translating policy objectives into a planning framework may pose particular problems for tourist transport systems since the political philosophy of national governments can lead to different approaches to funding and development of transport infrastructure projects. For example, many airports are now commercially operated enterprises, a principle established in the UK government's White Paper *Airports Policy* (HMSO 1978) which was mirrored in many other European countries (see Doganis 1992 and Chapter 7 for more detail). In this respect, infrastructure provision is now based in the private sector, with the UK government's market-led philosophy also applying to complementary infrastructure provision. For example, this was demonstrated by the public/private sector partnership between BR and the BAA's £300 million Heathrow Express rail-link project to connect Heathrow and Paddington in London (see Chapter 7).

Where countries do formulate a tourist transport policy, the following issues should be taken into account in its implementation through planning measures:

- the management of tourist traffic in large urban areas (see the London Tourist Board 1987 : 61–3, on managing transport used by visitors, particularly the problems of coach parking which is also discussed by the Bus and Coach Council 1991) and small historic cities (see Page 1992a for a case study of Canterbury). In 1997, the English Historic Towns Forum (1997) established a Code of Practice for coach tourism to address the issue among its members in line with growing concerns about managing tourists and non-tourists and traffic in town centres (Page and Hardyman 1996)
- the management of tourist and recreational traffic in rural areas (see Sharpley 1993 for an up-to-date assessment)
- the promotion of off-peak travel by tourists to spread the seasonal and geographical distribution of tourist travel and the resulting economic, social, cultural and economic impacts of tourism
- maximising the use of existing transport infrastructure and the use of more novel forms of tourist transport together with the provision of new infrastructure on the basis of long-term traffic forecasts
- integration of transport modes (Stubbs and Jegede 1998)

With these issues in mind, attention now turns to a case study of the state's role in tourist transport to assess which objectives are being pursued towards tourist rail travel in the UK.

CASE STUDY   The organisation and management of a tourist transport system: tourist rail travel in the United Kingdom

In this case study, the organisation and management of a tourist transport service – rail travel – is examined to illustrate the direct and indirect impact of government transport policy on a tourist transport system. The case study examines the implications of government policy prior to and after privatisation in 1994. It highlights how a systems approach, government regulation and policy issues need to be viewed in relation to the provision and consumption of a service for tourists which operates as an integrated system. In this system, the main input comprises the supply of a service and the demand for tourist travel, while the principal output comprises the tourist travel experience as a service encounter (see Figure 3.4), thereby introducing some of the ideas and concepts developed in Chapter 2. The nature and organisation of this system of tourist travel is indicated in Figure 3.4 which identifies the inter-relationships between the provider of the service and the purchaser (i.e. the tourist). As Figure 3.4 shows, a complex set of relationships affect the supply and demand for tourist travel by rail, with a variety of organisations directly affecting the supply via regulatory and financial controls (e.g. the Department of Transport and the Treasury). However, prior to discussing the role of the state in policy formulation and its implementation, it is useful to establish the dimensions of the supply and demand for tourist rail travel in the UK.

Tourist travel by rail in the UK pre-privatisation
Turton (1992a) examines the role of rail passenger transport in the UK, providing a broad overview of the various factors which affect the organisation and management of passenger services, the impact of government policy and the relative importance

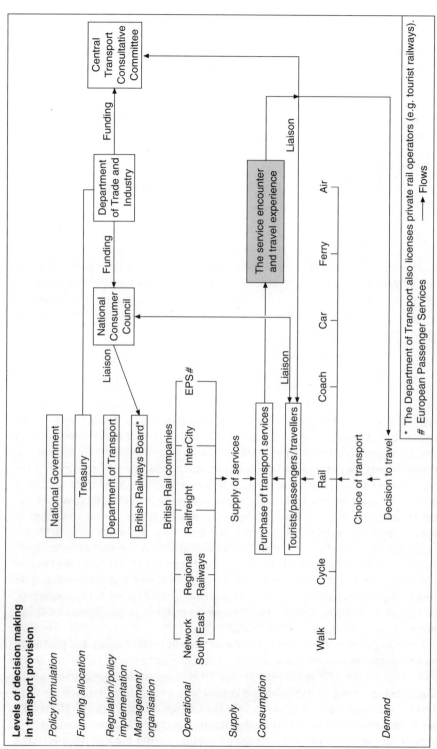

Figure 3.4   The tourist rail system in the UK prior to privatisation (redrawn from Page 1994)

of rail versus other modes of transport. The UK rail network comprises 16,585 route kms and some 719 million passenger journeys were undertaken in 1981/82 and 738 million journeys in 1996/97. Potter (1987) argues that the main determinants of the demand for rail travel are:

- speed
- cost
- comfort
- convenience
- access to stations
- the image of the service.

These affect the perceived quality of the service and are often judged in relation to competing modes of transport.

In terms of tourist demand, Turton (1991) reports that the Long Distance Travel Survey (1979–80) found that rail had a 20 per cent share of all long-distance trips (those over 40 km). According to Jefferson and Lickorish (1991), the number of domestic tourists using rail to reach their holiday destination in Great Britain dropped from 13 per cent in 1971 to 10 per cent in 1985 and 8 per cent in 1989. For journeys to the European mainland which included a sea crossing, 8 per cent of British tourists travelled by BR in 1986. In 1991/92, the business traveller using InterCity (long-distance high-speed rail services) comprised:

- 23 per cent of InterCity passenger volumes
- 35 per cent of receipts, and
- approximately 8 per cent of the UK long-distance travel market.

According to Green (1994) InterCity accounted for 8 per cent of all long-distance trips in the UK and 52 per cent of these trips were classified as 'leisure travel' (which includes tourist travel). A further 27 per cent of travellers were undertaking business trips which could also have an element of tourism since some of these trips would involve an overnight stay away from the travellers' normal place of residence. There is also a wide-ranging debate amongst transport analysts as to whether rail travel has experienced a drop in public support and patronage (Banister 1994), while Page (1993c) observed that there is evidence that in a European context rail travel is experiencing a renaissance. Public support is evident from the British Social Attitudes Survey reported in *Social Trends 1997* (HMSO 1997a), with 62 per cent of respondents disagreeing with the statement that rail services that do not pay for themselves should be closed down. A further 58 per cent of respondents agreed that 'Britain should do more to improve its public transport system even if its road system suffers'. However, the car still accounts for 88 per cent of passenger traffic in Britain, compared to rail with a 4.5 per cent share in 1995 (HMSO 1997a).

Rail travel in the UK has seen a significant drop in public support and patronage since the 1970s as competition with other modes of transport (e.g. air, car and coach travel) have eroded rail's market share. How far is this a result of government policy as opposed to changing attitudes and preferences among travellers, and to what extent is the decline in the use of rail for tourist trips a unique feature of the UK? (See Page 1993c for a discussion of the situation in other European countries.) For example, in the 1970s and early 1980s, BR penetrated the domestic long-distance

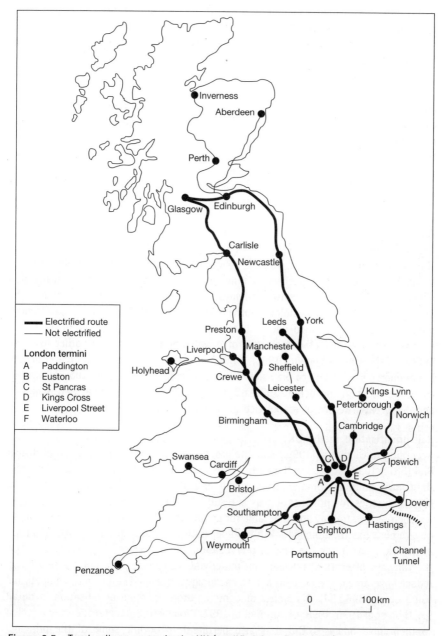

Figure 3.5 Trunk railway routes in the UK (modified from Page 1994)

holiday market, selling over 170,000 'Golden Rail Holidays' per annum in the late 1980s (Lavery 1989) but its share of the long-distance travel market continued to decline, despite attempts to diversify its product base. A significant proportion of tourists in the UK use trunk rail routes (Figure 3.5) and interurban travel dominates the traffic flows (Turton 1992b) since much of the demand emanates from the main towns and cities, which also contain many of the UK's main tourist gateways. Those rail routes which experienced the greatest volume of traffic were the electrified east and west coast main lines, linking London and Scotland with many of the UK's major towns and cities. Feeder services from rural areas to the main trunk routes and cross-country express services during periods of peak demand (e.g. summer services from inland towns and cities to coastal resorts) provide additional services connecting most destinations to the rail network.[2] Prideaux (1990) notes that the demand for InterCity rail travel in 1989 was 8.5 million passengers who travelled 77 million journeys. Within the overall demand, different segments are discernible (e.g. youth travel, business travellers, the elderly and disabled), each of which generates its own patterns of demand. The provision of such rail services is subject to the prevailing policy objectives of the national government, with the Department of Transport responsible for the implementation of the policy.

Government regulation of rail services in the UK prior to 1994: the Department of Transport
The Department of Transport was the government department responsible for regulating rail travel and it licensed a number of tourist railways until privatisation measures in 1994. It also licensed railways operated on a seasonal basis to meet the demand for scenic and nostalgic journeys (Halsall 1992). Since 1981, government policy in the UK has rolled back the role of the state in the regulation of public transport through selective measures of privatisation as discussed earlier (e.g. the 1985 Transport Act – Knowles 1985, 1989) in an attempt to develop a greater degree of commercialism in service provision. For example, BR's government subsidy was cut by 60 per cent in the period 1983–89. For BR, this has meant that it has had to adopt a greater degree of market-led planning towards rail services.

Rail privatisation in the UK
According to Gibb et al (1996 : 35), 'the process leading to the privatisation of British Rail (BR) actually started in the early 1980s and has been a more protracted process than most political commentators credit' and 'BR has been better prepared than most industries for privatisation as a result of a restructuring process that began well before the strategy and timetable for privatisation became clear'. Britain's Labour government has to launched a White Paper outlining the future for transport policy in the UK. However, in the case of privatisation, the government's policy to privatise the railways after 1992 is clear: an 'underlying neoliberal philosophy – seeking to reduce government involvement and to encourage a free market driven by competition' (Gibb et al 1996 : 37). A number of commentators examine privatisation in detail (i.e. Gibb et al 1996, 1998; Charlton et al 1997; Wolmar 1996) and readers are directed to their lucid analyses. But it is also pertinent to outline some of the principal changes that have occurred and the implications for tourist rail travel.

In 1992, the Department of Transport announced its plans to implement government policy objectives by introducing further measures to privatise the railway system (Department of Transport 1992a, 1992b, 1993). Whilst seeking to reduce government

---

[2] The recent recession in the UK, combined with falling passenger demand and profitability highlighted the need to be more market-led.

Table 3.5   Company groups within the rail industry according to their function

Function	Description
Central	Consultancies, research and telecommunications services
TOC	Train Operating Companies running passenger services
BRIS	British Rail Infrastructure Services (e.g. track renewal and maintenance)
BRML	British Rail Maintenance Limited (e.g. rolling stock maintenance and construction)
ROSLO	Rolling stock leasing companies
GoCo	Railtrack, Union Railways and European Passenger Services (EPS). The ownership of these businesses was transferred to central government prior to privatisation. Union Railways was charged with planning the high-speed rail link from London to the Channel Tunnel. Union Railways and EPS were combined as part of the deal with LCR in 1996 to develop the high-speed rail link.

Source: Modified from Charlton et al (1997).

subsidies for rail travel, a hybrid solution was sought to introduce private sector investment and involvement. Gibb et al (1996 : 38) show that 'up to 1982 the railways had been run as five geographical units. This was replaced with five business sectors – InterCity, Network South East, Regional Railways, Parcels and Freight.' Yet privatisation plans paradoxically were not necessary on financial grounds by the early 1990s, since InterCity and Freight were operating without subsidy and the Department of Transport's (1992a) *New Opportunities for the Railways* acknowledged that BR was the most efficient of any European railway. At the time of privatisation, the existing business units were not selected as the best option for sale. Instead the Department of Transport's (1992a) proposal intended to sell

> both freight and parcels to the private sector, the breaking up of passenger services into franchises to be operated by the private sector, the provision of rights of access to the rail network for private operators of freight and passenger services (overseen by a Regulator), the separation of track and train services with one part of BR becoming a track authority (Railtrack) with responsibility for running the track and infrastructure, and opportunities for the private sector to purchase or lease stations. (Gibb et al 1996 : 43)

The Department of Transport's (1994) *Britain's Railways: A New Era* outlined the final structure of the privatised railway industry, where BR was compartmentalised into independent businesses, each cooperating on a commercial and contractual basis. As Charlton et al (1997) argue, the government's desire to break BR's monopoly was enacted using the concept of a *common carrier*, which avoided the provision of new infrastructure. By separating ownership of infrastructure from service provision, the common carrier could provide customers with access to its resources – the companies who compete to offer rail services to the public. In this case, Railtrack is the common carrier. The rail industry now comprises a series of companies which operate rail services or support services (Table 3.5) and Figure 3.6 summarises the interactions in

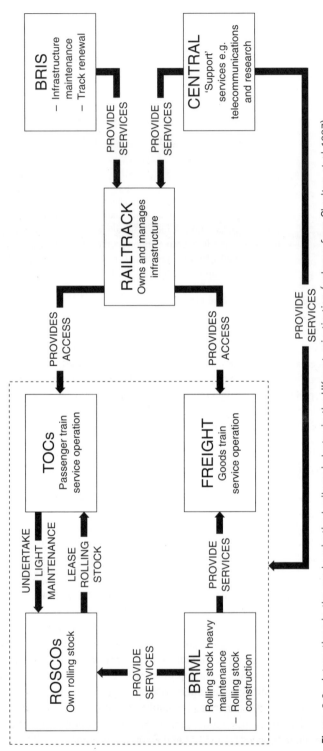

Figure 3.6 Interactions in the newly privatised railway industry in the UK post-privatisation (redrawn from Charlton et al 1997)

Table 3.6   Rail franchises announced in 1993 by former British Rail sectors

Date	InterCity	Network South East	Regional Railways
Feb 1993	East Coast Main Line (ECML) Gatwick Express Great Western Main Line	Isle of Wight Line[1] London–Tilbury–Southend South Western Division	ScotRail
May 1993	Anglia Cross Country Midland Main Line West Coast Main Line (WCML)	Chiltern Great Eastern Kent Services Thames Thameslink South London and Sussex Coast Northampton and North London West Anglia and Great Northern	Cardiff Valleys Line Central North East South Wales and West North West Mersey Rail Electric Services

[1] The Isle of Wight Line is a small railway and its management structure and operation are different from those which exist under privatisation. The franchise operates the line on a vertically integrated basis, where the operator is also responsible for track, infrastructure and operations.
Source: After Gibb et al (1996) based on the Department of Transport (1994).

this new system. Not only is this a complex operational structure (Charlton et al 1997), as a result of privatisation, 92 businesses emerged and 67 were offered for outright sale. The 25 train-operating companies (TOCs) were offered as regionally based franchises through a tendering process (Table 3.6). Railtrack is also responsible for central timetabling, train planning, signalling and the structural condition of the 2,500 stations leased to TOCs. Railtrack operates in 10 zones as illustrated in Figure 3.7. Charlton et al (1997) argue that government intervention and regulatory action have effectively prevented competition. Most TOCs are still supported by government subsidies to procure rolling stock and access to track. In fact, as Charlton et al (1997 : 151) indicates, 'in line with its [government] desire to break up BR's monopoly, the government also requires the Rail Regulator and Franchise Director ... to promote competition in the provision of railway services. However, so long as TOCs are not financially viable, the goals of protecting passengers' interests and promoting competition are largely incompatible'. In other words, the first round of rail franchises (Table 3.6) have been protected from competition so that no new competition will be allowed before 31 March 1999, and a wide range of restrictions will remain until 2002. Only in a few circumstances (e.g. where TOCs have separate routes between similar destinations or two or more operate the same route) does competition exist. Thus, 25 existing TOCs on balance have regional monopolies until 2002.

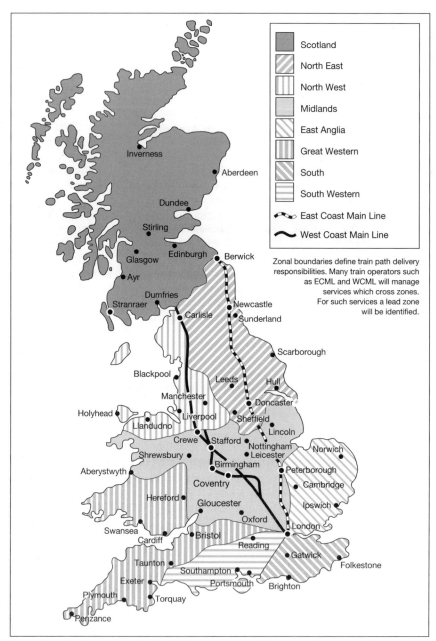

Figure 3.7   Railtrack zones in the UK (redrawn from Gibb et al 1996)

In a competitive context, Charlton et al (1997 : 152) observe that BR's monopoly has been replaced by an oligopoly due to horizontal and vertical integration (see Chapters 5 and 6 on supply issues) where corporate interests have acquired TOCs, franchises and rail businesses. For example, the coach operator National Express integrated horizontally after obtaining the Midland Main Line and Gatwick Express franchises. It also integrated vertically by purchasing the rolling stock company from which it leases the trains. Other examples also exist which have reduced competition. In terms of the outlook for tourists and other travellers, both Cole (1994) and Gibb et al (1996) observe that the future is uncertain. However, in a policy context the 'philosophical desire to promote competition may be dangerous both for the railways and for public transport more generally. Instead of a high-quality, responsive and affordable means of transport, the traveller may be left with a fragmented, confusing and expensive-to-use railway suffering from under investment and decline' (Gibb et al 1996 : 50–51). The new climate for rail services has effectively ended BR's 44-year-old role as the main provider of rail services, and reflects a political ideology dominated by a desire to improve efficiency and introduce competition (European Conference of Ministers of Transport 1992). The outcome is a more complex organisational framework for the tourist rail experience as outlined in Figure 3.8 (compare this with the pre-privatisation model in Figure 3.4). Recent studies have examined the utility of such an approach (e.g. Adamson et al 1991) and critics of privatisation have argued that lack of government investment is responsible for the poor performance of the nationalised railway company. Yet within the European railway industry, it is widely acknowledged that BR operated one of the most efficient railway networks in Europe, in view of the level of staffing and government subsidy provided per passenger kilometre travelled. Even so, the government encouraged public service providers such as BR to improve service quality and to establish a *Passenger's Charter* following the launch of the *Citizens' Charter* initiative in 1991 (National Consumer Council 1991).

It is too soon to assess the long-term impact of privatisation and the likely effect on tourist service provision. For this reason, the discussion focuses on BR and its long-distance passenger service – InterCity. This is because the case of InterCity illustrates how the establishment of a new business unit led to a customer-focused approach for tourist (leisure-traveller) and non-tourist travel. By emphasising how InterCity re-engineered its organisation to provide a travel experience, the study recognises the holistic nature of the components that affect the perception and evaluation of that experience. Although the InterCity example may now be less relevant following privatisation of rail services, the concepts and lessons it offers to other transport organisations seeking to become more customer-focused are informative in terms of competition with other modes of long-distance transport. Furthermore, if one accepts the Gibb et al (1996) argument that the privatisation of BR predates the 1994 legislation, then the formation of the InterCity brand was a measure of privatisation after 1982.

British Railways and rail service provision

Prior to the 1994 legislation (Department of Transport 1994), the majority of long-distance tourist trips by rail in the UK were undertaken on InterCity and to a lesser extent on regional railways, which competed with the express coach business (Page 1994a) and domestic airlines for business and leisure travel. In a climate of reducing government subsidy for rail travel, BR did achieve a complete reversal in InterCity's financial performance from a £157 million loss on £623 million passenger revenue in 1987/88 to a £324 million profit on passenger revenue of £725 million in 1988/89. In 1991/92, InterCity's income was £896.7 million, with £829.7 million from passenger

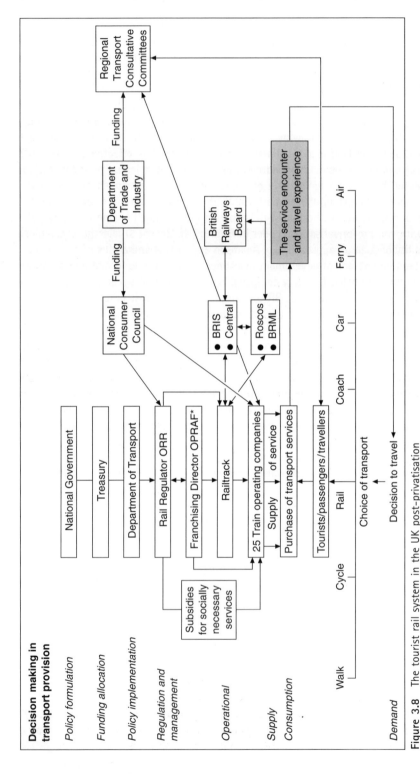

Figure 3.8 The tourist rail system in the UK post-privatisation

*Note: In July 1998, a government White Paper proposed the abolishing of OPRAF and the establishment of a strategic Rail Authority to create a more integrated system.

fares, £45.3 million from non-fare income (e.g. catering) and £21.7 million from ancillary services. In 1992/93, InterCity generated £889 million in revenue and produced a profit of £82 million, indicating the success achieved in its main business.

Prior to privatisation, in the medium-to-long term, the real potential for BR long-distance tourist travel was to expand its market share of the European passenger services running from London to mainland Europe through the Channel Tunnel (Page and Sinclair 1992a,b; Page 1994b, 1993d). Following privatisation, these services were awarded to a consortium – London and Continental Railways – in 1996. The consortium, led by National Express and Richard Branson's Virgin Group, was to use the profits from the service to fund the 68-mile Channel Tunnel high-speed rail link (Harper 1998a). However, in January 1998, LCR announced it was unable to raise sufficient capital to fund the scheme, and requested the government to increase its share of the development to £3 billion. In the event of deadlock, the government has the authority to take back the 18 Eurostar trains given to LCR to operate the service. This is because the agreement was that the assets were only transferred to LCR until it met its obligation to build the route. This is a further example of the problems associated with privatisation.

British Rail and the customer: the tourist travel experience and service encounter

For the tourist, the journey on a train represents the consumption of a travel service. From a tourism perspective, this is the culmination of the desire to travel and the selection of a mode of transport to fulfil that need. Researchers such as Poon (1989) acknowledge the growing sophistication of tourists and their higher expectations from travel and tourism services. For the supplier, it has led to a much greater focus on service quality. Within BR, those sectors which deal with leisure and business travellers on a regular basis (e.g. InterCity) have recognised the need to provide a consistent level of service which meets the expectations of customers.

InterCity rail services: delighting the customer

In a retrospective review of the development of InterCity in the 1980s and early 1990s, Vincent and Burley (1994 : 107–8) reveal that

> InterCity had to reorientate its thinking so that its activities reflected a market-led organ-isation. InterCity no longer sold train trips but aimed to provide a total travel experience. This meant thinking about customers' needs, not just when they arrived at stations to buy tickets, but when they were sitting at home planning to make a rail journey.

Key questions this posed for senior management included:

● How did customers want to make their first contact with InterCity?
● Were the telephone information points of contact welcoming?
● Was the route to the station adequately signposted?
● Did the station provide sufficient secure parking spaces?

The limitations on InterCity making further investment after 1982 meant the company turned to other solutions to improve market share in long-distance travel. Having become a profitable entity in 1988, InterCity sought at a time of severe recession

Customer needs

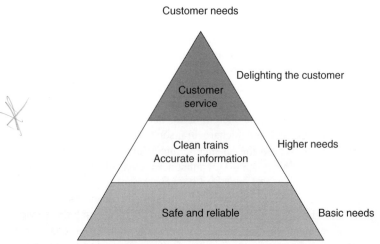

Figure 3.9   The hierarchy of rail travellers' needs (modified from Vincent and Green 1994)

to develop a business strategy to build the business. The vision for the 1990s which the company devised was one called 'Delighting the Customer' (Vincent and Burey 1994). Customer service and the travel experience were selected as avenues for development.

A radical change of culture was required to move from a product-led and operationally driven monolithic state enterprise to an organisation which was customer-focused. Within a two-year time frame, it required retraining and re-engineering an organisation of 30,000 employees. A guiding principle from the central management of BR, 'Quality Through People', this was the basis upon which changes would be built. Two driving forces in the process of change were empowerment of staff and teamwork to deliver tangible changes. This is a key component of total quality management (TQM) which is discussed in more detail in Chapter 9. One example of the culture change was the introduction in 1989 of a senior conductor role on board InterCity train services to replace the traditional guard function. Guards had been chiefly responsible for revenue protection and safety and they had not worked exclusively for InterCity. As senior conductors, retrained guards were empowered to deal with problems associated with delays (e.g. authorising customers to receive free refreshments if severe delays occurred) and to be more visible on board trains. They were selected to be ambassadors for the business. By listening to the problems experienced by the new senior conductors, senior managers were able to introduce innovations to keep them in contact with land-based service centres. This enabled conductors to pass problems to service centres for immediate resolution.

One major catalyst for change was the arrival of a new manager of InterCity in 1992. This led to the recognition of a hierarchy of customer needs modelled along the lines of Maslow's basic needs study (Hall and Page 1999). Figure 3.9 reveals three levels of needs which had to be addressed before the customer would be satisfied. The base of the pyramid illustrates the operational focus of most transport providers – to offer a safe and reliable service. In-house research conducted by InterCity in

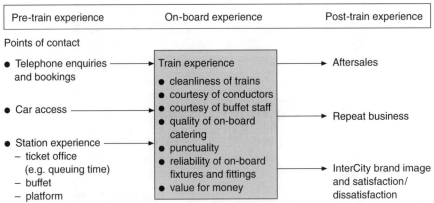

Pre-train experience	On-board experience	Post-train experience

Points of contact

● Telephone enquiries ⟶ Train experience ⟶ Aftersales
  and bookings
  ● cleanliness of trains
  ● courtesy of conductors
● Car access ⟶ ● courtesy of buffet staff
  ● quality of on-board ⟶ Repeat business
  catering
● Station experience ⟶ ● punctuality
  – ticket office   ● reliability of on-board
  (e.g. queuing time)   fixtures and fittings ⟶ InterCity brand image
  – buffet   ● value for money   and satisfaction/
  – platform   dissatisfaction

Figure 3.10   InterCity's 'trail to the train' model (modified from Vincent and Green 1994)

1991 recognised that it had a good product which was inconsistently delivered. As a result, to achieve the pinnacle of delighting the customer, a systematic review of service delivery and presentation was undertaken.

InterCity developed the concept of the 'trail to the train' (Figure 3.10) to identify difficulties travellers experienced and where improvements could be made. Three-monthly opinion polls were undertaken by external market research organisations to assess customer responses to service improvements and the ongoing problems. A number of practical responses to perceived problems included:

● Rationalisation of InterCity's telephone enquiry bureaus since 1,800,000 calls a month were made to the company but inconsistent answers were given to callers. By establishing 20 dedicated centres across the UK, introducing computerised timetables and telephone-answering technology, major service improvements were introduced, notably the conversion of calls to sales through a telesales service.
● Car parking facilities were enhanced at key stations, including the introduction of security surveillance equipment.
● A greater emphasis on customer welcome was pioneered to mirror the service reception provided by airlines. Since 90 per cent of InterCity travellers passed through 20 stations, 'individuals were selected because they wanted to work in a close customer environment' (Vincent and Burley 1994 : 113) at key stations to give InterCity a heart and positive customer interaction. One tangible innovation was the introduction of claret uniforms which were seen as synonymous with friendly service.
● Platforms were zoned to ensure travellers could identify the positioning of First Class and Standard Class carriages, following best practice in Germany.
● Improvements to engineering workshops to reduce on-board defects to trains improved the reliability of the on-board travelling environment.
● On high-frequency, middle-distance routes an InterCity Shuttle service was introduced to compete with motorway travel. These services eliminated ticket checks at barriers and replaced them with a trainside welcome. Trains were available to board 20 minutes before departure and buffet cars were opened and remained open throughout the journey and during their turnaround. Catering trolleys also moved throughout the train during the journey.

Vincent and Burley (1994 : 117) summarise the business principles which characterised the InterCity culture by 1994:

Plate 3.1 Interior of a refurbished GNER 225 carriage post railway privatisation (reproduced courtesy of GNER Railways, York, UK)

- High-speed passenger rail businesses must be market-led and customer-driven.
- Operations and production are clearly vital to the efficient and safe running of a railway, but they must be subordinate to the customer and the market if they are to survive.
- Customers' needs have to be thoroughly understood and acted upon by all staff.
- Good customer service is a key concern for all those who work for InterCity.
- Consistent standards of customer service are essential to a national brand under pressure from competition.
- Giving customers added value is paramount for repeat business.

This service encounter begins at the point when the ticket is purchased and includes the time spent waiting at the station through to the completion of the journey. The experience on board the train is also deemed to be one of the most memorable aspects of the journey and an integral part of the service encounter. Shilton (1982) suggests that while it is difficult to quantify the quality aspects of InterCity services, factors such as the condition of the rolling stock (Plate 3.1), the image of the high-speed trains (Plate 3.2) and additional advertising, accounted for half of the increased patronage of the long-distance travel market in the early 1980s. InterCity has also been conscious of the need to expand its share of tourist and non-tourist high-speed rail travel in the long-distance travel market by introducing new products (e.g. Family Railcards), cultivating niche markets such as off-peak premium travel (e.g. 'Weekend First' – discounted first-class travel) as well as budget advance purchase (Apex) tickets. As two-thirds of InterCity's business travellers chose second-class (standard) accommodation (Prideaux 1990), InterCity introduced the 'Silver Standard' service for full-fare standard-class passengers. This included free seat reservations,

Plate 3.2 A former British Rail InterCity 225 high-speed train repainted in GNER colours post railway privatisation (reproduced courtesy of GNER Railways, York, UK)

complimentary airline-style meals and steward service; it was a deliberate attempt to offer a competitive service equivalent to what was currently provided by domestic airlines.

The demand among more discerning travellers for enhanced service quality on InterCity services meant that new services were also provided for different market segments (Green 1994). In addition, the quality and ease of interchange facilities at major rail termini in the UK (e.g. at the ports, airports and in major cities – see Stubbs and Jagede 1998) are important factors affecting the tourist's travel experience and may influence the decision to travel by train.

Interchanges need to be integrated, with easy access to other modes of transport for onward travel, so that the commencement and completion of the journey are not marred by difficulties in making connections. One further development which has had a wide-ranging impact on the service quality of tourist use of rail, was the introduction of BR's *Passenger's Charter* which has now been a key feature post-privatisation for TOCs.

British Rail's *Passenger's Charter*: new standards for tourist rail travel?

It is widely acknowledged that BR was perceived by passengers in the 1970s and early 1980s as an archetypal state-owned enterprise, with a negative image in relation to service provision. Although there is no simple explanation for this image, factors such as limited public funding by successive governments since the 1960s and operational difficulties (e.g. service interruptions) have contributed to the poor public perception.

Table 3.7   The main features of the British Rail's *Passenger's Charter*

- Establishing performance standards and targets for groups of routes
- Setting punctuality standards (e.g. 90 per cent of InterCity trains should arrive within 10 minutes of their scheduled time)
- Setting reliability standards (e.g. 99 per cent of InterCity services should run)
- Waiting time at ticket offices should be a maximum of 5 minutes in peak periods and 3 minutes at other times
- The quality of staff communication is to be improved through training, particularly at retail outlets (e.g. ticket offices and enquiry bureaux)
- Managerial staff are to be placed on a performance-related pay scheme to ensure they reach predetermined targets
- Passengers should be kept informed prior to and during their journey of any delays
- Engineering work is to take place, wherever possible, at times which minimise inconvenience to passengers
- Those with special needs (e.g. the disabled) are to be given assistance
- Refunds are to be made for delayed or cancelled journeys within the confines of the new compensation scheme

---

Source: Based on the British Railways Board (1992) *The British Rail Passenger's Charter*, British Railways Board, London.

For example, the 'Consumer Concerns 1990' survey by the National Consumer Council (NCC) found that only 40 per cent of BR users were very satisfied to fairly satisfied with the service provided. This was confirmed by BR's acquisition of the Consumers' Association's 'Captive Consumer Award' in January 1992 for failing to adequately compensate consumers for poor service.

BR stated that some £7 million was paid to its customers in 1991 as cash refunds and travel vouchers through its Discretionary Compensation Scheme. However, to place service and performance on a more prominent footing, BR launched its *Passenger's Charter* in May 1992, with a view to having it fully operational by 1993. The funding of this new scheme was expected to cost £10–15 million and it aimed to ensure that rail travel for tourists and commuters meets certain standards. The charter set out to provide:

- a safe, punctual and reliable train service
- clean stations and clean trains
- friendly and efficient service
- clear and up-to-date information
- a fair and satisfactory response if things go wrong
  Source: British Railways Board (1992).

This required BR to establish standards for services and the most publicised aspect of the charter was the provision of compensation for delayed or cancelled rail services. From BR's point of view, it has meant publicising its internal procedure for measuring performance. This process required data generated through its Management Information Systems to monitor and evaluate different aspects of its business (Allen and Williams 1985). The main features of the *Passenger's Charter* are set out in Table 3.7 which shows that performance indicators and clear targets had to be met in terms of service provision. These were reviewed and published on a regular basis. In managerial terms, the process of producing a charter may have the advantage of encouraging the development of a corporate culture based on service quality, but staff training and commitment are required if this service ethos is to permeate staff consciousness in an organisation employing over 138,000 staff. Since the privatisation bill, the new TOCs have each introduced their own passenger charter and the Office of the Rail Regulator now has the power to impose fines on companies for non-compliance. TOCs are expected to maintain an operational environment as prescribed by the National Conditions of Carriage (British Railways Board 1996) and each must produce a passenger charter which is user-friendly and as favourable as that of its BR predecessor. In some cases, TOCs have sought to aspire to higher standards than their BR predecessor (Gibb et al 1998). Table 3.8 outlines the key features of one passenger charter for Great Eastern Railway Limited which operates services east of Liverpool Street station in Central London. Gibb et al (1998) observe that new passenger charters also have to set out TOCs' commitment to operating reliable and punctual train services, fares and facilities, to meeting the needs of disabled travellers and to putting measures in place to deal with delays. A useful comparison with the UK situation can be found in the Singapore Mass Rapid Transit Ltd (1997) *Customer Charter – At Your Service: SMRT's Commitment to Passengers*; it pledges:

- 94 per cent of trains to arrive within two minutes of schedule
- 96 per cent of trains to depart within two minutes of schedule
- train service availability to be at least 98 per cent
- a maximum of 6.7 failures per 100,000 uses of ticket vending machines
- a maximum of 5 failures per 100,000 uses of ticket gates

In contrast to the UK, each MRT station has customer feedback forms readily accessible, a World Wide Web page, a toll-free hotline and a voice-mail message system for out-of-hours contacts.

Critics, such as the Consumers' Association, have argued that passenger charters are a glossy public relations exercise with only a limited amount of finance devoted to the scheme, which may be better deployed on improving services rather than compensating travellers for poor service. Sceptics have also argued that the targets set by different rail services are too low. Nevertheless, such charters have publicised the importance of service quality and provided tourists and other travellers with clearer information on their role as consumers, and the standards and attributes to expect from a rail journey. However, the NCC is still seeking further improvements to the process of producing charters for public services to achieve long-term benefits for consumers (National Consumer Council 1992).

So what does the case study tell us about the tourist transport system?

- Marketing research is an important part of the strategic planning process for a tourist transport service, particularly when attempting to assess what tourist and non-tourist travellers require from a rail journey. This has a critical role in establishing a benchmark for service quality to ensure minimum standards of provision are met.
- Government policy has adopted a long-term policy favouring privatisation under the guise of improving consumer choice and service quality, citing the transformation in former state-owned enterprises which have been privatised and are more customer-led. Yet as Charlton et al (1997) indicate, turning policy into reality has been problematic due to internal contradictions associated with the logic of inducing effective competition in rail travel.

Table 3.8 The main features of Great Eastern Railway's passenger charter, January 1998

The purpose of the Charter

• Conditions of carriage

• Our standards: How are we doing?
                  – monthly reports of reliability to be published at key stations
                  – publication of consumer surveys about the service

• Being on time
Train times (actual arrival times) are measured against the advertised timetable on Monday to Friday departures between 0700–0959 and 1600–1859 (peak times arriving at Liverpool Street Station). The targets for services are:
– up to December 1997: 88 per cent of trains will arrive at their destination within 5 minutes of their scheduled time
– January to December 1998: 89 per cent of all trains will arrive at their destination within 5 minutes of their scheduled time
– after January 1999: 90 per cent of all trains will arrive at their destination within 5 minutes of their scheduled time

• Reliability (the percentage of trains that run instead of those cancelled)
A target of 99 per cent of all published services to run between Monday and Saturday

• Planning your journey
– details on a dedicated customer service centre
– details on buying a ticket
– how to make a national rail enquiry
– timetables, posters and leaflets provided to show how to contact Great Eastern Railway on its services
– train services and engineering works
– fares
– bicycles on trains
– policy on smoking on trains

• Looking after you
– sources of information which update the travel situation
– staffed and unstaffed stations
– train service updates
– on-board service announcements
– weather-related delays
– refreshments
– safety

• Customers with special needs

• Asking your views
– at the station
– during the journey
– contacting the Customer Relations Team
– reviewing your comments

• Paying compensation for providing a poor service
– season ticket holders
– all ticket holders
– journeys you do not make

Table 3.8   Cont'd

- Penalty fares

- Our commitment to customer service

- Other commitments

- Summary of customer contacts

- Route maps

Source: Great Eastern Railways Limited (1997).

- Since the 1970s, tourist use of railways in the UK has declined but there is potential for increasing the number of journeys involving air and rail for distances of 500–1,000 km now the Channel Tunnel has opened. In 1995, 1.9 million outbound UK residents used the Channel Tunnel as a travel mode and 1.7 million inbound tourists chose it as a means of transit (HMSO 1997a).
- Further investment is needed in the UK's aged railway rolling stock and infrastructure to improve journey times. This is likely to be limited within the context of existing policy which favours a greater role for the private sector. It is unlikely to yield major investment in new rolling stock as the franchises available to private operators are only offered on a short-term basis, though some TOCs such as Richard Branson's Virgin Railways have made massive investments on the west coast main line.
- The introduction of the passenge charter is a direct consequence of government policy towards public service provision and it has led to a greater emphasis on service quality in the tourist transport system.
- Since privatisation, the unified InterCity brand and image built up since 1982 has been supplanted by the TOCs with their narrow focus on their own services rather than a broader rail travel perspective.
- For international tourists visiting the UK, the changes since privatisation are complex, confusing and not visitor-friendly. Even long-distance booking through services which cross several TOCs is a deterrent to tourists' choice of rail if it is not simple to make and readily available at local rail stations rather than at regional centres.
- The long-term implications for rail service provision outside of the main conurbations are less clear. The integrity of the existing network is in question as many of the TOCs operate under a subsidised regime and a series of privately owned monopolies have replaced BR's monopoly. Even with the new government White Paper of mid-1998, the situation has changed dramatically. However, it is worth considering Whitelegg's (1998 : 21) comment on the London Underground which received no government subsidy operating grant. It is an indication of what can happen if a private sector ideology dominates public transport as 'two and a half million journeys are made . . . every day . . . Travelling on it is also a very unpleasant experience . . . A poor-quality public transport system sends out strong signals to those seeking work, tourists and inward investors. They are less likely to choose London as a destination if their experience of moving around, is one of delay, overcrowding, inconvenience and misery.'
- Government policy can radically affect the way in which transport services are delivered to tourists and non-tourists as well as ensuring a simplistic or complex operational structure exists.
- The degree of privatisation and model chosen in transport policy determine the style of competition and likely benefits which will result for consumers. It is too soon to assess the

long-term competition which will result for rail travel, with many franchises protected from competition and receiving subsidies. If effective competition were to arise even in a modified form akin to bus deregulation in the UK, tourist and leisure travel would assume an even more complex form that might exacerbate the role of the car as a preferred mode of transport.

• Speakman (1996 : 47) argues that 'the current process of rail privatisation is . . . a serious threat to UK tourism. The recent controversy about the proposed withdrawal of the West Highland Sleeper Service focuses on the degree to which rail travel is still . . . a vital part of the region's tourism product.' The privatisation process may also mean a loss of discounted travel and accommodation arrangements for tour operators. This is serious for rail companies since Speakman (1996) estimates that 40 per cent of rail travel is leisure-based. One welcome sign is that one of the former elements of BR's regional railways division – ScotRail – has launched a short breaks brochure and discounted rail tickets and a Freedom of Scotland Travel Pass (with discounts on other forms of public transport) and discounted entry to local tourist attractions. In fact now some rail companies (TOCs) are owned by bus companies, such ventures should be a natural development. This will assume an even greater significance as inbound rail travellers from mainland Europe choose to use services beyond London to the UK regions. Speakman (1996 : 48) is optimistic that 'a more entrepreneurial approach . . . [by TOCs] . . . together with a trend towards more environmentally sustainable forms of tourism . . . could open the way to some exciting and creative partnerships with local authorities and tour operators to rediscover the country railways'.

Summary

It is evident from the analysis of government policy towards tourist transport that it is often subsumed under the wider remit of national transport policy making, rather than being seen within an integrated planning context where the relationship between transport and tourism is recognised. Different political ideologies affect the continuity in investment decisions for transport planning and development, and this often leaves future governments with a legacy in view of the timescale involved in major tourist transport infrastructure projects coming to fruition. This is certainly the case with rail travel following privatisation and the election of a Labour government with a greater concern for public transport to supplant the insatiable appetite for car use in the UK. The influence which governments exercise on tourist transport systems to regulate the efficient movement of people cannot be viewed in a vacuum: it is not isolated from the operation, management, provision and consumption of tourist transport services. The example of the tourist rail travel system in the UK highlights this very point, emphasising the importance of a systems approach to analyse the multidisciplinary issues associated with the transport service which InterCity recognised in the late 1980s. The situation is made more complex because the EC (now the EU) has also tried to influence European rail travel with its attempt to develop a Common Transport Policy.

Clearly, the situation in any tourist transport system is both dynamic and in a constant state of flux as government policy, the business environment and the requirements of the consumer are forever changing. For the innovative transport provider, staying abreast of these developments is a major challenge,

although day-to-day operational and management issues assume great importance in an economic activity where the logistics of moving large numbers of
people from origin to destination areas for business and pleasure require much
skill and organisation. Not surprisingly, the formulation and implementation
of government transport policy is often viewed as a long-term issue for any
tourist transport operator as their concern for the efficient management and
operation of profitable services consumes the time of their operations staff. It
is the corporate strategists and planners who have a long-term view of their
future position in the marketplace and the means of achieving strategic planning objectives. Even so, where governments pursue private sector solutions
based on tendering and short-term franchises, long-term planning and policy
issues are replaced by an overriding concern for the bottom line – profitability
and retaining the contract. This runs contrary to wider strategic goals such as
those espoused by the Wales Tourist Board (1992 : i) which argues that

> the provision of the basic tourism infrastructure upon which the industry depends
> . . . covers [the] fields of roads, public transport, information provision and visitor
> facilities. Accessibility from the principal visitor markets is crucial in determin-
> ing the number of visitors . . . The quality of . . . facilities [and infrastructure] will
> undoubtedly influence the visitor's perception of Wales.

Such issues are of less concern for a fragmented and diverse transport sector
which often fails to recognise the vital contribution tourism and recreational
traffic makes to its business. Even so, an understanding of consumer demand
for tourist travel services is a fundamental requirement for tourist transport
operators wishing to plan their supply of services in the short to medium term.
For this reason, the next chapter examines the demand for tourist transport
services and some of the data sources available to assess this issue.

Questions

1. How do governments ensure that transport policies are formulated to accommodate the tourist and leisure traveller?
2. Is the privatisation process now affecting many components of tourist transport
   services at a global scale?
3. How will the rail privatisation process affect the travel experience for tourists and
   leisure travellers?
4. Why do governments of less developed countries still retain state ownership of
   tourist transport services?

Further reading

Department of Transport (1994) *Britain's Railways: A New Era*, London: Department
of Transport.
Findlay, C., Chia, L. and Singh, K. (eds) (1997) *Asia Pacific Air Transport*, Singapore:
Institute of South East Asian Studies.
(World Wide Web address: http://www.Iseas.edu.sg/pub.html)

Gibb, R., Lowndes, T. and Charlton, C. (1996) 'The privatisation of British Rail', *Applied Geography* 16 (1): 35–51.

Gibb, R., Shaw, J. and Charlton, C. (1998) 'Competition, regulation and the privatisation of British Rail', *Environment and Planning C: Government and Policy*, in press.

Hall, C.M. and Jenkins, J. (1995) *Tourism and Public Policy*, London: Routledge.

Knight, S. and Johnston, H. (1998) *The Comprehensive Guide to Britain's Railways*, Peterborough: EMAP.
This contains a wealth of detail on each aspect of the newly privatised railway system. It also sets out the TOCs plans for each franchise it operates.

Page, S.J. (1993) 'European rail travel', *Travel and Tourism Analyst*, 1: 5–30.

Vincent, M. (1994) *The InterCity Story*, Somerset: Oxford Publishing Company.

Wheatcroft, S. (1994) *Aviation and Tourism Policies*, London: Routledge/World Tourism Organisation.

See also the special issue of the *Journal of Air Transport Management*, 3/4 (1997) on aviation deregulation in Europe.

Internet sites

Singapore Mass Rapid Transport is at:

> http//www.smrt.com.sg

Two examples of TOCs can be found at:

> http://www.firstbus.co.uk
> http://www.gre.co.uk

For the privatised rail infrastructure company, Railtrack, visit:

> http://www.railtrack.co.uk/whatsnew

Railtrack launched a major investment campaign in March 1998 following public criticism of its falling standards of maintenance in the national press after a number of accidents and delays to rail services.

Chapter 4

Analysing the demand for tourist travel

Introduction

Leisure travel has become a key feature of the leisure society which now characterises many developed countries. One corollary of this is that tourist travel has become a global activity and it is assuming a much greater role in the leisure habits of developed societies now that holidays and overseas travel have become more accessible to all sections of the population. This growth in travel also poses many challenges for the transport industry since understanding the demand for tourist transport is a critical part of the strategic planning process for transport operators and organisations associated with the management and marketing of transport services for tourists. At government level, accurate information on the use of tourist transport infrastructure is critical when formulating transport policies and particularly in assessing the future demand. At the level of individual transport operators, it is necessary to have a clear understanding of the existing and likely patterns of demand for tourist transport, to ensure that they are able to meet the requirements of tourists. This means that for transport providers, high quality market intelligence and statistical information are vital in the strategic planning process and day-to-day management of transport, so that the services offered are responsive and carefully targeted at demand, cost-effective and efficient. Ultimately, transport companies seek to operate services on a commercial basis so that supply matches demand as closely as possible but there are also situations in which such services are subsidised to meet social objectives not related to tourism. In such situations, tourism is really an added bonus for subsidised services.

The types of information required by decision makers associated with tourist transport provision are usually gathered through the marketing research process (see Chapter 2) and are likely to include the following:

- the geographical origin and spatial distribution of demand in the generating region
- the demographic and socioeconomic characteristics of tourist travel demand (e.g. age, sex, family status, social class, income and expenditure)
- the geographical preferences, consumer behaviour and images of tourists for holiday destinations and tourist travel habits, including the duration of visit
- when it is likely to occur (e.g. temporal and seasonal distributions of use)
- who is likely to organise the holiday (e.g. independently or as part of a package)

- the choice of transport likely to be used in the tourist transport system
- future patterns of demand (e.g. short- and long-term forecasts of tourist travel)
- government policy towards tourist transport operations
- the implications of tourist travel demand for infrastructure provision and investment in tourist modes of transport (e.g. aircraft, airports, passenger liners, ferries and ports)

The purpose of this chapter is to examine a range of the main types of data sources available to assess the demand for tourist transport at different spatial scales, from the world scale down to individual countries. International and domestic sources of data are introduced, with the emphasis on the relative merits and weaknesses of each source. As most tourism textbooks tend to focus on definitions of tourism, tourists and ways of measuring tourism per se, rather than on the implications of tourism statistics for assessing the demand for tourist transport, the discussion here is more focused on transporting the tourist. Building on the previous chapter, a case study of international outbound tourist travel from Japan shows how a government's tourist travel policy can be used to encourage a growth in demand for outbound travel, and the consequences for tourist transport provision. The implications for managing international and tourist transport systems are emphasised. Lastly, the role of forecasting the demand for tourist transport is discussed in relation to the assessment of future growth scenarios for tourist travel. This is important because it enables travel organisations (e.g. tour operators) and transport providers to plan ahead to remain competitive and anticipate tourist travel requirements.

The international demand for tourist travel

Ryan (1991) discusses the economic determinants of tourism demand which are associated with the purchase of an intangible service, usually a holiday or transport service, which comprises an experience for the tourist (see Chapter 2). The consumption of tourist transport services as part of a package holiday, or as a separate service to meet a specific need (e.g. a business trip or a visit to see friends and relatives), has manifested itself at a global scale in terms of the worldwide growth in international tourist travel. Among the economic determinants of the growth in international tourism are rising disposable incomes and increased holiday entitlement in developed countries. Transport operators have stimulated demand by more competitive pricing of air travel and other forms of travel for international tourists. This has been accompanied by the 'internationalisation' of tourism as a business activity (see Witt et al 1991), as global tourism operators emerge through mergers, takeovers, strategic alliances (e.g. airlines cooperating and code sharing on routes), investment in overseas destinations and diversification into other tourism services. One consequence is that tourist transport operators view the determinants of tourist travel as crucial to their short- and long-term plans for service provision.

Aside from the economic determinants of the demand for travel, Ryan (1991) emphasises the significance of psychological determinants of demand in explaining some of the reasons why tourists travel. Although there is no theory of tourist travel, a range of tourist motivators exists (Pearce 1992; Ross 1994). Ryan's (1991 : 25–29) analysis of tourist travel motivators (excluding business travel) identifies reasons commonly cited to explain why people travel to tourist destinations for holidays. These include:

- a desire to escape from a mundane environment
- the pursuit of relaxation and recuperation functions
- an opportunity for play
- the strengthening of family bonds
- prestige, since different destinations can enable one to gain social enhancement among peers
- social interaction
- educational opportunities
- wish fulfilment
- shopping

Although it is possible to identify a range of motivators, it is possible to classify tourists according to the type of holiday they are seeking and the travel experience they desire. For example, Cohen (1972) distinguished between four types of tourist travellers:

- *the organised mass tourist* on a package holiday; they are highly organised and their contact with the host community in a destination is minimal
- *the individual mass tourist*, who uses similar facilities to the organised mass tourist but also desires to visit other sights not covered on organised tours in the destination
- *the explorers*, who arrange their travel independently and who wish to experience the social and cultural lifestyle of the destination
- *the drifter*, who does not seek any contact with other tourists or their accommodation, seeking to live with the host community (see Smith 1992).

Clearly, such a classification is fraught with problems, since it does not take into account the increasing diversity of holidays undertaken and inconsistencies in tourist behaviour (Pearce 1982). Other researchers suggest that one way of overcoming this difficulty is to consider the different destinations tourists choose to visit, and then establish a sliding scale similar to Cohen's (1972) typology, but which does not have such an absolute classification. Pearce (1992) produces a convincing argument which highlights the importance of considering the tourists' destination choice. By establishing a blueprint for tourist motivation, Pearce (1992) argues that

> Tourism demand should not be equated with tourism motivation. Tourism demand is the outcome of tourists' motivation as well as marketing, destination features and contingency factors such as money, health and time relating to the travellers' choice behaviour . . . Tourism demand can be expressed as the sum of realistic behavioural intentions to visit a specific location . . . [which is] . . . reduced to existing travel statistics and forecasts of future traveller numbers. Tourist motivation is then a part rather than the equivalent of tourism demand. (Pearce 1992 : 113)

In other words, transport providers need to recognise the travellers' choice, behaviour and travel intentions at destinations to understand fully the wider transport requirements beyond simple aggregate patterns of travel statistics.

The implication of research on motivation and demand is that governments and transport operators need to recognise what economic, social and psychological factors are stimulating tourist travel. This may help in establishing the different types of travellers and their preferences for various destinations and specific activity patterns on holiday. Tour operators selling holidays need to recognise the complexity of tourist motivation to travel and airlines need to understand the precise effect on the availability of aircraft. In particular, they must be able to rotate and interchange different aircraft in a fleet to meet daily and seasonal travel requirements through complex logistical exercises. For airports, the expected number of passengers, use of airspace and runways need to be planned in advance (see Chapter 7) and they need to take account of where there is going to be a long-term growth in demand. This also means that infrastructure such as airports has to consider future investment and development plans. More specifically, transport operators will need to understand the range of motives and expectations of certain types of traveller since the level of service they provide will need to match the market and the requirements of travellers. Operators need to understand not only the dimensions of demand (see Plate 4.1), but the market segments and the behaviour and expectations of consumers which they will need to accommodate in providing a highquality tourist experience.

Throughout much of the chapter, the emphasis is on establishing patterns of tourism demand. Yet one should not forget the significance of domestic tourism which Pearce (1995a) argues is ten times greater than international tourism in numerical terms. However, all too often data sources on domestic tourism are poorly documented and irregular in terms of survey timing. Therefore prior to discussing international tourism, one example of domestic tourism in China is used to illustrate the type of patterns of tourist activity that developed in spite of a limited transport system.

Domestic tourism patterns in China: development amid infrastructure constraints

There is a growing research literature on tourism in China (e.g. Lew and Yu 1995; C.M. Hall 1997) and a wide range of articles have been published in academic journals, particularly those with an Asia-Pacific focus (e.g. *Pacific Tourism Review*). Much of the interest in China has been focused on its post-1979 opening up to international tourism and the effect on the demand and supply for tourism infrastructure (e.g. transport). As reviews of tourism research in China show (e.g. C.M. Hall 1997), much of the effort has been expended on the urban and coastal areas where rapid tourism development has occurred and on the implications for the management of ancient heritage sites. Only recently has a new genre of studies emerged that begins to deal with more

Plate 4.1   A former London Transport routemaster (now operated by FirstBus) at Trafalgar Square, London. Tourists make extensive use of the 1960s buses as they are icons of the London tourist transport product (Source: S.J. Page)

diverse aspects of tourism such as rural tourism (e.g. Getz and Page 1997) and domestic tourism (Zhang 1997).

In terms of the growth of domestic tourism in China and the history of travel, recent reviews (Sofield and Li 1997) highlight the significance of ideological traits in shaping tourism in post-revolutionary China. Prior to the 1970s, the large proportion of domestic travel was business-oriented, VFR (visiting friends and relatives) or health-related (i.e. to visit a spa). Zhang (1997) highlights the problem of documenting the volume, patterns and activities of domestic tourism prior to 1970 since 'there are no statistics on tourism during this period' (Zhang 1997 : 565). Low living standards and shortages of accommodation and transport acted as constraints to domestic travel. Economic reform in China stimulated the growth of inbound tourism to help raise much-needed foreign revenue and yet as Zhang (1997 : 566) argues 'not much attention was given to the development of domestic tourism [for Chinese]. The improvement of facilities and supplies appropriate for domestic tourists was ignored. As the lack of adequate passenger transportation was the 'bottleneck' in the development of the national economy, demands were far from satisfied, consequently curbing the growth of domestic tourism.'

One of the principal drivers of domestic tourism demand in China in the late 1980s was rising living standards, which saw the national average per capita income rise by 3.44 times that of 1979 to RMB 1189 (RMB = renminbi

Table 4.1   Growth in domestic tourism in China

Year	Domestic tourist arrivals (millions)
1986	270
1987	290
1988	300
1989	240
1990	280
1991	290
1992	330
1993	410
1994	520
1995	630
1996	640

Source: Modified from Zhang (1997).

yuan). However, in the coastal areas and large cities, rates of income were higher (e.g. Shanghai RMB 4501; Beijing RMB 3035). Using statistics from China's National Tourism Administration, Zhang (1997) documents the post-1985 growth in arrivals and receipts from domestic tourism. As Table 4.1 shows, the 1990s experienced a significant growth in arrivals. Zhang (1997) explains this growth in terms of both the spillover effects of developing international tourism and expanding the opportunities for growth. In addition, the private sector began to recognise the market opportunities for domestic tourism as government policies progressively moved to the simultaneous promotion of international and domestic tourism. This was also facilitated by economic growth which enabled greater investment in domestic tourism infrastructure. The development of a socialist market economy system after 1992 has generated economic growth of 9 per cent in China's GNP per annum. Patterns of consumption by China's population have risen in parallel with this and domestic tourism has been a major beneficiary. For example, between 1990 and 1995, domestic arrivals grew by 17.62 per cent while average receipts increased by nearly 52 per cent. Domestic tourism arrivals reached 640 million in 1996, an increase of 2.286 times the arrivals in 1990. Zhang (1997) points to the China Domestic Tourism Sample Survey which documents the growth in domestic package holidays from 8.2 million in 1993 to 34.65 million in 1995.

To explain such a growth in domestic travel, one needs to examine a greater coordinated level of development among the constituent parts of the tourism industry (especially in transport). For example, by 1995 China had 1.39 billion m² of road in its urban areas, a 500 million m² increase on 1990. Other major infrastructure projects such as the Beijing–Jiujiang–Jiulongand Nanning–Kumming railway projects plus other infrastructure projects such as the Beijing–Shijiazhuang and Shanghai–Hangzhou expressway projects have

Table 4.2   Expenditure (RMB) on domestic tourism in China

Category	1993	%	1995	%
Transport	125.50	28.2	157.46	26.6
Accommodation	81.30	18.3	104.82	17.7
Food & beverage	81.60	18.3	109.09	18.4
Urban transport	–	–	19.63	3.3
Communication	4.00	0.9	3.94	0.7
Sightseeing	23.10	5.2	34.76	5.9
Entertainment	9.70	2.2	16.59	2.8
Shopping	86.20	19.4	102.94	17.4
Other categories	33.80	7.6	42.86	7.2

Source: Modified from Zhang (1997).

assisted in intercity tourist travel. Greater mobility resulting from the growth in other forms of urban transport has favoured domestic travel. Yet as Hilling (1996 : 225) explains 'the relatively flat terrain of a number of the larger cities (Beijing, Shanghai, Tianjin) undoubtedly favours cycling, as does the serious inadequacy of public transport in the face of rapidly rising demand and urban roads which cannot easily accommodate larger vehicles – in Beijing an estimated 12 per cent of the urban road network cannot take buses'. This is supported by Zhang (1997 : 568) who argues that 'the transportation capacity is still not satisfactory, and the basic infrastructure needs further improvement'. Against these constraints, one also has to recognise that in the early 1990s the Chinese government began to introduce a national holiday system allowing employees 7–15 days' annual leave based on length of service. This combined with the gradual introduction of a five-day week to generate opportunities for long-distance domestic travel.

In terms of tourist travel in China, Table 4.2 illustrates the proportion of expenditure on various aspects of domestic tourism. Transport remains the dominant element of expenditure and it increased in 1996 when the use of urban transport was included. If one also adds the use of transport at destinations for sightseeing, then transport accounts for a significant element of tourism expenditure. Therefore, the 1990s can be characterised as the decade in which domestic travel for pleasure began to develop. As Zhang (1997) explains, travel is associated with the acquisition of new knowledge and understanding in Chinese society, but he also adds that most Chinese people are not affluent enough to travel extensively. Yet even though the cost of domestic travel by rail and coach increased twice in 1996, and again in early 1997 and despite the fact that 60 million Chinese are estimated to live in poverty, Zhang (1997) argues that 1 billion people are able to meet their basic needs and the market for domestic travel is around 300 million trips per annum. Much of the demand is urban-based and Zhang believes it could exceed 400 million trips per annum by the year 2000.

Table 4.3   Air travel in China 1990–2015

Year	Revenue passenger kilometres (billions)
1990	11.6
1995	32.8
2000	71.6
2005	124.6
2010	193.4
2015	291.9

Source: Boeing Commercial Airplane Group (1996).

Constraints on further tourism growth are related to shortage of supply of budget hotels aimed at the domestic market. But 'passenger transportation is the bottleneck of development' (Zhang 1997 : 571) and could act as a major constraint on the country's ability to meet the growing demand for domestic travel. According to Ballantyne (1996), air travel in China is characterised by infrastructure problems and a poor image of safety. This is compounded by the scale and demand for domestic, and to a lesser degree, international air travel in China. The Boeing Commercial Airplane Group (1996) assessed the existing and expected growth in air travel in China as indicated in Table 4.3. Such growth rates highlight the importance of demand management (Taplin 1993) to accommodate the volumes involved. While China has a five-year plan to upgrade 40 airports, Beijing, Shanghai and Guangzhou currently handle 42 per cent of the country's total air passenger volume. To address such shortcomings, the Chinese government invested US$1.42 billion in aviation infrastructure in 1996, a 30 per cent increase on 1995 levels of investment. China's airport authorities are seeking overseas capital to assist in the investment programme, and Ballantyne (1996 : 20) summarises the situation for domestic air travel since 'China's airlines have become vibrant and yet while the scale of tourism demand in China is of a scale many people find hard to recognise and comprehend', much of the data is dependent upon state statistical sources which have not been extensively used by researchers. The example of domestic tourism demand in China also raises a further debate: whether tourism demand can be stimulated by infrastructure in developing countries. Hilling (1996) outlines two schools of thought which are pertinent here.

Firstly, the supply-led school of thought has traditionally viewed demand as following the provision of infrastructure (White and Senior 1983). A more recent explanation and now widely used model used in transport provision is that it should be a response to demand. Hilling (1996) explains that in supply-led models, development would occur where a latent demand existed. In the demand-led model now coming increasingly into vogue due to the enormous cost of transport infrastructure investment, such as in China, tourism use of infrastructure is seen as a new market to support the demand for infrastructure investment. But it is not the case that infrastructure is being provided in advance of demand. In China, a demand-led model probably characterises the situation affecting the domestic tourism market, with economic cost-benefit criteria assuming a higher priority in infrastructure development decisions, given the scarcity of resources to meet all transport needs.

For the remainder of this chapter the emphasis is on the dimensions of international tourism demand, due to its global significance. A focus on international tourism will illustrate the interrelationship between government policy and tourism demand as well as the implications for developing transport systems which both domestic and international travellers can use. However, the 'analysis [of international tourism] is complicated by the paucity of appropriate data' (Pearce 1987 : 35). So, what sources of data are available to assess the demand and use of different modes of tourist transport?

Data sources on international tourist travel

The analysis of tourism, tourists and their propensity to travel and previous travel patterns is 'a complex process . . . involving not only the visitor and his movements but also the destination and host community' (Latham 1989 : 55). Tourist transport providers will often have statistical information relating to their own organisation's services and tourist use. But how can a new entrant into the tourist transport business examine the feasibility of providing a transport service? What statistical information on tourist transport is available? How is it gathered? And who publishes it?

At a global scale, Hilling (1996 : 1) acknowledges that 'there are vast differences in the availability of transport, indeed there is a stark contrast between a relatively immobile Third World and the highly mobile advanced economies . . . much of the infrastructure is poorly maintained and in disrepair and is inadequate for present needs without the complication of growth of demand in the future'. The global discrepancies which exist in transport provision obviously have a major impact on the tourism-generating potential and patterns of demand which result. As Table 4.4 shows, the developed regions of North America, Europe and Oceania comprise 26 per cent of the earth's land area, 15 per cent of the population and contain 60 per cent of the world's commercial vehicles. Africa is notable for its underprovision and Japan is responsible for 64 and 67 per cent respectively of Asia's commercial and passenger vehicles. While newly industrialised countries (NICs) such as Thailand, South Korea and Brazil have experienced increased mobility, the gap between developing and developed countries is certainly great and reflected in terms of patterns of international tourism demand.

One immediate problem which confronts the researcher interested in tourist transport is the absence of international statistics which monitor every mode of tourist travel. For example, organisations such as the International Civil Aviation Organisation (ICAO) publish annual statistics on international air travel for their members' airline operations. Table 4.5 provides a very

Table 4.4   World transport provision in 1990

	Percentage of world total			
	Area	Population	Commercial vehicles	Passenger vehicles
North America	16.1	5.2	40.1	35.2
Europe	3.6	9.4	18.2	37.4
Oceania	6.3	0.5	1.9	2.0
Former USSR	16.4	5.7	0.2	4.8
Central & South America	15.1	8.5	8.7	6.7
Africa	22.2	12.1	3.5	2.0
Asia (excluding Japan)	20.2	58.8	27.4 (9.7)	11.7 (3.8)

Source: Modified from Hilling (1996 : 2) based on the *United Nations Statistical Yearbook*.

general indication of the scale of passenger activity at the world's major air-ports (based on ICAO data). The data is unable to identify tourist and leisure trips, being a broad measure of terminal passengers and international travel. Such data is dependent upon the willingness of airports to report statistical information on passenger flows and sometimes ICAO estimates the flows on the basis of part-year data. In this respect, Table 4.5 is only indicative of the scale of travel at an aggregate level. Not surprisingly, the data in Table 4.5 confirms Hillings' (1996) identification of a world transport gap. Table 4.5 is dominated by North America and Europe, and in the case of Singapore, Japan, Thailand and Hong Kong it represents the rapid expansion in air travel in Asia-Pacific discussed in Chapter 3. In contrast, aviation statistics collected by national aviation organisations such as the UK's *Civil Aviation Authority UK Airports* outline the origin and destination of all terminal passengers. This data relates to air transport movements at UK airports and is based on oper-ators' requirement to report each service they operate, passengers uplifted and discharged. Only data on direct flights is recorded. Table 4.6 is only an indica-tion of the relative importance of passenger flows between the UK and other countries. Like Table 4.5, it is not possible to distinguish between arrivals and departures or the regional breakdown of flights. However, Table 4.6 does highlight the dominance of Western European travel flows, which comprised nearly two-thirds of the total passenger movements to and from the UK in 1996. This was followed by the USA and a range of other destinations throughout the rest of the world. Such industry-specific studies do provide an indication of the volume of travel. For example, in the case of cruising, the following case study outlines the sources, patterns of demand and significance of single industry-based studies.

Table 4.5 Traffic movements and passenger flows at world airports in 1995

Country	Location	Name	Terminal passengers (millions)	International terminal passengers (millions)	Commercial air transport movements (thousands)	International commercial air transport movements (thousands)
USA	Chicago	O'Hare	67	7	828	N/A
USA	Atlanta	Hartsfield	57	3	578	N/A
USA	Dallas	Fort Worth	56	3	864	N/A
UK	London	Heathrow	54	47	421	348
USA	Los Angeles	International	54	13	703	N/A
Germany	Frankfurt	Frankfurt-Main	37	30	366	287
USA	San Francisco	International	36	6	396	N/A
USA	Miami	International	33	14	499	163
Rep.Korea	Seoul	Kimpo International	31	13	197	81
USA	New York	John F. Kennedy	30	17	324	97
USA	Detroit	Metropolitan	28	3	432	N/A
France	Paris	Charles de Gaulle	28	26	325	286
Hong Kong	Hong Kong	International	27	27	150	150
USA	New York	Newark	27	4	400	N/A
France	Paris	Orly	27	11	233	92
Netherlands	Amsterdam	Schiphol	25	25	291	280
USA	Houston	Houston International	25	3	354	N/A
USA	Boston	Logan	24	3	442	N/A
USA	Orlando	Orlando International	22	3	300	N/A
UK	London	Gatwick	22	21	193	159
USA	Honolulu	International	22	6	250	N/A
Singapore	Singapore	Changi	22	22	156	156
Japan	Tokyo	New Tokyo International	22	21	120	116
Thailand	Bangkok	Bangkok International	21	15	149	107
Italy	Rome	Fiumicino	21	12	209	111

[1] All commercial movements including positioning and local movements.

Source: HMSO (1997 : 135) based on ICAO (1995) *Civil Aviation Statistics of the World*.

Table 4.6 United Kingdom international passenger movements by air: arrivals plus departures by country of embarkation or landing 1986–96

	1986	1988	1990	1992	1994	1995	1996
European Union[1]							
Austria	381	707	904	966	1,143	1,173	1,161
Belgium	986	1,229	1,406	1,528	1,880	1,865	1,925
Denmark	622	696	862	1,047	1,283	1,390	1,598
France	3,667	4,882	6,202	6,566	7,260	6,567	6,399
Finland	206	306	377	321	415	493	618
Germany	3,820	4,492	5,572	5,537	6,190	6,539	6,939
Greece	3,349	3,796	3,556	4,470	4,963	4,578	3,589
Irish Republic	2,107	3,521	4,429	4,323	5,126	6,017	6,938
Italy	2,714	3,095	3,392	3,586	4,188	4,693	4,941
Luxembourg	106	136	144	160	160	161	161
Netherlands	2,312	2,896	3,251	3,536	4,013	4,285	4,941
Portugal & Madeira[2]	2,044	2,157	2,247	2,559	2,692	2,761	2,674
Spain & Canary Islands	12,678	14,865	11,782	13,076	17,645	18,267	17,789
Sweden	566	762	947	956	1,122	1,213	1,306
Total EU	35,558	43,541	45,072	48,641	58,080	160,003	160,979
Other Western Europe							
Norway	706	821	858	916	1,074	1,166	1,283
Switzerland	2,112	2,489	2,730	2,595	2,647	2,717	2,870
Yugoslavia[3]	899	1,047	1,066	68	93	132	173
Gibraltar	174	315	290	208	147	167	158
Cyprus	617	812	1,325	2,151	1,940	1,834	1,521
Turkey	257	913	669	885	1,613	2,166	2,368
Malta	746	1,062	1,014	1,181	1,186	1,063	948
Other[4]	72	85	89	99	112	128	161
Total Other Western Europe	5,583	7,544	8,040	8,102	8,812	9,372	9,481
Total Western Europe	41,142	51,084	53,112	56,744	66,892	69,375	70,460
Rest of world							
USSR[3]	168	250	288	297	384	397	399
Rest of Eastern Europe	375	501	687	915	1,207	1,365	1,565
North Africa	734	932	845	988	1,079	1,149	1,148
South Africa[5]	395	493	613	654	772	942	1,043
Rest of Africa	859	757	843	913	962	899	969
Israel	445	492	464	537	762	862	885
Persian Gulf States	314	290	295	337	329	359	363
Saudi Arabia	296	262	255	300	338	337	357
UAE	190	261	301	420	557	644	698
Rest of Near East & Middle East	364	317	332	442	493	564	605
USA	6,335	8,574	10,143	11,430	12,173	13,249	14,402
Canada	1,501	1,877	2,057	1,979	2,096	2,292	2,543
South America	106	151	250	321	391	409	466
Central America	0	31	64	29	237	306	471
Caribbean	419	611	645	717	933	1,003	1,116

Table 4.6   Cont'd

	1986	1988	1990	1992	1994	1995	1996
Australia	409	476	522	661	714	834	714
New Zealand	85	107	121	123	153	135	158
India	312	426	561	520	751	855	970
Pakistan	211	248	261	235	280	309	340
Rest of Indian sub-continent[6]	196	223	294	314	394	424	449
Japan	357	552	838	1,035	1,186	1,241	1,348
Hong Kong	516	644	731	809	947	939	1,023
Singapore	429	452	538	515	585	604	765
Thailand	144	223	306	316	350	380	441
Rest of Asia	252	305	405	589	824	1,031	1,162
Total Rest of world	15,411	19,454	22,661	25,397	28,899	31,528	34,400
All air passenger movements	56,554	70,540	75,772	82,141	95,790	100,902	104,861

[1] Austria, Finland and Sweden joined the EU in 1995, but are included in the EU section for all years to show the time series on a consistent basis.
[2] Includes Azores and Cape Verde Islands.
[3] Or former constituent states.
[4] Includes Iceland and the Faroe Islands.
[5] Includes Botswana, Zimbabwe and the Republic of South Africa.
[6] Includes Indian Ocean islands.
Source: HMSO (1997) *Transport Statistics*, based on the Civil Aviation Authority's *United Kingdom Airports*.

CASE STUDY   The demand for specific forms of tourist transport – cruising

There has been a growing academic interest in cruising and research studies have considered its historical development (Douglas and Douglas 1996) and its impact on Pacific Island destinations (Table 4.7). More social psychologically based studies (e.g. Morrison et al 1996; Moscardo et al 1996) and a range of research concerned with the economics of cruising (Bull 1996) and the economic impacts of cruise tourism (Dwyer and Forsyth 1996) provide an up-to-date series of reviews of cruising and tourism. More specific industry-derived studies and commentaries (Cockerell 1997; Burt 1998; Dickinson and Vladmir 1997) indicate the scale and volume of growth in individual contexts. These complement the studies by industry analysts such as Peisley (1995) which outline the prospects for the cruise ship industry.

There is a growing recognition that cruising has been capacity-led over the last two decades. As Burt (1998 : 4) argues 'capacity is filled whenever new ships are brought into service' and the state subsidies used in France, Finland, Italy and Asia to maintain their shipbuilding industry (often up to 30 per cent of production costs) have made cruise ships a competitive capital purchase. This is fuelled by what Burt

Table 4.7 Cruise ship passengers for selected Pacific Island ports as a percentage of total tourist arrivals, 1989–93

Destination	1988	%	1989	%	1990	%	1991	%	1992	%	1993	%
Fiji	19,991	9	30,392	11	27,874	9	27,332	10	29,835	10	7,933	3
New Caledonia	42,762	41	33,169	29	42,158	33	35,330	30	49,802	38	38,742	31
Solomon Islands	4,547	30	2,981	23	2,616	23	2,649	19	4,886	28	2,656	19
Tonga	7,536	25	7,120	23	5,789	20	5,760	19	7,338	22	4,442	13
Vanuatu	50,932	74	41,311	63	41,867	54	37,023	48	59,346	58	43,059	49

Source: Modified from Douglas and Douglas (1996) and World Tourism Organisation *Compendium of Tourism Statistics* 1988–93.

Table 4.8   Profile of existing and future North American cruise ship passengers, 1994

	Cruise passenger of 1990s (%)	Future projections (%)
Male	54	53
Female	46	47
Age profile		
25–39	29	50
40–59	36	39
60 or over	35	11
Average	50	42
Income profile (US$)		
20,000–39,999	31	36
40,000–59,999	30	40
60,000–99,999	28	20
100,000 and over	114	–
Marital status		
Married	76	70
Single	24	30
Household structure		
Children in the household	27	46
Vacation with children	15	29
Vacation without children	12	17
No children in the household	72	54

Source: Modified from Peisley (1995) based on Cruise Lines International Association Survey.

(1998) indicates is an average of 9 per cent growth per annum in passenger demand, a figure confirmed by Peisley's (1995) analysis of data from the industry body – the Cruise Lines International Association (CLIA). For example, in the USA, the demand increased from 3.3 million passengers in 1989 to 4.6 million in 1994. Morrison et al (1996) argue that cruising has outperformed any other vacation category in the USA over the last 25 years. North America accounts for 80 per cent of the world cruising market, much of which has focused on the Caribbean. In 1994, CLIA examined the demographic profile of cruise ship passengers, which can be summarised thus:

● A quarter of passengers were under 40 years of age, with a third over 60 years of age.
● Less than two-thirds of clients earn less than US$60,000 per annum.
● A quarter of passengers still have children at home.

As a result, CLIA identified the profile of existing and the 'hot prospect' market segments (Table 4.8) where the future cruise ship passengers are likely to seek shorter and cheaper products. In fact, Peisley (1995 : 8) argues that 'a younger lower income clientele with more family commitments would certainly suit the mass or middle-market lines which believe that the mega-ships with all the entertainment facilities . . . [may have] the capacity to keep prices down'. Table 4.9 outlines the ranking of cruise ship companies in terms of capacity, highlighting the market dominance by the Carnival

Table 4.9   Cruise ship companies and capacity 1995 and 1998

	Number of berths		Rank	
Company	1995	1998	1995	1998
Carnival Corporation	36,649	56,512	1	1
Royal Caribbean Cruise Line	16,530	28,969	2	2
P&O Cruises Division	14,531	22,306	3	3
Kloster Cruise	11,253	11,731	4	4
Cunard	7,983	6,127	5	8
Costa Crociere	7,565	9,765	6	6
Chandris	6,434	11,684	7	5
Regency Cruises	5,008	6,408	8	7

Source: Modified from Peisley (1995).

Corporation, Royal Caribbean and P&O Cruises Division. The destinations in which the North American Cruise Ship industry plied its trade in 1994 were:

● the Caribbean (52 per cent)
● the Mediterranean (9.2 per cent)
● Alaska (7.8 per cent)
● Trans-Panama Canal (5.9 per cent)
● Mexicana Riveria (5.3 per cent)
● Northern Europe (3.8 per cent)
● Bermuda (3 per cent)
● the South Pacific
● other regions (10.8 per cent)
   Source: Peisley (1995).

Peisley (1995) documents the problems which various cruise ship lines have faced in terms of market segmentation and failed attempts to position certain products in niche markets. However, critics believe the cruise ship industry is not sufficiently mature to support niche marketing, though Carnival carried 90,000 child passengers in 1994, which is a new development to attract families. Peisley (1995) also documents regional initiatives in Far East Asia and Oceania to develop the cruise passenger market with its sought-after tourist spending. The main markets appear to be firmly developed in North America, the UK, Europe and Asia and it is predicted that the passenger business will grow to reach 8 million a year by the new millenium.

At a smaller scale, Cockerell (1997) discussed the development of river cruising in Europe. In the UK, the Passenger Shipping Association (PSA) collates data on river cruises. In 1990 Germany dominated both ocean and river cruises with 184,000 ocean cruise sales in the UK compared to 218,000 in Germany. By 1996, however, the figures were 416,000 in the UK compared to 255,000 ocean cruise sales in Germany. Table 4.10 outlines the popular destinations, assisted by the UK tour operator Thomson Holidays becoming a member of the PSA, thereby providing data on volume and destinations chosen. While the Nile was a major growth market in 1996, the terrorist attack in Egypt in 1997 may well have severely affected this market. New river cruise destinations, facilitated by the opening up of the former Eastern Europe, have meant

Table 4.10 River cruise passengers booked in the UK

Cruise location	1992 (000s)	1996 (000s)
Nile	15.0	70.0
Rhine and tributaries	20.0	27.0
Neva/Volga	–	15.0
Danube	1.0	10.0
Yangtse	0.1	5.0
Rhone	1.0	5.0
Elbe	0.6	4.0
Mississippi	0.5	2.0
Other rivers (e.g. Amazon, Irrawaddy and Po)	5.0	4.0
Total	43.2	142.0

Source: Passenger Shipping Association and Cockerell (1997).

the Neva–Volga cruise is one of the fastest-growing segments of the UK cruise market which travels between Moscow and St Petersburg. Although the Rhine retains a prominent position, new options such as cruises on the Elbe now compete with those on the Rhine. Likewise, the new canal between the Rhine and the Danube has also led to a major growth in Danube cruises. The construction of a new dam on the Yangtse river in China also prompted a major growth in cruises prior to the flooding of the scenic Three Gorges region in 1997. Therefore this case study illustrates how a range of cruise industry data sources can be combined to analyse the demand for tourist activities based on the marine environment (Orams 1998). However, it also illustrates how dependent the researcher is upon the industry sources to generate reliable, consistent and regular data which documents tourist use of such transport. For this reason, one is forced to refer to international tourism statistics as they are more consistent and comprehensive in their treatment of tourist travel, particularly in relation to transport issues.

Tourism statistics

Within an international context, tourism statistics provide an invaluable insight into:

- tourist arrivals in different regions of the world and for specific countries
- the volume of tourist trips
- types of tourism (e.g. holidaymaking, visiting friends and relatives and business travel)
- the number of nights spent in different countries by tourists
- tourist expenditure on transport-related services.

Such information may indicate the order of magnitude of tourist use of transport systems and their significance in different locations. In this context, Latham's (1989) seminal study on tourism statistics is essential reading, since it provides a useful insight into the complex process of assessing the demand for international tourist transport. Studies by Latham (1989), Jefferson and Lickorish (1991) and Veal (1992) document the procedures associated with the generation of tourism statistics, which often use social survey techniques such as questionnaire-based interviews with tourists at departure and arrival points.

Unlike respondents in other forms of social survey work, tourists are a transient and mobile population. This raises problems related to which social survey method and sampling technique one should use to generate reliable and accurate statistical information that is representative of the real world. Due to the cost involved in undertaking social surveys, it is impossible and impractical to interview all tourists travelling on a specific mode of transport. A sampling framework is normally used to select a range of respondents who are representative of the population being examined. While there are a number of good sources which deal with this technical issue (see S.L.J. Smith 1989 and Veal 1992), it is clear that no one survey technique and sampling framework is going to provide all the information necessary to enable decision makers to make strategic planning decisions. It is possible to discern three common types of tourism survey:

- pre-travel studies of tourists' intended travel habits and likely use of tourist transport
- studies of tourists in transit or at their destination, to provide information on their actual behaviour and plans for the remainder of their holiday or journey
- post-travel studies of tourists once they have returned to their place of residence.

Clearly there are advantages and disadvantages with each approach. For example, pre-travel studies may indicate the potential destinations which tourists would like to visit on their next holiday but it is difficult to assess the extent to which holiday intentions are converted to actual travel. In contrast, surveys of tourists in transit or at a destination can only provide a snapshot of their experiences to date rather than a broader evaluation of their holiday experience. One interesting survey instrument which, in part, addresses the potential weaknesses associated with surveys of holiday experience would seek to assess the tourists' previous travel behaviour and their changing travel behaviour through time. Where retrospective post-travel studies are used, they often incur the problem of actually locating and eliciting responses from tourists which accurately record a previous event or experience. Each approach has a valuable role and generally individual transport operators and tourism organisations use the approach appropriate to their information needs.

The most comprehensive and widely used sources of tourism statistics that directly and indirectly examine international tourist travel are produced by the World Tourism Organisation (WTO) and OECD (Pearce 1987, 1995a). National governments also compile statistics on international tourism for their own country (inbound travel) and the destinations chosen by outbound travellers. These are normally commissioned by national tourism organisations as a specialist research function, though it depends upon each country and how the management of public sector tourism is organised. WTO publishes a number of annual publications which include the following: the *Yearbook of Tourism Statistics* (published since 1947 as *International Travel Statistics*, then as *World*

Travel Statistics, and now as *World Travel and Tourism Statistics*). This has a summary of tourism statistics for almost 150 countries and areas, with key data on tourist transport, and includes statistical information in the following order:

- world summary of international tourism statistics
- tourist arrivals
- accommodation capacity by regions
- trends in world international tourism arrivals, receipts and exports
- arrivals of cruise passengers
- domestic tourism
- tourism payments (including international tourism receipts by countries calculated in US$ millions, excluding international fare receipts)
- tourism motivations (arrivals from abroad and purpose of visit)
- tourism accommodation
- country studies which examine the detailed breakdown of tourism statistics collected for each area, including tourism seasonality

In addition to WTO, OECD produces *Tourism Policy and International Tourism in OECD Member Countries*. Although the data collected is restricted to 27 countries (BaRon 1997), it deals with other issues such as government policy and barriers to international tourism.

International tourist travel: trends and patterns

Both WTO and OECD statistics are compiled by each agency from returns and data which national governments supply according to the criteria laid down by each organisation. International tourist arrivals have risen from 25 million in 1950 to 593 million in 1996. Figure 4.1 illustrates the growth in arrivals by WTO region since 1986; it shows there was a 75 per cent growth in total arrivals between 1986 and 1996. This is a good rate of growth given the prevailing global economic conditions which had a bearing on different tourism regions.

The key feature of Figure 4.1 is the continued dominance of Europe, with 352 million arrivals in 1996, although the relative position of Europe in world arrivals has declined from a 64 per cent market share in 1986 to 59 per cent in 1996. Europe recorded one of the lowest regional growth rates for the period 1986–96, with an average of 5 per cent per annum. However, Latham (1992) argues that Europe's continued dominance as an origin and destination area for tourism is due to the fact that its population contains segments with a high disposable income and a highly developed tourism infrastructure. The Americas received nearly 115 million arrivals in 1996, which represents a 19 per cent market share of world arrivals. The Americas recorded a decline in their market share from 21 per cent in 1986.

The East Asian Pacific region experienced a 47 per cent growth in arrivals between 1986 and 1996, with 87 million arrivals. This comprised 15 per cent of world arrivals in 1996, a significant growth from the 10 per cent share in 1986. Africa (excluding Egypt and Libya) received 20.6 million arrivals in 1996, which comprised 3.5 per cent of the world share of arrivals; that increased by an average of 8.3 per cent per annum in the decade 1986–96.

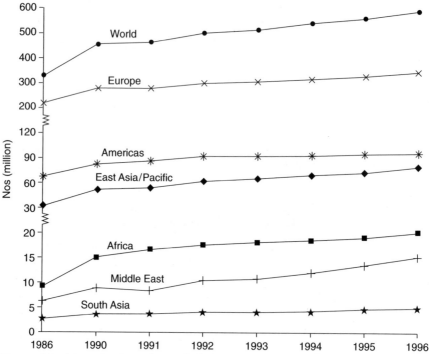

Figure 4.1 International tourism arrivals by region 1986–96 (Source: World Tourism Organisation)

South Asia, however, received 4.5 million arrivals which represented 0.8 per cent of world arrivals and the Middle East received 15.3 million arrivals in 1996, which comprised 2.6 per cent share of world arrivals.

As arrivals have grown, tourist receipts increased from $352 billion in 1994 to $423 billion in 1996. Tourist arrivals are not distributed evenly in each region, with some countries and their destinations performing better than others. In terms of the specific countries and their position in world arrivals, Table 4.11 provides a detailed breakdown by each WTO region (see BaRon 1997 for more detail). What Table 4.11 highlights is the significance of global tourism flows in contributing to such trends. Both Hoivik and Heiberg (1980) argue that up to four-fifths of all international flows were between Europe, North America and Japan, and one-twentieth between other countries in the world. Pearce (1995a) has examined these flows in more detail and Rowe (1994) lists the country-to-country flows in order of magnitude:

- Canada to USA, a flow of 15 million
- USA to Mexico, a flow of 14 million
- USA to Canada, a flow of 12 million
- France to Spain, a flow of 12 million
- Germany to Italy, a flow of 10 million
- Germany to Austria, a flow of 10 million
- Hong Kong to China, a flow of 10 million

Table 4.11   World tourism arrivals and key destinations

1996 ranking	WTO region main destinations	Tourist arrivals	
		1996 (000s)[1]	% of world total
	World	**593,638**	**100.0**
	Top Ten	289,535	48.8
	Africa (CAF)	20,562	3.5
39	Morocco	2,693	0.5
27	South Africa	4,944	0.8
31	Tunisia	3,885	0.7
	Other Africa	9,040	1.5
	Americas (CAM)	**114,706**	**19.3**
	Northern America	83,605	14.1
10	Canada	17,386	2.9
7	Mexico	21,428	3.6
2	USA	44,791	7.5
	Other America:	31,704	5.3
29	Argentina	4,286	0.7
35	Puerto Rico	3,065	0.5
	Other America	24,353	4.1
	E.Asia/Pacific (EAP)	**87,025**	**14.7**
30	Australia	4,167	0.7
6	China	22,765	3.8
15	Hong Kong	11,703 V	2.0
26	Indonesia	5,034	0.8
32	Korea	3,684	0.6
28	Macau	4,890	0.8
21	Malaysia	7,138	1.2
22	Singapore	6,608	1.1
20	Thailand	7,192	1.2
	Other East Asia/Pacific	13,884	2.3
	Europe (CEU)	**351,612**	**59.2**
	Central/Eastern Europe	83,479	14.1
37	Bulgaria	2,795	0.5
12[2]	Czech Republic	17,000	2.9
8	Hungary	20,670	3.5
9	Poland	19,410	3.3
36	Romania	2,834	0.5
14[3]	Russian Federation	14,587	2.5
	Other Central/Eastern Europe	6,173	1.0
	Northern Europe	39,314	6.6
25	Ireland	5,280	0.9
38	Norway	2,746 H	0.5
5	UK	26,025 V	4.4
	Other Northern Europe	5,263	0.9
	Southern Europe	98,507	16.6
18	Greece	8,987	1.5

Table 4.11   Cont'd

1996 ranking	WTO region main destinations	Tourist arrivals	
		1996 (000s)	% of world total
4	Italy	32,853	5.5
17	Portugal	9,900	1.7
3	Spain	41,295	7.0
	Other Southern Europe	5,472	0.9
	Western Europe	118,244	19.9
11	Austria	17,090 A	2.9
24	Belgium	5,753 A	1.0
1	France	61,500	10.4
13	Germany	15,205 A	2.6
23	Netherlands	6,546 A	1.1
16	Switzerland	11,097	1.9
	Other Western Europe	1,053	0.2
	East Mediterranean	12,068	2.0
19	Turkey	7,966	1.3
	Other East Mediterranean	4,102	0.7
	Middle East (CME)	**15,256**	**2.6**
40	Bahrain	2,669	0.4
33	Egypt	3,675	0.6
34	Saudi Arabia	3,458	0.6
	Other Middle East	5,454	0.9
	South Asia (CSA)	**4,477**	**0.8**

[1] A = arrivals at accommodation establishments reporting; H = arrivals at hotels reporting; V = visitor arrivals (includes day visits).
[2] Former Czechoslovakia.
[3] Former USSR.
Sources: WTO, National Reports (some preliminary); modified from BaRon (1997).

- Switzerland to Italy, a flow of 10 million
- France to Italy, a flow of 9 million
- Germany to Spain, a flow of 7 million
- Mexico to USA, a flow of 7 million
- UK to Spain, a flow of 7 million
- Austria to Italy, a flow of 6 million

While such patterns are well established in international tourist travel, the growing significance of EAP and outbound/intraregional travel needs to be recognised. Although the existing Europe–Americas dominance of flow remains marked, the growth of arrivals in other regions is a noticeable feature of the 1980s and 1990s. Although WTO (1992) did provide a summary of tourist transport, WTO stopped publishing tourist transport statistics in their summary and full compendium of tourism statistics in 1996, so it is unfortunately no longer possible to investigate the role of transport and arrivals. This

Table 4.12   Mode of transport used by European holidaymakers

	Car (%)	Train (%)	Plane (%)	Boat (%)	Motorbike or bike (%)	Coach (%)
By origin						
Belgium	77	6	10	1	8	7
Denmark	59	14	18	11	3	4
France	81	15	6	2	2	7
Germany	61	16	17	3	1	7
Greece	78	4	13	25	1	0
Ireland	51	11	31	18	1	6
Italy	73	15	5	5	2	11
Luxembourg	62	10	19	4	0	15
Netherlands	70	8	14	5	6	14
Portugal	76	17	3	3	1	16
Spain	70	16	5	2	0	12
United Kingdom	59	11	24	8	0	14
EU member states	68	14	13	5	1	10
By destination						
Own country	78	14	1	8	1	8
Other country	52	11	32	6	2	13
Non-EU Europe	53	15	29	17	0	18
Outside Europe	35	19	86	1	0	15

Note: Totals exceed 100 owing to multiple responses.
Source: EC Omnibus Survey, cited in Page (1993d : 12).

exemplifies the problem facing tourism researchers in developed countries and particularly in developing countries (Weaver and Elliot 1996; O'Hare and Barrett 1997), where inadequate data exists beyond general aggregate statistics that do not investigate the tourism–transport interface. However, occasional market studies published in *Travel and Tourism Analyst* do focus on particular tourism flows between WTO regions (e.g. see Cockerell 1994 on flows from Europe to the Asia-Pacific region.)

Other statistical sources on international tourist travel

At a transnational level, the EC (now the EU) is also an important source of statistical information on tourist transport (Jefferson and Lickorish 1991). In the past, the EC has undertaken various studies of tourist travel, most notably the 1986 Omnibus Survey *The Europeans and their Holidays*, which examined the travel patterns in member states and recognised the overwhelming importance of carborne travel. Although Table 4.12 is dated, it remains one of the comprehensive sources of data on tourist modes of travel between member states. More recently, the EC publication *Transport, Communications, Tourism Statistical Yearbook* (Eurostat 1987) recorded generalised time series statistics on tourist travel in relation to:

- transport infrastructure in member states
- distances covered by various modes of transport
- definitions and explanations of tourist travel on railways, roads, waterways and aircraft by country
- summary information on tourism demand.

However, it is now too dated to provide detailed insights into the high level of intraregional tourist travel (i.e. international tourist travel within the EC). For example, in 1989 there were 220,779,000 tourist trips in the Europe/Mediterranean region, the majority undertaken by West Germans (25.5 per cent), French (9.5 per cent), British (9 per cent), Dutch (5.4 per cent), Swiss (5.3 per cent) and Italians (4.9 per cent). For these trips, air and road travel assumes an important role.

For more detailed analyses of tourist use of transport systems in the EC, such dated figures have to be complemented by international travel surveys undertaken by member states. These are well documented in accessible sources such as Latham (1989) and Jefferson and Lickorish (1991) and the technical issues associated with these surveys need not be reiterated here. In the UK, the most important government survey of international tourist travel is the International Passenger Survey (IPS) which began in 1961 and is a stratified random sample of tourists arriving/departing at the UK's main ports of entry. Latham (1989 : 64) identifies its principal aims as the collection of:

- data on the government's travel account (expenditure by incoming tourists and outgoing visitors) which is used to calculate tourism's contribution
- detailed information on visitors to the UK and outbound travellers to overseas destinations
- information on international migration
- data on the routes used by tourists to assist shipping and aviation authorities.

These statistics are published annually in a revised format as the *International Passenger Survey* (HMSO 1990), although for a commercial annual subscription (now in excess of £20,000), companies can purchase the raw data from the IPS survey and additional questions can be added to the survey at a cost to transport operators. In addition, other organisations publish summaries of transport statistics (e.g. Potter 1997) of travel to and from the UK and the Department of the Environment, Transport and the Regions (DETR) produces the UK transport statistics, although they do not distinguish between tourist and non-tourist use of transport. Ad hoc studies, such as the former UK *Long Distance Travel Survey* and the *National Travel Survey 1993/95* (see HMSO 1997a) also provide more detailed insights into the use of tourist transport. For example, the *National Travel Survey 1993/95* indicated that on average, each person in the UK travelled 871 miles a year, comprising 29.6 journeys a year (excluding commuting). Although few tourism surveys specifically target tourist transport as the primary focus, these sources of data can be used to understand the nature of international tourist use of different forms of transport.

From this discussion of data sources on tourism demand, it should be evident that up-to-date market intelligence is essential for tourist transport operators, particularly in a situation where demand is increasing at a rapid rate. For this reason, attention now turns to a case study of outbound travel from Japan to illustrate the type of data sources used by researchers, and the need to consider the extent to which demand is shaped by a variety of factors (e.g. government policy, cultural traits and the perception of tourism and leisure time).

CASE STUDY   The demand for outbound tourist travel and the transport system in Japan

One of the main areas of expansion in outbound travel in the 1980s and 1990s occurred in the Asia-Pacific region, which comprises the established industrial and newly industrialised countries of South East Asia, most notably Japan (Edwards 1992; C.M. Hall 1997). The continued growth in outbound travel from these markets has attracted a great deal of interest from tourist transport operators, especially global airlines, since these markets have been growing at a time of relative stagnation in many of the mature outbound travel markets in North America and Western Europe. Japan (Figure 4.2) is an interesting example as it was the third-largest generator of international tourism expenditure in 1996, after North America and Germany. Japanese outbound travellers spent US$37,977 million in 1996 (BaRon 1997). It is a lucrative market which many international tourist destinations have nurtured, given that Japan's 123 million population constitutes a dominant source market for international tourism. In 1996, the Japanese took 16.7 million overseas trips. According to Nozawa (1992), Japanese travel to destinations in the Asia-Pacific regions is set to expand in the late 1990s, with Australia, Canada, New Zealand and Switzerland ranked as the most desirable destinations (see Page 1989a for a discussion of Japanese outbound travel to New Zealand). Within Western Europe, the UK, France and Italy were also 'desirable destinations for historical . . . sightseeing and folklore tours for Japanese tourists' (Nozawa 1992 : 232) and in the Asia-Pacific region, destinations such as Thailand, Singapore and Hong Kong have remained popular (Morris 1990). In addition, Pacific Island destinations have proved popular locations for tourism development and invest-ment (Page 1996; Hall and Page 1996). This case study examines the situation in Japan in the 1980s and 1990s and the significance of a government policy designed to expand outbound tourist travel. Emphasis is placed on the factors associated with a boom in outbound travel and the constraints imposed by the existing tourist transport system, which is inhibiting further growth. This is followed by a critical discussion of the situation in the 1990s and the significance of government policy and its long-term effect on Japanese travel. It is pertinent, however, to commence with a discus-sion of sources of data on international tourism demand in Japan and the historical evolution of outbound travel to place the case study in a broader context.

Data sources on Japanese outbound travel
According to Pearce (1995), one of the main sources of data available to assess international travel by Japanese tourists is the passport applications generated by the Ministry of Justice's Immigration Bureau prior to 1970. Since there were restric-tions on outbound travel from Japan prior to 1970, this data provides evidence of

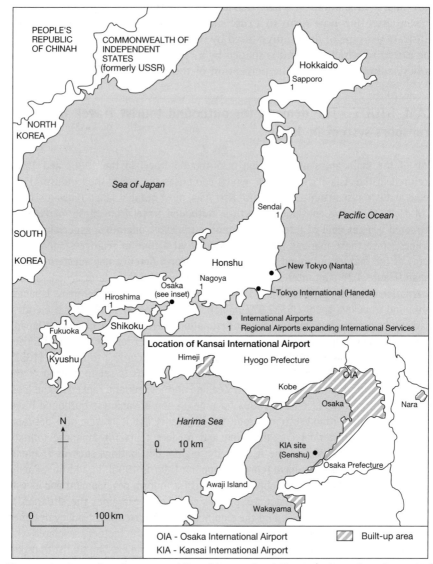

Figure 4.2   Japan: location map and Kansai International Airport (redrawn from Page 1994)

the 'purpose of visit and main country visited' (Pearce 1995). After 1970, restrictions on travel and multiple visa applications were gradually lifted, making the data more difficult to analyse, and therefore recent trends are calculated using a number of statistical sources. The Ministry of Justice still collates passport applications, but the Japanese National Tourist Organisation reanalyses the data to ensure they are consistent with the methodology employed by organisations such as WTO. The Ministry of Transport, responsible for the formulation and implementation of tourism policy, collects tourism data for policy purposes. It is the main source of data for the transport and travel trade. Other bodies which produce useful research data include:

- The Japan Travel Bureau (JTB), which is Japan's premier travel company and produces regular reports on tourism market trends together with an annual report on the Japanese travel market
- The Market Intelligence Corporation, which surveys air travellers quarterly at key airports, generating data for reports and other bodies such as JTB (Morris 1994)

The evolution of tourism demand

Pearce (1995) documents the evolution of Japanese outbound travel between 1964 and 1981, when the number of tourists travelling increased from 100,000 to over 4 million (Tokuhisa 1980). A number of factors can be attributed to this rapid expansion in demand:

- a continued growth in international trade and the generation of wealth from activities in the Japanese economy
- increased personal affluence for the population and additional disposable income to spend on luxury items like travel
- the liberalisation of currency restrictions in 1964
- the easing of administrative procedures to obtain a passport and visa
- improved aircraft technology which facilitated easier access to overseas destinations
- the development and marketing of package holidays which reduced the price of travel.

Since 1981, outbound tourist travel has expanded at an even greater rate than in the 1970s, with the number of outbound trips more than doubling from 5 million to 11 million between 1986 and 1991. And in 1996, outbound travel reached 16.1 million trips. This growth has caused a great deal of debate among analysts over the extent to which it was stimulated by government policy rather than the result of an underlying trend towards international travel in Japanese society. For example, Polunin (1989) interprets this expansion as a result of an underlying growth within Japanese society which was only marginally influenced by official policy. The significance of Polunin's (1989) analysis is that whilst encouraging a growth in demand, the government has failed to match this with an expansion in the supply of tourist transport infrastructure to accommodate growth. In contrast, a recent synthesis by Burns (1996) questions the validity of such arguments. For this reason, it is useful to examine the role of government policy in stimulating the demand for outbound travel, followed by the range of underlying factors which Polunin (1989), Nozawa (1992) and Burns (1996) cite as influential in shaping demand.

Government policy and outbound travel in Japan

In 1986, the Japanese government announced its Ten Million Programme which sought to encourage a doubling of outbound travel between 1986 and 1991. The rationale for this programme has been interpreted in various ways, including:

- the promotion of international citizenship among Japan's population
- a response to growing international pressure for Japan to reduce its trade surplus generated through its balance of payments
- a strong export-led economy in the 1970s and 1980s.

Polunin (1989) suggests that this policy was little more than a public relations exercise since outbound travel was already expanding by over 11 per cent in 1986 and this trend continued with expansion in excess of 20 per cent in 1987 and 1988.

However, the significance of this policy statement was that the 'Japanese tend to regard government bureaucrats as well-intentioned partners in progress... They respond positively to being told what to do. The Ten Million Programme says that travel is desirable and indeed expected' (Polunin 1989 : 6). In this context, the Ten Million Programme may be viewed as legitimising outbound travel in Japanese society where there have been cultural barriers to travel, since the work ethic has traditionally militated against long holidays. Government policy endorsed outbound travel and positively encouraged travel as a way of addressing international criticism of the accumulating trade surplus.

More recently, the government's Ten Million Programme has been expanded to incorporate the Two Way Tourism 21 Programme, designed to nurture incoming tourism to Japan. This has focused on attracting the growing Asian outbound markets, by emphasising the unique culture, history and traditions of Japanese society. This policy aimed to boost the existing $3,435 million receipts from inbound tourists in 1991 to offset the $23,983 million spent by outbound Japanese travellers. Yet for the tourist transport operators in Japan it is interesting to ascertain what factors facilitated the Japanese travel boom in the late 1980s and the effect of expansion in demand for the tourist transport system.

Evaluating the Ten Million Programme and the expansion of outbound travel from Japan

Polunin (1989) argues that the growing internationalism of Japanese society has encouraged a growth in outbound tourism, as Japan developed a more outward-looking position in the world economy, accompanied by increased international investment in overseas countries. Such a change in outlook led to a greater awareness in Japanese society of the desirability of overseas travel. Nozawa (1992) argues that the 1989 annual White Paper on leisure activity in Japan found that 32 per cent of respondents would like to travel abroad if given an extra day off a week. This reflects a shift in emphasis within Japanese society from the work ethic, as in 1987 'average annual Japanese working hours were 2150 hours compared with typical US hours of 1924' (Nozawa 1992 : 228). Hard work and loyalty to one's company have traditionally resulted in long working hours and few official overseas holidays. However, a number of specific factors appear to have stimulated a Japanese travel boom in the 1980s:

- Japan's gross national product outperformed the world average in the 1980s, creating additional disposable income available for discretionary spending on tourism and leisure services in Japan.
- In 1986, the state ended Japan Airline's (JAL) monopoly of international scheduled services (Polunin 1989) and some of the capacity constraints of the late 1970s; the 1980s were characterised by additional airlines offering extra scheduled services. In 1991, there were 52 scheduled airlines operating to 32 destinations and a further 43 seeking traffic rights. Yamauchi (1997) examines the gradual liberalisation of air transport policy in Japan and the effect on international travel, particularly the consequences for Japanese airlines.
- In 1987, the International Tourism Institute of Japan was formed to assist the government in the promotion of outbound travel, providing tourism consultants to developing countries on projects assigned by the Ministry of Transport (Nozawa 1992).

- The Japanese government encouraged the Ministry of Labour to reduce the working week from 48 to 40 hours in 1988 to generate additional leisure time for overseas travel. It was followed by the promotion of a 'Long Holiday Week' and a 'Tourism Week' to encourage outbound rather than domestic travel.
- The year 1988 was deemed 'the year of liberalization for Japanese tourism' (Nozawa 1992 : 227), with a streamlining of emigration procedures and tax privileges for outbound travel (see Morris 1990).
- The Japanese government has negotiated bilateral visa waivers for short-stay visitors in other countries, most notably the USA in 1988, which stimulated an increase in demand for visits to this destination.
- Overseas investment by Japanese tourist transport enterprises has been characterised by diversification into overseas hotels (Morris 1990), as Japan's international carriers (e.g. JAL and All Nippon Airways) secured a stake in the major destinations served by these airlines (see Go and Murakami 1990). For example, Morris (1990) noted that 60 per cent of hotel rooms in Hawaii were owned by Japanese companies.
- A 'Working Holiday Scheme' for young Japanese travellers was developed for Australia and subsequently expanded to New Zealand and Canada.
- State investment in international airports and the promotion of regional airports for the use of charter flights on inclusive tours has only had a marginal impact on capacity constraints at Narita and Osaka international airports.

However, Burns' (1996) seminal study argues that while the Ten Million Programme achieved its numerical objectives ten months ahead of its target, the programme was an abject failure in social terms. Burns (1996) analysed the Ten Million Programme and identified four elements:

- a publicity campaign by the government and tourism industry to encourage Japanese people to take more holidays abroad
- the promotion of overseas school tours to develop the idea of international citizenship
- assistance with public relations for foreign national tourism associations based in Japan
- a campaign to promote longer holidays

In his assessment, Burns (1996) examines the mechanisms employed by the Japanese government to achieve its objectives. He concludes that no underlying cultural change occurred in Japanese society in terms of holiday-taking. The amount of time which the Japanese population spend at work has remained high. Their perception of holidays remains a barrier, with the 9–15 days' paid leave to which employees are entitled is often viewed as emergency paid sick leave or to use for attending events such as funerals. In the 1990s, just over 10 per cent of the Japanese population travelled overseas. Burns (1996) highlights the paradox of the Ten Million Programme as statistically significant, but in policy terms it was a smokescreen to give the illusion that the country was reducing its trade surplus with its trading parties.

Unfortunately, the demand for outbound travel and its continued expansion in the late 1980s and 1990s has not been matched by the supply of airport capacity. One simple explanation is that airport development requires a long-term programme which is phased in over a number of years (see Doganis 1992). The expansion in demand of over 20 per cent per annum in the late 1980s would be difficult to accommodate in any tourist transport system. Hunt (1988) notes that a Bank of Japan study on Japanese airports found 1.5 million tourist trips were deterred by capacity constraints (e.g. Narita, Tokyo, Osaka and Itami airports only had one runway). According to Nozawa (1992) this was compounded by a curfew on night flying, constraining existing capacity by 30 per cent. In addition, the expansion of Narita airport, involving the construction of two extra runways and a new terminal, was delayed until 1995. Among Japan's regional airports which operate as international gateways, an absence of long runways, terminal facilities and associated services has constrained expansion. In Osaka an innovative solution to the problem has been the construction of the New Kansai International Airport on a floating island.

Kansai International Airport

Therivel and Barrett (1990) examine the construction of this new airport, focusing on the environmental impact of developing a 24-hour service in a country where land is a scarce commodity. Kansai International Airport (KIA) is located in Central Japan, and provides much-needed access to Asia (Morris 1990). The KIA project was conceived to avoid the complaints and problems associated with Osaka International Airport, which was built near a densely populated area and attracted considerable public criticism due to noise pollution (see Chapter 6). The planning of KIA can be dated back to the 1960s. Based on a feasibility study of sites in 1968, an Aviation Deliberation Council (ADC) examined seven sites in the Osaka region, with a view to their being able to:

- handle domestic and international flights
- offer 24-hour airport operation
- provide comprehensive environmental protection
- offer the possibility of further expansion
  Source: Therivel and Barrett (1990 : 82).

It was not until 1974 that the Japanese Ministry of Transport released the ADC report and this identified a site 5 km off Senshu as the preferred location (Figure 4.2, page 134). Despite opposition from the local authorities concerning the environmental impact of such a project, a decline in the local economy in the mid-1970s prompted a greater consideration of local planning gains from the project (Therivel and Barrett 1990). In 1982, following environmental reports, the project was approved and the KIA Company was formed. The Company's plans sought:

- 511 ha of reclaimed land
- a 3,500 m runway
- terminals and associated buildings
- capacity for 16,000 take-offs and landings per annum
- a second phase to expand the site by 1,200 ha with two additional runways of 3,000 m and 3,500 m
  Source: Therivel and Barrett (1990 : 83).

The airport received planning permission after a complex environmental review process (see Chapter 6 for a more detailed discussion of environmental assessment) and opened in 1994, fifteen months behind schedule due to landfill problems associated with reclamation works. This has meant that the project was over budget at ¥1.4 trillion, which may jeopardise the second phase expansion with its estimated cost of ¥1.8 trillion. But what are the prospects for improving international airport capacity in other regions of Japan in the 1990s?

The Ministry of Transport's Five-Year Airport Development Plan (1991–95) advocated that a further Y3.19 trillion was needed for investment in airport infrastructure. However, almost 60 per cent of the budget has already been allocated to projects at Narita, Haneda and the KIA, and to the upgrading of regional airports, which suffer from a lack of long runways, terminal facilities and associated services (see Figure 4.2, page 134). For example, in the early 1990s new airports were opened at Fukushima and Hiroshima.

Lessons from the Japanese experience

● Facilitating an expansion in the demand for outbound travel requires appropriate investment in tourist transport infrastructure so that inadequate supply does not constrain demand.
● The rhetoric and actions of government policy need to be set in a rational framework where the demand for outbound travel is sensitively managed to avoid congestion and a decline in the 'tourist travel experience'. Deregulating and liberalising air travel in Japan was not a short-term solution to expanding capacity, as airports are seriously constrained in terms of the volume of traffic they can accommodate. Morris (1994 : 63) argues that 'the airport log jam that has long hindered the development of the outbound market will be eased by the opening of the new airport in Kansai in 1994'.
● Measures to facilitate outbound travel need to be phased in as new airport capacity comes on stream, but Burns (1996) is critical of existing measures which are viewed as 'window dressing'.
● Despite the lead time and cost of airport investment, an expansion in outbound travel was achieved in the timespan of the five years envisaged in the Ten Million Programme.

Prospects for the Japanese outbound market: implications for tourist transport systems

It is evident from the number of operators seeking to develop airline operations to serve Japan and the Asia-Pacific region that the Japanese outbound market offers considerable potential for the new millenium. Morris (1990) forecast that the outbound market would double by the year 2000, reaching 20 million trips and averaging annual growth rates of approximately 10 per cent. Nozawa (1992) argued that market research in Japan suggests outbound travel would expand in the following market segments:

● outbound travel by the female traveller market, especially in the 20–39 age group among office professionals (single females with a high discretionary income)
● Japanese housewives using package tours
● older married couples aged 45 or more (full mooners) with the 'greying' of its population
● the 'silver market' (the over 60s)

and these are markets which were discussed by Page (1989a) within the context of New Zealand.

Outbound Japanese tourists are likely to require more sophisticated package tours and independent travel on scheduled airlines. Since the Japanese tourist is preoccupied with quality in service provision, airlines and providers of tourism services will need to be conscious of the needs and 'tourist experience' sought by these travellers. Repeat travel among the Japanese to destinations they visited on package tours is growing, particularly in New Zealand. As sightseeing and touring dominate their itineraries, these offer potential niche markets for transport service providers in destination areas. The tendency among Japanese tourists for shorter stays abroad and the complex motives that influence their travel remain important factors which tour operators and tourist transport systems must try to take into account if they are to nurture the Japanese market (Morean 1983; Moore 1985). An understanding of the objectives and values of such travellers (e.g. the desire for service quality and safety) and compliance with these demands will influence the perception and image of specific tourist transport systems in the 1990s and the airlines tourists choose to fly on. Transport operators serving Japan not only need to recognise the dimensions of demand in inbound and outbound tourism markets, but also to consider how future demand will be accommodated in tourist transport systems. The Japanese case study shows that infrastructure was unable to balance the demand with adequate supply.

Yamauchi (1997) identifies the difficulty which foreign airlines have in entering the Japanese market due to government policy and airport congestion problems. New proposals to allocate airport slots at Hareda airport, after the completion of a new runway, may mark a shift to a more transparent and competitive approach to air travel. But a very traditionalist approach to international aviation could continue to add to constraints on making travel more available and cost-driven for the benefit of consumers.

Attention now turns to the significance of the demand expressed as market segments before considering tourism forecasting to understand future patterns of tourist demand for travel. This will help to develop a sophisticated analysis to enable transport planners and policy makers to recognise the opportunities for tourist transport systems so that adequate infrastructure is provided to meet demand.

Analysing tourism demand: a market segment approach

In the previous case study, the discussion of growth in the Japanese outbound market focused on the prospects for specific market segments. This approach has found a great deal of support among research companies and tourist transport operators who seek to understand how they may be able to target niche or broadly defined markets to understand their travel requirements. In a growing global marketplace with transport operators seeking to offer new products to customers, research adopting a market segment approach can provide another source of data to assess the demand for travel. For this reason, three markets (business travel, VFR and the senior market) are now outlined to illustrate the value of such specialist studies in assembling data on a specialist form of demand.

Business travel in the 1990s

Within the literature on tourism demand, comparatively little attention has been devoted to business travel. Much of the research has been in the form of specialist reports by agencies such as American Express in the 1980s and a range of studies by market intelligence organisations and analysts. Rowe (1994)

Table 4.13   WTO estimates of business travel by region

Region	Business-related arrivals (000s)
Europe	47,108
Americas	8,082
East Asia-Pacific	7,079
South Asia	405
Africa	2,524

Source: WTO (1992) cited in Rowe (1994).

is an excellent example of an in-depth market intelligence report combining secondary sources of travel data and a blend of primary research commissioned to examine the business issues and prospects for this activity. Rowe (1994 : 1) defines business travel as 'non-discretionary trips made for a work-related purpose'. Rowe's key finding is that the global volume business travel is probably underestimated in WTO statistics. The data in Table 4.13 is based on estimates from WTO in 1992 and illustrates the poorly documented nature of the subject. While Rowe's (1994) study is instrumental in explaining the extent of business travel by industry sector and the average company travel expenditure, it also provides detailed insights into the nature of demand. Macroeconomic influences have a major impact on the demand for this form of travel, although other factors also influence the rate of growth, such as the role of trading blocs. According to Rowe the fastest growth in global business travel is likely to occur in Central and South America, the Middle East and all parts of the Asia-Pacific region. Table 4.14 highlights the regions of forecast growth in business travel within the East Asia-Pacific (EAP) region.

The UK VFR market

The VFR market segment has only recently attracted attention. According to Beioley (1997), VFR constituted 29 per cent of UK domestic tourist trips in 1995, which equates to 39 million trips. This generated £3.25 billion of revenue, equivalent to 13 per cent of the UK's total tourism spend. The interesting feature which Beioley identifies is the dominance of such trips by young people (under 30 years of age) who have a strong reliance upon public transport. Overseas tourists visiting the UK for VFR purposes tend to stay longer than non-VFR visitors. In fact, within the domestic VFR market, 33.1 million trips were spent in the homes of friends and relatives in the UK. In terms of mode of transport used, rail companies in the UK estimate that between 30 and 50 per cent of traffic is leisure-oriented (Beioley 1997), with some airlines (e.g. Ryanair) viewing VFR leisure travel to/from Ireland as comprising 70 per cent of their business. Similarly, scheduled carriers such as Qantas view VFR business as comprising 50 per cent of their market. In geographical terms, much of the VFR demand in the UK is located in the

Table 4.14 Future growth circles for business travel in the East Asia-Pacific region

Growth circle	Participating countries
• Golden quadrangle	S.W. China, Burma, Thailand and Laos
• Japan Sea economic zone	Japan, North and South Korea, N.E. China and the Russian Far East
• Growth circle	Taiwan and the neighbouring Chinese province of Fujian
• Growth circle	Hong Kong, Guangdong and Macau
• Growth circle	South China, Vietnam, Cambodia, North East Thailand and Laos
• Yellow Sea economic zone	Maritime China, Zhanjiang and Liaoning province
• The Straits growth triangle	Singapore, Malaysia's Johor and Indonesia's Riau province
• Northern growth triangle	North Malaysia, South Thailand and Northern Sumatra
• Growth circle	Malaysia and Indonesian-governed area in Borneo, the South Philippines and North Sulawesi
• Growth circle	Southern Sulawesi and Australia's Northern Territory

Source: Modified from Rowe (1994) based on *Asia Inc. Business Magazine*.

larger cities (14.8 million trips) compared to small towns (9.7 million trips), countryside areas (6.8 million trips), London (4.8 million trips) and coastal towns (3.6 million trips). It is evident from Beioley's (1997) study that there are a group of market segments and understanding the demand for VFR travel is a more complex task than was hitherto the case.

The senior market

In recent years, researchers have begun to focus on the increasing significance of one market segment that has exhibited many growth characteristics, notably the elderly or senior market (Viant 1993). The senior market is invariably defined as those travellers in middle age (55 years and over). Their significance is illustrated by Smith and Jenner (1997) who argue that international arrivals by this group will exceed 100 million in 1997, accounting for one-sixth of global trips. This is a recognition of the 'greying' of the population base of many Western nations, with the 55+ age group growing steadily in the USA and Europe and also in Japan. As a market segment, this group is normally described as possessing a significant disposable income and an expectation of increased mobility reflected in greater international travel habits. Smith and Jenner (1997 : 44) cite a US magazine, *Seniors Travel Tips*, which notes that seniors account for '80 per cent of luxury travel, 70 per cent of coach tours, 65 per cent of cruises, 32 per cent of overnight stays in hotels and 28 per cent of foreign travel, control 50 per cent of discretionary income, and 44 per cent of passports'. This is a clear indication of the market's significance in North America and its contribution to domestic and international tourism demand.

Table **4.15**   Senior population aged 55 or over for main tourism-generating countries, 1997

Country	Total population (000)	Population 55+ (000)	Senior population (% share)
Austria	8,054	2,101	26.1
Belgium	10,204	2,725	26.7
Denmark	5,269	1,349	25.6
France	58,470	14,678	25.1
Finland	5,109	1,254	24.5
Germany	84,068	23,846	28.4
Greece	10,583	2,988	28.2
Ireland	3,556	695	19.5
Italy	57,534	16,559	28.8
Japan	125,717	35,221	80.0
Luxembourg	422	104	24.6
Mexico	97,563	9,043	9.3
Netherlands	15,653	3,603	23.0
Portugal	9,868	2,535	25.7
Spain	39,244	10,257	26.1
Saudi Arabia	20,088	1,331	6.6
Sweden	8,946	2,436	27.2
Taiwan	21,656	3,288	15.2
UK	58,610	14,971	25.5
USA	267,955	55,913	20.9

Source: US Bureau of the Census, International Data Base, cited in Smith and Jenner (1997 : 47).

Smith and Jenner (1997) indicate that there is a dearth of research on this topic following a series of ad hoc reports by the European Travel Data Centre and the Aviation and Tourism Consultancy report on the subject in 1992. WTO do not collate any comprehensive data on this market, which means researchers have to look to other sources, and in some cases make estimates of demand. While Smith and Jenner (1997) debate the semantics of what is a senior traveller, they provide a useful overview of the senior population by main tourism-generating countries (Table 4.15). This highlights the volume and variable proportion of the senior population in each country (see Smith and Jenner 1997 for individual profiles of the senior travel market by country). On the basis of estimates derived from their research, Smith and Jenner (1997) consider that:

- On average the senior population comprises 25 per cent of the population in most countries, with Germany, Italy, France, the UK and Spain having the largest populations in this age group in the EU.
- The older people become, the less inclined they are to travel. For example, in the EU, the senior market (aged 55–59 years) made 2.4 domestic trips a year and 0.8 international trips on average. By age 75, the number of domestic trips drops to 1.3 domestic and 0.5 international trips.
- In 1993, there were an estimated 49 million senior arrivals in the EU, many of which were intraregional and by 1996 this had risen to 60 million trips.

Table 4.16 Transport used by seniors on international
trips in the EU, 1996

Mode of transport	Share of market (%)
Air	
Scheduled	17
Charter	14
Car (private/rental)	27
Coach/Bus	24
Train	7
Cruise/Ferry	10
Other	1
Total	100

Source: Estimates by Smith and Jenner (1997).

- The senior market is far less seasonal than other tourism markets, given its flexibility to travel in the non-peak period.
- almost one-third of trips made by the senior market in Europe were by air and they also favour other land-based forms of transport (e.g. coach travel) which they perceive to be safe. Table 4.16 outlines the use of transport by this market and also highlights the interest in cruising.

A range of specialist senior travel organisations exists which nurture such travellers (e.g. the Saga group in the UK) while outbound North American educational tours have also seen significant growth in recent years (e.g. the Elderhost group). Smith and Jenner (1997) rightly point to the future growth prospects of this market, expecting a 30 per cent growth in the EU to 83 million arrivals by the year 2000. At a global scale, they forecast a growth to 136 million arrivals in the year 2000.

It is apparent from a consideration of the significance of three types of demand expressed through market segments that researchers often need to move beyond aggregate data sources produced by WTO and OECD to derive more specific data and information. This often involves developing matching data sets to produce estimates of tourism demand, as in the case of the senior market. Transport operators and planners therefore need to have a detailed understanding of tourism demand and the difficulties associated with data sources used to understand demand. The implications here are significant when attempting to predict future patterns of demand through forecasting.

Forecasting the demand for tourist transport

According to Jefferson and Lickorish (1991 : 101), forecasting the demand for tourist transport is essential for commercial operators, 'whether in the public or private sector . . . [as they] . . . will seek to maximise revenue and profits in moving towards maximum efficiency in [their] use of resources'. Archer (1987) argues that:

no manager can avoid the need for some form of forecasting: a manager must plan for the future in order to minimise the risk of failure or, more optimistically, to maximise the possibilities of success. In order to plan, he must use forecasts. Forecasts will always be made, whether by guesswork, teamwork or the use of complex models, and the accuracy of the forecasts will affect the quality of the management decision. (Archer 1987 : 77)

Reliable forecasts are essential for managers and decision makers involved in service provision within the tourist transport system to try and ensure adequate supply is available to meet demand, while avoiding oversupply, since this can erode the profitability of their operation. In essence, 'forecasts of tourism demand are essential for efficient planning by airlines, shipping companies, railways, coach operators, hoteliers, tour operators . . .' (Witt et al 1991 : 52).

Forecasting is the process associated with an assessment of future changes in the demand for tourist transport. It must be stressed that 'forecasting is not an exact science' (Jefferson and Lickorish 1991 : 102), as it attempts to make estimations of future traffic potential and a range of possible scenarios, which provide an indication of the likely scale of change in demand. Consequently, forecasting is a technique used to suggest the future pattern of demand and associated marketing activity is required to exploit the market for tourist transport services.

According to Jefferson and Lickorish (1991 : 102) the principal methods of forecasting are:

- 'the projection by extrapolation, of historic trends' (i.e. how the previous performance of demand may shape future patterns)
- 'extrapolation, subject to the application of . . . [statistical analysis using] . . . weights or variables'
- and structured group discussions among a panel of tourism transport experts may be used to assess factors determining future traffic forecasts (known as the Delphi method).

Bull (1991) recognises that the range of tourism forecasting techniques are determined by the methods of analysis they employ. There are two basic types of forecasting method: those based on qualitative techniques, such as the Delphi method, which Archer (1987) argues are considerably less rigorous than quantitative forecasting methods, using techniques developed from statistics and economic theory.

Bull (1991 : 127–28) classifies the quantitative techniques forecasters use in terms of the degree of statistical and mathematical complexity based on:

- time-series analysis of trends (e.g. seasonality in travel) which involve simple statistical calculations to consider how past trends may be replicated in the future
- economic theory models, used in econometrics (see Witt and Martin 1992).

Clearly, in an introductory book such as this, there is insufficient space to consider the detailed technical aspects of forecasting which are reviewed in a number of other good sources (e.g. Archer 1987; Witt and Martin 1992; Witt and Witt 1995; Frechtling 1996; Sinclair and Stabler 1997). The important issue to recognise here is that in forecasting, a number of variables are examined which relate to factors directly and indirectly influencing tourist travel. These variables are considered according to their statistical relationship with each other. Bull (1991 : 127) notes that the most common variables used are:

- number of tourist trips
- total tourist expenditure and expenditure per capita
- market shares of tourism
- the tourism sector's share of gross domestic product.

Depending on the complexity of the methodology employed, the forecasting model may examine one dependent variable (e.g. tourist trips) and how other independent variables (e.g. the state of the national and international economy, leisure time, levels of disposable income, inflation and foreign exchange rates) affect the demand for tourist trips.

Approaches to forecasting can also be classified according to what they are attempting to do. For example, time-series models of tourism, using statistical techniques such as moving averages (see Witt and Martin 1989 : 6) may be easy and relatively inexpensive to undertake, but they are 'non-causal'. This means that they do not explain what specific factors are shaping the trends: they only indicate what is happening in terms of observed trends. In fact, there is also evidence that non-causal techniques have been more accurate than more complex econometric models (Witt et al 1991). Econometric models are termed 'causal' since they are searching for statistical relationships to suggest what is causing tourist trips to take a certain form, thereby producing particular trends. Thus, the level of complexity involved in causal modelling is considerably greater.

Usyal and Crompton (1985) provide a good overview of different methods used to forecast tourism demand, concluding that:

> Qualitative approaches when combined with quantitative approaches enable forecasts to be amended to incorporate relevant consumer demand data. When used alone quantitative models have conceptual limitations. Typically they are philosophically blind . . . Lack of appropriate data means that they are unable to incorporate an understanding of consumer motivations and behaviour which explain tourism demand and may cause it to shift unpredictably in the future. (Usyal and Crompton 1985 : 14)

Ultimately, forecasting attempts to establish how consumer demand for tourist transport has shaped previous trends and how these may change in the future, often over a five- to ten-year period. At a world scale, the detailed study by Edwards (1992) *Long Term Tourism Forecasts to the Year 2005* and subsequent updates remains an invaluable and widely cited source which examines the future demand for tourism. This can be used to assess how the demand for tourist transport will change on a global basis and within different countries over the next decade. A useful discussion of the significance of using forecasts of tourist use of a new transport system can be found in Page and Sinclair's (1992b) analysis of the demand for Channel Tunnel rail services. Nevertheless, such reports remain highly technical and only of particular use to economists who understand the techniques and methods used to forecast tourism (see Frechtling 1996). However, industry organisations such as the Air Transport Action Group commission specialist agencies to produce forecasts. For this reason, some of their key findings from their North America Traffic Forecasts 1980–2010 (Air Transport Action Group 1994) are now examined.

Air transport traffic forecasts to the year 2010

The Air Transport Action Group (ATAG) was established in 1990 as an industry coalition aimed at developing 'economically beneficial aviation capacity improvements in an environmentally responsible manner'. It is concerned to lobby bodies to ensure the future growth in air travel can be accommodated through appropriate investment in airports and air traffic management. The organisation recognises that its impact has been greatest in Europe, North America and Asia through a range of publications it produces. Most of its reports are also available on the Internet (http://www/atag.org/atag/Index.htm). Given ATAG's values and objectives, the report it commissioned from IATA in 1993 provides a detailed analysis of North American traffic forecasts to 2010 (http://www.atag.org/NATF/Index.htm). While there is not space here to examine the report in detail, as a source of data it derives forecasts based on total passenger numbers in North America between 1980 and 1993. The report is extremely detailed and differentiates between various elements of existing and future demand by US and non-US citizens and also between US and non-US carriers. Some of the key findings are interesting given the implications that ATAG identifies for its members, government agencies and private business interests:

- Air traffic in North America is heavily dominated by domestic travel and the USA accounts for approximately 80 per cent of passenger traffic, followed by Canada (4 per cent) and Mexico (3 per cent). Intra-NAFTA (North American Free Trade Area) traffic constituted 5 per cent and international traffic 8 per cent.
- Future growth rates for international traffic to and from the USA on routes to the year 2010 are forecast to be 8.2 per cent on Pacific routes, 5.6 per cent on Atlantic routes and 6 per cent on Latin American routes.
- Domestic travel in the USA is expected to rise from 450 million passengers in 1993 to over 700 million in 2010.
- IATA forecasts that the number of international passengers travelling to and from the USA will reach 226 million by 2010, a 187 per cent increase on 1993 figures of 78.8 million passengers.
- As a consequence of these growth rates, ATAG identifies the following supply-side issue: delays at US airports will increase from 23 to 33 airports by the year 2002, based on cumulative annual delays of over 20,000 hours a year.

Thus ATAG provides a range of reports on air travel in key markets which utilise IATA, ICAO and other data sources to assess demand. Its reports are easily accessed on the Internet and the detailed analysis of the North American market by domestic and international routes is a useful starting point to consider the relationship between demand, supply and the role of forecasting. Such research also highlights the significance of using various data sources and techniques to assess future demand. As Witt and Witt (1995) conclude in their seminal review of tourism forecasting methods, future developments in forecasting need to make improvements in model building and to use a range of methods to produce forecasts.

Summary

This chapter has shown that the demand for tourist transport can be examined from a range of accessible data sources published by the WTO and OECD, which may be complemented by more detailed studies by individual researchers, consultancies and transport operators interested in understanding the needs of their customers. As both Chapters 3 and 4 emphasised, the demand for tourist transport can be directly influenced by government policy to expand both incoming tourism and outbound tourism (e.g. Japan), although this has important consequences for infrastructure provision and the supply issues in tourist transport which are dealt with in subsequent chapters. In the case of China, the demand for domestic travel is of a scale which many planners, operators and researchers would find difficult to visualise and conceptualise. Growth in demand has major implications for infrastructure and the use of transport modes. Even a minor growth in demand in China has major consequences for tourist travel, given the large base population and distances involved in domestic trips. While data on domestic tourism in China is notoriously difficult to access outside of China, existing studies do point to the significance of diverse data sources in reconstructing patterns of demand and expected future growth. In this respect, WTO data can only be a rudimentary starting point for any analysis of demand. As the case study of Japan and the discussion of market segments reveal, a variety of other data sources (which do not always correspond with each other) are vital in understanding the human dimension of tourism demand.

The Ten Million Programme promoted by the Japanese government appeared to overlook the constraints imposed by airport capacity, despite the licensing of new carriers and routes to serve the Japanese market. Many airline companies have set their sights on penetrating this emerging market for outbound tourism in the 1990s, at a time of relative stagnation in other mature travel markets. Such strategies are based on forecasts of the likely growth in the Japanese outbound market since companies need to establish how future investment programmes in new routes and infrastructure can be justified. Yet as Burns (1996) argues, statistical growth in tourism is not necessarily an indication of widespread cultural and cognitive changes towards holiday and overseas travel. The Japanese outbound market would have considerable growth potential if it was able to develop a holiday-taking culture commensurate with its disposable income. It is not surprising that large tourist transport operators employ forecasting experts in-house and use a variety of transport analysts to ensure that they receive the best sources of market intelligence to understand how tourism markets are performing and changing to remain competitive. Such information inevitably remains confidential, even though forecasting is not an exact science which is able to offer definitive answers on how particular tourist markets will perform over a five- to ten-year time span.

Yet the demand for tourist transport should not be viewed in isolation from supply issues. The tourist transport system operates through the interplay of government policy, consumer demand and the supply of transport services. For this reason, the next two chapters consider the supply of tourist transport services from two perspectives: how to understand supply issues and how they are subsequently managed.

Questions

1. What are the main published sources of data available to researchers to assess the demand for domestic and international travel?
2. Why do private and public sector tourist transport providers commission research on the demand for their services?
3. Why do tourist transport organisations undertake forecasting?
4. What are the main types of forecasting technique which researchers use when assessing the future market potential for tourist transport services?

Further reading

In terms of international tourism statistics, the most accessible synthesis of trends is:

BaRon, R. (1997) 'Global tourism trends to 1996', *Tourism Economics* 3 (3): 289–300.

On forecasting and tourism data consult:

Archer, B.H. (1987) 'Demand forecasting and estimation' in J.R.B. Ritchie and C.R. Goeldner (eds) *Travel, Tourism and Hospitality Research*, New York: Wiley, 77–85.
Withyman, W. (1985) 'The ins and outs of international travel and tourism data', *International Tourism Quarterly*, Special Report No. 55.
Witt, S. and Martin C. (1992) *Modelling and Forecasting Demand in Tourism*, London: Academic Press.

A good article which examines China's rapidly expanding transport system and the problems which infrastructure constraints pose can be found in:

Shen, Q. (1997) 'Urban transportation in Shanghai, China: Problems and planning implications', *International Journal of Urban and Regional Research* 21 (4): 589–606.

Internet sites

The best source to consult on the internet sites for the Pacific Rim region is:

Hall, C.M. (1997) *Tourism in the Pacific Rim*, 2nd edition. Melbourne: Longman Cheshire as it contains the World Wide Web sites for national tourism organisations (where available) which can be used as a statistical data source on tourism demand.

The World Tourism Organisation can be accessed at:

http://www.world-tourism.org

The Cruise Lines International Association is at:

http://www.ten-io.com/clia/

Chapter 5

Analysing supply issues in tourist transport

Introduction

In Chapter 4 the demand for tourist transport was discussed in terms of the commercial opportunities it affords operators who are able to understand and harness it. To meet the demand for tourist transport, businesses and operators can employ a range of concepts to analyse what they need to do to match supply to demand. For this reason, the next two chapters examine supply issues in two discrete and yet interrelated ways. First, this chapter considers some of the broader issues affecting the supply of tourist transport, particularly a conceptual framework in which both the traveller's and the transport provider's perspectives are considered to try and maximise the commercial and non-commercial opportunities within the tourism industry. Secondly, the next chapter discusses how operators and the transport sector employ particular management tools (e.g. logistics and information technology applications) and business strategies to provide the tourist as a user and consumer of the transport system with a range of opportunities to enhance their overall tourist experience. While the two chapters examine supply issues in different ways – looking at the complexity of analytical and operational issues separated into discrete sections – in reality one needs to view these two elements in a holistic manner as they are intertwined in the real world.

It is also important to emphasise the broader tourism context in which tourist transport exists. The recent study by Harris and Masberg (1997) that examined vintage tram operations in 26 North American cities indicates how the supply of a mode of transport can be harnessed for tourism and yet is not able to generate tourism without appropriate infrastructure to support it. Harris and Masberg (1997) argue that while trams may constitute a cultural icon influencing tourists to visit a destination, a pool of attractions is needed to get visitors to use trams to tour within the destination. In other words, the supply of transport itself is not sufficient to stimulate tourism development but can be a catalyst if it is integrated as part of a wider strategy to develop attractions, accommodation and an urban tourism experience (Page 1995b). This example serves to illustrate the significance of developing various approaches and concepts which promote a broader understanding of transport supply issues. A similar argument can also be developed for transport modes such as car hire which support tourist activities (see Loverseed 1996), although in a number of cases where transport is the tourist experience (e.g. cruising and vintage railways), supply issues may assume an even greater significance as the entire experience is directly dependent upon the service the operator provides.

There is a relative paucity of research on supply issues in tourism and transport studies (Eadington and Redman 1991). Witt et al (1991) consider that the:

> subset of transport studies that directly relates to tourism is relatively neglected . . . [and] . . . it is a major task of research to bring together the work done in transport studies with that more specialised work on tourism . . . [as] . . . many of the relevant studies are privately commissioned and often not widely disseminated. (Witt et al 1991 : 155–6)

The absence of any synthesis of supply-related research which integrates tourism and transport into a more cohesive framework led Sinclair (1991 : 6) to argue that the 'literature on transportation and its implications for domestic and international tourism merits separate analysis'. The efficient management and operation of transport systems for tourists require that demand issues are analysed in relation to supply since the two issues coexist and they determine the future pattern of use and activities within the tourist transport system.

The supply of tourist transport has been dealt with in various popular tourism textbooks (e.g. Holloway 1989; Lavery 1989), which consider the characteristics, principles and organisation associated with each mode of tourist transport. In a book such as this, it is inappropriate to reiterate the empirical discussion of different modes of transport in these publications since it would inevitably result in a descriptive listing which is documented elsewhere (Collier 1994). Furthermore, a variety of good accessible publications also document many of the issues associated with different modes of tourist transport, using up-to-date market intelligence compiled by transport analysts. (The former Economist Intelligence publications and the subsequent travel and tourism intelligence journals such as *Travel and Tourism Analyst, International Tourism Reports* and their occasional reports are essential reading.) The chapter commences with a discussion of theory related to supply issues; this is followed by a framework which develops the context in which both the traveller's and the transport provider's perspectives are considered in relation to an underlying concern for service quality. The concept of transaction analysis is introduced as a method of understanding the central role of transport in the supply of tourist travel. Transaction analysis assists in assessing the transport supplier's involvement in the distribution chain and the ways in which they may influence and control the chain and service quality in the supply of transport services. To illustrate the extent of one transport operator's involvement in the distribution chain, a case study of Singapore Airlines is examined.

Theoretical perspectives on tourism and transport supply issues —

Despite the rapid growth in research studies in tourism in the 1980s and 1990s, those publications which make a contribution to the advancement of knowledge and our understanding of the subject are still comparatively few.

This is certainly the case in relation to supply issues and one explanation may be related to the fact that

> Tourism supply is a complex phenomenon because of both the nature of the product and the process of delivery. Principally, it cannot be stored, cannot be examined prior to purchase, it is necessary to travel to consume it, heavy reliance is placed on both natural and human-made resources and a number of components are required, which may be separately or jointly purchased and which are consumed in sequence. It is a composite product involving transport, accommodation, catering, natural resources, entertainment, and other facilities and services, such as shops and banks, travel agents and tour operators. (Sinclair and Stabler 1997 : 58)

Thus many businesses supply components which are combined to form the tourism product, and because they operate in different markets, it makes it difficult to analyse supply issues. In fact, it proves even more complex when seeking to separate out one element of the tourism product (i.e. transport) to identify the range of supply issues affecting an individual element.

Probably the most influential and pertinent publication to date which assists in addressing supply issues in a theoretical framework is a recent synthesis of *The Economics of Tourism* (Sinclair and Stabler 1997). Sinclair and Stabler infer that one can explain how firms operate under different conditions and therefore it may be possible to identify factors which affect supply issues in relation to tourism in general, and transport in particular. While it is not possible to present a detailed analysis here, the main principles outlined by Sinclair and Stabler are discussed as they focus on four market situations:

- perfect competition
- contestable markets
- monopoly
- oligopoly

In this discussion, attention is directed to the transport sector.

Perfect competition

In economic models of conditions of perfect competition, a number of assumptions exist:

- there are a substantial number of consumers and firms, implying that neither can affect the price of an undifferentiated product
- there is free entry to and exit from the market, assuming that there are no barriers.

Sinclair and Stabler (1997) explain how a perfect market operates and how prices are derived, with the tendency towards a break-even price in a situation where consumers derive a benefit. However, in the real world, many economists believe that markets are not perfectly competitive.

Contestable markets

In this situation, there are 'insignificant entry and exit costs, so that there are negligible entry and exit barriers. Sunk costs which a firm incurs in order to produce and which would not be recoupable if the firm left the industry, are not significant' (Sinclair and Stabler 1997 : 61). Due to technology, information and supply conditions are available to all producers and while producers cannot change prices instantaneously, consumers can react immediately. The key principle here is that new and established firms are able to challenge rival businesses through pricing strategies. Firms in contestable markets are seen to operate in a similar way to those in perfect markets, since they charge similar prices for a product; existing operators cannot charge more than average cost because more competitors would enter the market. Due to low sunk costs and low entry/exit barriers, rivals establish to compete.

Monopoly

This is probably best described as the opposite of perfect competition, where a major business or firm is able to exercise a high level of control over the price of the product and level of output. The implications are that firms operating in a monopolistic market charge prices above the average cost of production to generate high profit levels, so consumers pay a price higher than that which would exist in a competitive market. In many countries, domestic air and rail networks operate under monopoly conditions even though it can be against the interests of consumers.

In some cases, a monopoly condition may be more beneficial than competition, as in the case of deregulation in the transport industry. In such situations, an influx of new entrants following deregulation may lead to smaller firms being taken over by larger businesses (see Chapter 6 on airline deregulation in the USA). On monopoly routes this results in higher prices which are detrimental to consumers' interests. Sinclair and Stabler (1997) also examine other scenarios where monopoly conditions may be beneficial to the wider public good in tourist transport operations. Where governments have privatised state transport interests, one outcome has been a greater degree of concentration of operations among a small number of operators. However, where monopoly situations exist, regulation by the state is normally imposed to prevent higher prices and supernormal profits.

Oligopoly

An oligopoly exists where a limited number of producers dominate the transport sector. Williams (1995 : 163) highlights the situation in relation to tourism and transport since 'tourism has a highly dualistic industrial structure which is polarised between large numbers of small firms (typically in retailing, accommodation services) and a small number of large companies (for example, in air transport)'.

In an oligopoly, each firm controls its price and output levels and there are entry and exit barriers. An oligopoly market situation is characterised by supply conditions dependent in part upon the output and pricing decisions of competitors. In an ideal world, oligopolies prefer prices to be set at levels where the profits are maximised for all producers in that industry sector. If the firms colluded to set prices, it could lead to a monopoly and higher profits for producers if they restricted the supply. Sinclair and Stabler (1997) point to the impact of inter-airline pricing and route-sharing agreements to achieve joint profits in an oligopolistic situation. In an oligopolistic market, producers can alter output and prices while taking account of their competitors' likely reactions.

In their overview of the air travel market, Sinclair and Stabler (1997) argue that

> Although a domestic monopoly or oligopoly structure has been common, with a single state-supported airline or a small number of competing airlines, deregulation has made some markets competitive in the short run. In the international market some routes are competitive, being served by many carriers. Most of the others are served by at least two carriers, indicating an oligopolist market, although a few routes are served by a single carrier which may be tempted to exercise monopoly powers. (Sinclair and Stabler 1997 : 81)

In terms of the remaining transport sectors, Sinclair and Stabler conclude that

> The structures of the bus, coach and rail sectors are similar to that of air travel in that they too, experience the problems of high capital costs, fixed capacity, peaked demand, the need for feeder routes to sustain profitable ones. Some state support and regulation characterise these modes. (Sinclair and Stabler 1997 : 81)

It is apparent that where a large number of small firms operate in the transport sector a competitive market exists. In contrast, where a limited number of firms operate, akin to an oligopoly or at the extreme, a monopoly, different conditions affect the supply of transport services for tourists. What emerges from the excellent analysis of supply conditions in tourism is that various criteria influence the competitive conditions which exist in different markets, and factors such as the degree of market concentration or price leadership affect the extent and nature of inter-firm competition. For example, French (1996) examines the advent of 'no frills' airlines in Europe and the effect on the market. In a similar vein, Peisley (1997) examines the case of the Channel Tunnel as a new market entrant and its impact on the cross-Channel ferry market. Here the competition led to a reduction in yields with a competitive price war. The result has been a greater degree of concentration in the ferry market with the merger of P&O and Stena Line. Thus, in any analysis of transport supply issues, a range of criteria need to be investigated in different market conditions. These are based on Sinclair and Stabler (1997 : 83):

- the number and size of firms
- the extent of market concentration
- entry and exit barriers
- economies/diseconomies of scale and economics of scope
- costs of capital, fixed capital and costs of operation
- price discrimination and product differentiation
- pricing policies (e.g. price leadership, price wars and market-share strategies)

The final two points are examined in more detail in Chapter 6 since they are concerned with the strategies businesses pursue in competitive markets. In the case of the first five points, which pertain to market structures, Sinclair and Stabler identify a range of data sources available to researchers. In the airline sector, some data sources on demand also contain evidence on supply such as ICAO, IATA and industry bodies. Other sources, such as company annual reports, also contain interesting data. Therefore, having outlined some of the principal theoretical considerations affecting the supply of transport for tourism, attention now turns to a conceptual framework – the supply chain. This enables one to recognise the theoretical issues and to appreciate how the criteria associated with competitive markets may affect the organisation and delivery of supply.

The supply chain in tourist transport services

Prior to the innovative synthesis by Sinclair and Stabler (1997), there was an absence of detailed research on the supply of tourism and transport services. This has acted as a major constraint on the development of literature in this area, and with the exception of research by transport economists, economists' interest in supply issues has been limited, if not peripheral to the main studies in the area (Sinclair and Stabler 1997). The situation has been compounded by the image of supply research in tourism and transport studies, which is sometimes perceived as descriptive, lacking intellectual rigour and sophistic-ated methods of study, since 'generally there is little research on the tourism [and transport] industry and its operation which is analytical in emphasis' (Sinclair and Stabler 1991 : 2). This is perpetuated by the treatment of supply issues in many general tourism texts which broadly discuss 'passenger trans-portation', since there are methodological problems in differentiating between the supply and use of transport services by the local population for travel to work, leisure and recreational travel purposes and more specific tourist use. In fact, Sinclair and Stabler (1997 : 70) argue that 'categories [such] as transport . . . are very broad and benefit from disaggregation into sub-markets with different structures and modes of operation'. Thus it is not surprising to find that research has focused on established areas of tourism and transport supply, notably:

- descriptions of the industry and its operation, management and marketing
- the spatial development and interactions which characterise the industry at differ-ent geographical scales
  Source: Sinclair and Stabler (1991 : 2)

Studies of transport systems within tourism research have been character-ised by a preoccupation with how their operations are organised to provide a service to travellers and how the international nature of transport facilitates tourism activities and development. This approach to research on tourist trans-port systems is rooted in economics, as emphasised in Chapter 2, based on the concept of the 'firm', developed by Coase (1937) and discussed further by

Buckley and Casson (1985). In the context of tourism and transport supply, Buckley (1987) notes that the analysis of a firm or company is characterised by certain relationships within the organisation and with its purchasers or consumers. The external process of selling a product or service involves a transaction between two parties following an agreement to purchase, often though not exclusively, using a monetary transaction. Commercial transactions are based on agreed conditions and enforced within a framework of contractual obligations between the parties. Therefore, transaction chains develop to link the tourist with the suppliers of services in tourism and the 'tourism product or service' is defined as the sum of these transactions (Witt et al 1991 : 81). Such research highlights the significance of the 'chain of distribution' for transport and tourism services, which is the method of distribution of the service from production through to its eventual consumption by tourists. A more general discussion of the distribution chain in tourism can be found in Holloway (1995).

Transaction analysis

Buckley (1987) describes some typical transaction chains for tourism which identify the integral role of transport services in linking origin and destination areas (see Figure 5.1). The nature of the specific supply chain depends upon a wide range of factors which are internal and external to individual firms in the transport sector. For example, what is the primary force driving the supply system? Is it driven by pull factors, where a tourist destination may market a region and supply transport services on a state-owned airline to stimulate demand for tourism? Or is it driven by push factors, where the tourist generates the demand, and the transport and accommodation sectors respond to this as a commercial opportunity? The overall business environment, government predisposition to tourism and planning constraints may have a moderating influence on the supply system. In addition, transaction analysis illustrates the significance of 'agents' in the system, corporate policy in transport provision and contractual arrangements in the supply chain.

Much of the existing knowledge available on these issues has been generated through interviews with managers in each sector of the transport industry about their commercial practices (e.g. contracting arrangements, profit margins and global strategies). It is rare to find researchers being given access to commercial information on supply (and demand) issues, due to the confidential and sensitive nature of the data, and the perceived threat it might pose to a company's competitive advantage if rival operators obtained such information. In some cases this amounts to paranoia among companies, as media coverage of the British Airways and Virgin Atlantic libel case in 1992–93 highlighted. The result is that the relationship between transport supply and tourist use remains poorly understood, with commercial research primarily concerned with the effect of pricing transport services, the behaviour of consumers (see Gilbert 1990) and the outcome in terms of use and profitability for producers. It is within this context that Buckley's (1987) research proves useful in understanding the nature of relationships which may exist in the supply chain.

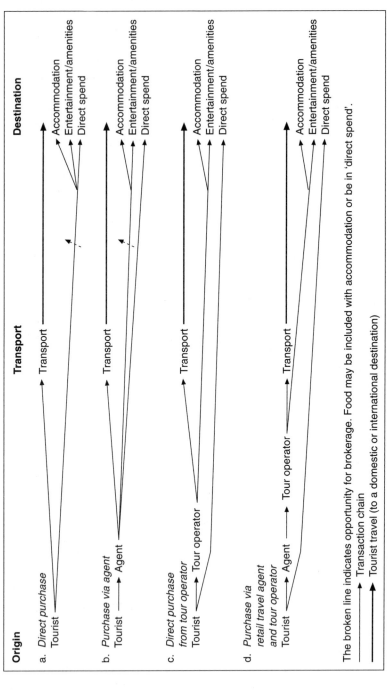

Figure 5.1 Four types of tourism transaction chain (redrawn from Page 1994, based on Witt et al 1991)

From Buckley's four chains (Figure 5.1), it is evident that a variety of distribution systems exist for the sale and consumption of transport services by tourists (O'Brien 1990; Barnes 1989). One of the critical issues in the distribution system for the seller is access to superior information on available services, so that these can be sold to the consumer. There are various studies which document tourism and transport retailing (e.g. Holloway and Robinson 1995), where the agent or broker is normally paid a commission on their sales. The travel agent comprises a convenient one-stop location for tourists to buy tourism services as an inclusive package, which includes transport and accommodation, usually marketed through the medium of a brochure. The packaging of these products or services (much of the literature interchanges these terms) by wholesalers (e.g. tour operators) reduces the transaction costs to the tourist of purchasing each element independently (Laws 1995). Thus a travel agent normally receives around 10 per cent commission on a sale of a holiday marketed by a tour operator, but the overall cost to the consumer is markedly lower than arranging the same components independently. However, in the case of air tickets, commissions to agents are declining as airlines seek to reduce this expense in competitive market conditions. The tour operator is able to reduce the number of transactions involved by packaging a holiday, thereby making economies in the supply through wholesale purchasing and by entering into long-term contracts with the suppliers of accommodation and transport services. Not only does this have benefits for the price charged to the purchaser, but it has more beneficial effects for the supplier as a number of intermediaries or brokers in the chain are eliminated by large tour operators and airlines which control a significant part of the distribution system. This has been the focus of innovative research by Bote Gómez and Sinclair (1991) who discuss the nature of corporations controlling the transaction chain to:

- maximise profit by eliminating costs
- reduce the price to the consumer to boost market share
- increase their level of concentration in the tourism industry.

Company strategies often pursue horizontal and vertical integration in the tourism and transport sectors, not only to control the production process of tourism services, but also to improve efficiency through economies of scale and long-term profitability. Although there are various economic theories to explain integration (see the excellent review by Bote Gómez and Sinclair 1991), the two terms – vertical and horizontal integration – have received little detailed analysis (Sinclair 1991a).

Integration in the tourism sector: implications for the supply of tourist transport

According to Bote Gómez and Sinclair (1991):

- Integration is based on the concept of common ownership which may involve the coordination or control of the production process or may have no direct effect on it.

Table 5.1   Integration in the UK package holiday business in 1998[1]

	Operator/parent company				
	Thomson	**Cooks Sunworld**[2]	**Airtours**	**Inspirations**	**First Choice**
Retail outlets	Lunn Poly 800 shops	Thomas Cook 385 shops	Going Places 713 shops	A.T. Mays 409 shops	Independent
Airline	Britannia 28 aircraft	Airworld 5 aircraft	International 21 aircraft	Caledonian 11 aircraft	Air 2000 no data
Cruise ships	MS Sapphire		MS Carousel MS Seawing MS Sundream		

[1] In 1997, the UK package holiday industry had four organisations selling over a million packages a year. Thomson led the market with 4 million customers; Airtours sold 2.9 million, First Choice sold 2 million and Thomas Cook sold 1.05 million holidays. Inspirations was ranked fifth with 411,600 sales.
[2] Thomas Cook has a 14 per cent holding in First Choice.
Source: Modified from Bullock (1998) 'The UK Package holiday industry', *Management Accounting*, 74 (4): 36–38.

- Horizontal integration occurs where two enterprises with the same output combine to increase the companies' control over output. It can occur through mergers, acquisitions, collaboration, franchising agreements and more complex contractual arrangements and may induce concentration in the same business.
- Vertical integration occurs when an enterprise with different interests and involvement in the supply chain acquires or merges with companies contributing inputs to its activities, or where output purchasers provide a ready market for the service. This has the advantage of decreasing economic uncertainty in the supply system and the avoidance of problems related to contract breaking.

The significance of integration in the UK tourism industry is documented in Table 5.1 and the implications for tourist transport have been well documented by Sinclair (1991). It is evident that:

> the transport function is an important point in the exchange of rights in the tourism transaction chain. If this function is subcontracted to an independent operator this delicate and central function can go out of control; hence the close integration of transport with other facilities in the integrated multinational company to ensure a degree of control in the distribution system in the supply of transport services to tourists. (Witt et al 1991 : 83)

Buckley (1987) notes that integration in tourist transport operations, especially vertical ownership, may help to reduce costs where higher load factors can be guaranteed for associated companies. Transaction analysis highlights not only the driving force in the supply system but also raises questions which researchers may wish to address in relation to specific companies and their role in tourist transport.

Transaction analysis also provides an opportunity to consider the changing patterns and processes shaping the tourist transport system and the growing internationalisation of the supply chain. One of the growing issues in the globalisation of tourism supply is a reflection of the development of mass leisure tourism markets, which have been subject to the process of internationalisation. Williams argues that the 'internationalisation of tourism activity and investment has to be seen in the more general and increasingly rapid process of globalisation of international investment' (Williams 1995 : 163). As a result, transnational business interests have developed in the tourism industry, epitomised by some of the larger airlines such as BA, which aims to become a global carrier. Williams (1995 : 164) suggests that 'firms' competitive strategies are based on seeking cost leadership, product differentiation, and focusing on market niches; under certain circumstances these may dictate the internationalisation of tourist activity and investment'. Such research is underpinned by the earlier findings of Dunning (1977) that transnational development in sectors of the tourism industry such as transport is due to firms entering international markets for offensive and defensive reasons when using recognised brand names. They aim to establish location-specific advantages and to reduce risk by using the existing business advantages of an established company. Williams (1995) provides an interesting assessment of transnationalism by seeking to apply theoretical constructs to the international aviation industry.

Transnationalism in the supply chain also raises important issues for service quality and service delivery. The implications for service quality are notable because integration raises a fundamental problem for the tourist transport operator: if an independent airline becomes vertically integrated, should the parent company manage its operation even if it has no experience of this specialised activity? If not, how can the parent company ensure it controls both continuity and quality in the service encounter throughout the supply chain? This has been referred to as *Total Supply Management*. In the area of corporate strategy (Tribe 1996), there is growing evidence that contractual arrangements are being used to ensure that service quality is a continuous process throughout the supply chain. For example, large companies such as Thomson Holidays are operated as independent organisations in the International Thomson group of companies which includes the sister company Britannia Airlines and their own travel agency chain, Lunn Poly. One option available to the group is to have a corporately administered system to ensure that a particular benchmark in service quality for the tourist's experience is achieved throughout the supply chain. This would commence with the sale of a service via Lunn Poly, or direct through Thomson Holidays, and would include the flight and arrival in the destination, where a company representative meets tourists and travels with them on the airport transfer to their accommodation. This is then followed by a welcome where general information and a sales function is undertaken by promoting tours with coach companies endorsed by Thomson. During the holiday, the representative is empowered to deal with consumer issues in situ and to organise the return airport transfer for the departure. During the return flight, a corporate customer satisfaction questionnaire is administered by

Thomson Holidays to assess the level of satisfaction with the package holiday, especially the transport element. This provides a cheap form of market intelligence on corporate quality standards and may highlight areas for attention to reduce service interruptions and customer dissatisfaction.

Introducing a corporate quality control system may help to minimise customer dissatisfaction, particularly if the organisation employs and trains staff to deal with the service encounter as an ongoing process, rather than viewing the services as a series of discrete elements over which they have only limited control. Yet even in a corporate quality control system, employees have to recognise the limit of their responsibilities and be able to refer customers to the relevant personnel empowered to deal with an interruption in the service requested.

Where the tour operator and purchaser of the transport service is unable to directly control the inputs and outputs throughout the system, one option may be to develop a contractually administered quality control system. Here all parties involved in the supply chain may make a contractual commitment to supply services to a certain standard to avoid weak links in the system (e.g. poor quality food and service on board an aircraft) which can affect the tourist's impression of the entire service. All parties involved need to agree on a particular quality principle (e.g. the British Standard 5750 for service systems which some transport operators, such as P&O Stena already employ) to implement throughout the supply chain, using performance indicators to ensure that the necessary standards are being reached. One way of examining the supply chain in the case of tourist transport operators which are public companies is to examine their annual report. For this reason, it is pertinent to consider the role of such data sources, their purpose and how they may be used.

Analysing annual reports: company accounts

An annual report is used by companies to provide a review of the year's activities and it contains company accounts which are prepared within accounting guidelines in force in the country where the head office is based. For example, in the case of SIA (Singapore International Airlines), the accounts must be deposited in Singapore.

Bird and Rutherford (1989) argue that company accounts contain messages which use specialist jargon to deal with a complex situation. Once the specialist jargon is decoded by the reader, company accounts provide an insight into the financial performance of businesses. There are two key elements within company accounts:

- a balance sheet
- a profit and loss account.

Within the balance sheet, items of value (assets) are listed and any claims against them are set out. A claim is a liability, such as an unpaid bill. Assets are divided into fixed assets, which are those acquired for use within the business, and current assets, comprising cash and other items which are to be converted into cash. Liabilities within the accounts are also divided into current liabilities, where settlement will be made within one year, and long-term liabilities to be settled after one year. It should be recognised that a balance sheet only provides a snapshot of an organisation's activities at one point in time. Therefore, analysts tend to consider company accounts over a three- to five-year period to give a more realistic assessment of an organisation's business performance.

There is a great deal of debate among accountants over the reliability of such documents, due to the degree of creative accounting which characterises them. In other words, critics argue that company accounts only record what a company wants to state publicly about its activities. In fact one can argue that the flexibility and vagueness of accounting rules in relation to the preparation of company accounts and financial results means that they are no longer an absolute measure of success. Manipulation of the profit and loss account and the balance sheet to report flattering results which meet corporate investors' and stock markets' expectations have contributed to the pressure for creative accounting. Notwithstanding these limitations associated with company accounts within published annual reports, they provide an important public relations function for companies and offer an insight into the organisation, operation and scale of a company's activities.

CASE STUDY   Singapore Airlines and the supply of tourist transport services

An interesting way to identify the extent to which integration exists within the tourist transport system is to consider one company and its activities. The case of Singapore Airlines (SIA) provides a useful example since it is based within one of the fastest-growing tourism markets in the 1990s – the Pacific Rim (Edwards 1992). SIA was also one of the most profitable international airlines in 1990/91 (Humphries 1992) and this has continued throughout the mid-1990s. Its transformation from a state-owned enterprise (SOE) to a privatised company has been the focus of recent research (Sikorski 1990). As a tourist transport operator, SIA has been widely regarded as Singapore's premier enterprise, formed in 1947, developing as the national flag carrier in 1972 and privatised after 1985. Its performance has been consistently profitable (Bowen 1997) and its close relationship with the government has meant it has developed a synergy with the Singapore Tourism Promotion Board to assist with the development of inbound tourism.

Sikorski (1990) provides a detailed analysis of how SIA was developed and financed as well as its route to privatisation. This case study focuses on SIA as an international transport operator, documenting its growth and then focusing on one data source – SIA's annual report to illustrate the scope and involvement of one company in the supply chain for transport and tourism services.

The development of SIA and aviation in Singapore
A recent study by Raguraman (1997) examines how civil aviation has contributed to the development of the nation and national identity in Malaysia and Singapore.

Raguraman (1997) outlines the development of civil aviation in each country in three phases; 1930–57, during the era of British colonial rule; 1957–71, which was characterised by the development of regional air services; and post-1971, when each nation developed its own international airline service and national carrier. Prior to 1930, Imperial Airways, a British-subsidised airline, developed services and all-weather landing strips in Malaya. In April 1931, Imperial Airways made its maiden flight to Singapore on its mail service, *City of Cairo*. In 1934, Singapore established itself as a hub, linking the Imperial Airways flight from London to the flight operated by Qantas Empire Airways (QEA) to Darwin. The passage to Australia could be completed in $12\frac{1}{2}$ days (Raguraman 1997 : 241). In 1937 Imperial Airways carried 70,000 passengers and flew in excess of 6 million miles; while in 1938, it was providing an extensive mail service using Empire flying boats to South Africa, New Zealand and Hong Kong. The airline fulfilled geographical objectives for the British government, seeking to sustain its sovereignty over the empire. In 1939, the British government nationalised and merged the newly formed British Airways with Imperial Airways to establish the British Overseas Airways Corporation (BOAC). The Second World War disrupted aviation development in Singapore; services resumed in 1946 with BOAC and Qantas flights between England and Australia.

In 1947, Malaysian Airways Limited (MAL) began scheduled services on a Singapore–Kuala Lumpur–Ipoh–Penang route. The company was formed as a local joint venture between the Ocean Steamship Company, the Straits Steamship Company and Mansfield Company. In 1948, BOAC acquired a 10 per cent shareholding in the company. During this era of aviation development in Singapore, arrivals grew from 2,735 in 1937 at Kallang Airport to 183,023 in 1955 when the airport was closed and relocated to Paya Lebar. In 1957, Malaya became an independent nation while Singapore remained a colony and MAL became a public company with exclusive operation rights of air services between Malaya, Singapore and Borneo for a decade. Qantas and BOAC each acquired a 33 per cent share of the new MAL since it removed the need for state investment in capital and human resources. Yet it also allowed outside interests to retain air rights.

In 1965 Singapore was separated from the Federation of Malaysia, and in 1966 an injection of S$15 million of capital from Malaysia and Singapore's governments provided them with majority control of MAL. This provided the national governments with a 33.74 per cent equity each, with BOAC and Qantas retaining a 13.2 per cent share and 6.12 per cent being held by the Brunei government, private business interests and the public. In 1967, MAL underwent a name change to Malaysia–Singapore Airlines (MSA) and by 1970, the MSA network covered urban destinations in Asia, Oceania and Europe. Increased frequencies resulted from the introduction of Boeing 707 jet aircraft while the domestic network saw considerable development. MSA developed a profitable business.

A divergence of interests between the national objectives of Malaysia and Singapore towards MSA led to public and national dissatisfaction. While Malaysia privatised domestic services, Singapore was primarily concerned with international links. Despite the obvious commercial advantages of MSA (it made a small loss in 1966/67 and a pre-tax profit of M$42.46 million in 1971), the respective governments decided to

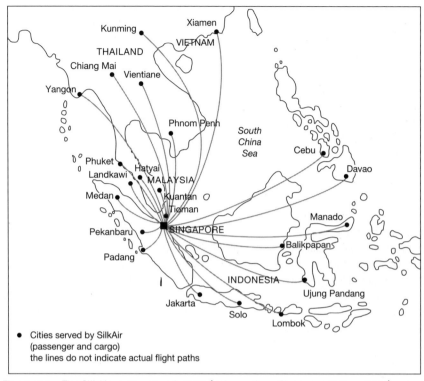

Figure 5.2   The SilkAir route network 1996 (redrawn from Singapore Airlines 1997)

split the airline in 1972. The assets were divided, minority shareholders were compensated, and new airlines (Singapore Airlines and Malaysian Airline System) were established.

During the early years of SIA, it did not receive the same level of government support as other Asian airlines, but being in a city-state removed its social obligation to provide unprofitable domestic services. Raguraman (1986) documents the early years of growth at SIA and acknowledges the numerous barriers it faced in gaining foreign market access. Through lobbying governments directly in the UK and Australia, SIA gained rights to fly to Manchester, Brisbane and Adelaide. The airline also recognised the growth potential of tourist flows to secondary cities in South East Asia at a time when Asian governments were allowing selective liberalisation of routes (Bowen 1997). As a result, SIA established SilkAir as a wholly owned subsidiary in April 1992. Graham (1995) examined the launch of regional scheduled services from the Singapore hub and Figure 5.2 shows the SilkAir route network in 1996. This route network reflects the decision of the Indonesian government to allow Singapore carriers to fly on a range of routes, acknowledged as secondary gateways for business and leisure travel. Competition from Singapore-based Region Air, flying Singapore to Vietnam and Thailand in 1994–95 ceased less than a year after scheduled operations commenced.

However, as Raguraman (1997 : 247) acknowledges, 'in negotiating for more liberal bilateral agreements, the Singapore government and SIA have also increasingly found it necessary to move away from seeking absolute reciprocity in the exchange of traffic rights and to accept some restrictions in route access and capacity provisions'. In 1985, SIA was privatised and although the government holding company, Temasek Holdings, still owns a majority share, it is without a doubt a commercially oriented organisation with a clear profit objective, as the company's mission statement in 1997 states

> Singapore Airlines is engaged in air transportation and related businesses. It operates worldwide as the flag carrier of the Republic of Singapore, aiming to provide services of the highest quality at reasonable prices for customers and a profit for the company.

Raguraman (1997 : 247) outlined the historical growth in traffic for SIA, where it achieved an annual growth rate of 10 per cent between 1974 and 1994. Its network expanded from services to 24 cities in 1972 to 68 cities in 1994 and 76 cities in 41 countries in 1998 (including 3 serviced only by freighters) (Figure 5.3). One immediate outcome of the success of SIA was the spillover effect for the development of Changi Airport as a hub and gateway to South East Asia (Hanlon 1996).

Thus SIA is among Singapore's top three enterprises. By developing a large international route network, SIA can also achieve economies of scope (Sinclair and Stabler 1997), by increasing frequencies or points in its network through the use of SilkAir in South East Asia. Bowen (1997 : 136) recognises that companies such as SIA can gain benefits from economies of scope by:

- developing a denser network of connections, through increasing efficiency in the use of resources, such as hub facilities at Changi, thereby reducing costs
- having a larger network gives an opportunity for price leadership
- deriving benefits from large-scale marketing that can offer a wide range of destinations.

Bowen (1997) also examines policy issues that could affect the future liberalisation of air travel in Asia, something which may yield additional benefits for established market leaders such as SIA. Having considered the growth of SIA and a range of issues affecting the airline in the 1980s and 1990s, we now turn our attention to the 1996/97 annual report.

The SIA annual report

The SIA annual report for 1996/97 comprises 80 pages compared to 139 pages in 1991/92) and some of the key themes are listed in Table 5.2. As a source of information on tourist transport, it contains a significant amount of public relations material which complements the statistical information on the airline. As an accessible data source for researchers, it is a baseline of information from which further research can be undertaken. The initial statistical highlights within the report provide an executive summary of the company accounts, but it is the discussion of the SIA group activities that is the most informative in terms of airline operations and integration within their business activities.

The general operational review discusses the company's performance as SIA's group activities saw its operating profit decline by 14.3 per cent in 1996/97 to

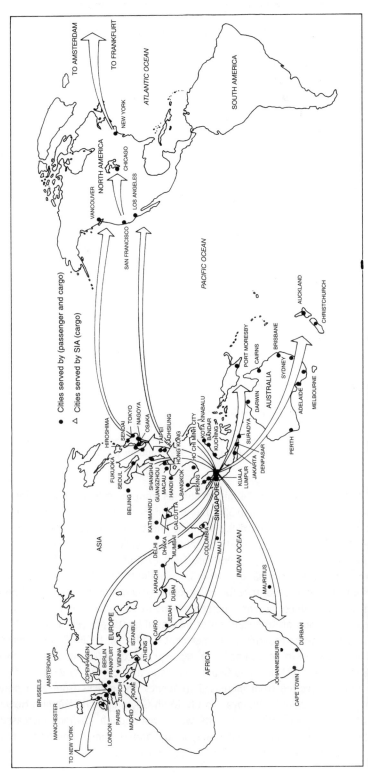

Figure 5.3   The Singapore Airlines route network 1998 (Source: Singapore Airlines 1997)

Table 5.2 Key features of SIA annual report 1996/97

- Mission statement
- Statistical highlights
- Chairman's statement
- Calendar of significant events
- The 50th anniversary
- General operational review
- Fleet and route development
- Cargo
- Our people
- Subsidiaries
- Finances
- Financial review
- Report by the board of directors
- Statement of the directors' profit and loss accounts
- Auditors' report
- Profit and loss accounts
- Balance sheets
- Cash flow statements
- Notes to the accounts
- Half-yearly reports of the group
- Five-year statistical record of the company
- SIA's fleet
- Group corporate structure
- Information on shareholdings
- Share prices and turnover

http://www.singaporeair.com

Source: Modified from SIA (1997) contents page.

S$896 million (including a S$174 million surplus from the sale of aircraft). The profit before tax dropped 0.6 per cent to S$1,076 million. Passenger traffic increased by 9.8 per cent on a capacity growth of 8.2 per cent and the overall load factor increased by 1.1 per cent to 70.5 per cent. At the same time, yields declined by 4.6 per cent. This meant that the break-even point rose from 62.6 per cent to 65.9 per cent.

In terms of route performance, Table 5.3 highlights that North and South East Asia remains the major contributor to revenue generation by route in 1996/97, although it had declined by half a percentage point compared to 1995/96. In contrast, the contribution of American routes increased by 0.7 per cent on the 1995/96 result. European routes reported a 0.4 per cent decline in 1996/97 compared with 1995/96 while the South West Pacific routes experienced a 0.2 per cent increase in 1996/97. In the case of North and South East Asia, traffic grew by 8.9 per cent and capacity growth resulted from increased frequency on services from Singapore to Hong Kong, Beijing, Manila, Nagoya, Ho Chi Minh City and Kuala Lumpur. However, SIA reported a 5.4 per cent drop in its yield even though loadings increased by 0.8 per cent, due to a highly valued Singapore dollar and competition on the routes.

Table 5.3　SIA route performance 1995/96 and 1996/97

	Revenue ($ million)		Overall load factor (%)		Passenger seat factor (%)	
	1966/97	1995/96	1996/97	1995/96	1996/97	1995/96
North and South East Asia	2,167	2,103	61.5	60.7	70.5	69.6
Americas	1,482	1,372	73.4	71.7	77.8	75.8
Europe	1,342	1,305	76.6	74.6	77.7	73.7
South West Pacific	743	699	70.8	71.3	74.1	74.4
West Asia and Africa	589	559	63.6	64.3	68.3	70.8
Systemwide	6,323	6,038	70.5	69.4	74.4	73.0
Non-scheduled services and incidental revenue	196	214				
	6,519	6,252				

Source: SIA (1997 : 44).

In the Americas, route performance saw traffic grow by 14.2 per cent while capacity increased by 11.7 per cent. An improvement in loadings by 1.7 per cent in 1996/97 was a result of the provision of additional passenger services to San Francisco and additional freighter services to New York and Chicago. However, the overall yield dropped by 5.4 per cent, due to greater competition. In Europe, traffic grew by 7.3 per cent in 1996/97, while capacity increased by 4.4 per cent and the load factor rose by 2 per cent. The rise in capacity was a result of greater frequency of services to Paris and Rome although the yield dropped by 4 per cent, due to the strong Singapore dollar and competition. The South West Pacific, in contrast, saw traffic rise by 5.9 per cent and capacity grew by 6.6 per cent. At the same time, the load factor dropped by 0.5 per cent in 1996/97. The main capacity increases resulted from additional service frequencies from Singapore to Adelaide, Sydney and Cairns, established in 1995/96. The overall yield increased by 0.3 per cent, assisted by the strengthening of the Australian and New Zealand dollar against the Singapore dollar. In West Asia and Africa, traffic increased by 11.8 per cent while capacity rose by 18.1 per cent. Frequency increases in 1996/97 resulted from a growth in services to Male, Johannesburg and Cape Town. However, the overall yield declined by 5.9 per cent, due to the strength of the Singapore dollar.

Aside from these operational issues, the annual report also provides information on new route developments, which are summarised in Figure 5.4. When this is compared with the existing route network (Figure 5.3 on page 167), it is evident that SIA is an expansionist airline which seeks to grow its global influence as a major carrier. These developments comprise a strategy which includes code-sharing agreements, new air service agreements and an expansion of existing air service agreements.

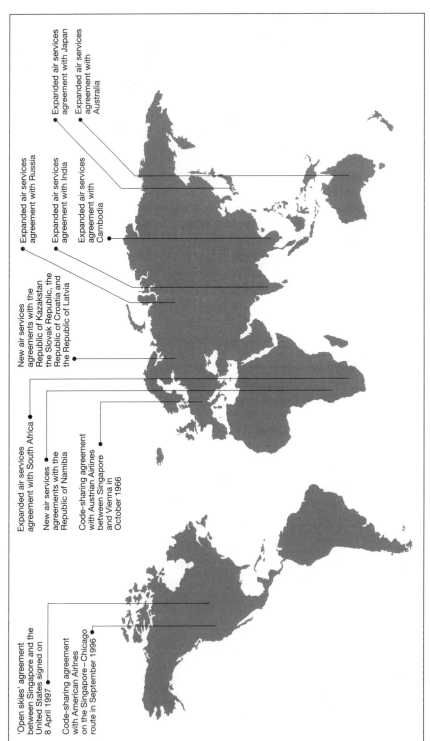

Figure 5.4 Singapore Airlines route expansion plans (redrawn from Singapore Airlines 1997)

Plate 5.1   Singapore Airlines Internet page (reproduced courtesy of Singapore Airlines)

Product and service development
SIA's reputation as a global leader in the airline industry for product and service development saw the introduction in 1996/97 of the Krisworld in-flight entertainment system. This trend-setting innovation led to the introduction of an individual entertainment system with 22 video channels at the seat of every passenger. While other airlines have attempted to introduce such technology, it has been characterised by in-flight service breakdowns. Krisworld is a state-of-the-art technology which has proved popular with passengers, allowing customer choice in relieving the tedium of long-haul air travel. In 1996/97, 32 of the SIA fleet had this technology installed and the company's intention is to install it on all its Megatop-747s (Plate 5.2), Jubilee 777s and Celestar 340s in all classes. In contrast, airlines such as Air New Zealand have only introduced such technology in first class and business class, preferring to emphasise the comfort of their new ergonomically designed seating in economy class. SIA, however, has not stopped at Krisworld. For first class, it introduced active noise reducing (ANR) headphones in 1996, providing a high-quality sound system. For 1997/98, the airline is also planning the introduction of audio and video on demand (AVOD) in first class and business class (Raffles Class). This will allow travellers to choose a wide range of movies and music from SIA's CD music library.

Other innovations include Internet check-in for first class, business class and priority passengers departing from Singapore. Check-in formalities via e-mail or SIAs World Wide Web site (Plate 5.1) will be followed by SIA confirming details by return e-mail.

Plate 5.2 A Singapore Airlines 747-400 megatop with stretched upper deck to accommodate an enlarged carrying capacity for its Raffles Class (business class) (reproduced courtesy of
Singapore Airlines)

At the airport, passengers can then collect their boarding passes from SIA check-in
counters up to 45 minutes before departure and check in any luggage. Since 1997,
first-class passengers departing from Changi Airport in Singapore on flights leaving
between 2300 and 0000 hours can have their in-flight meal served at the Silver Kris
(SIA lounge) prior to embarkation. This then enables passengers to rest upon boarding. While SIA continues to receive accolades for innovation in airline travel, its
policy of continuous improvement to services and product development (including
regular customer satisfaction surveys) indicates the competitiveness of international
travel, particularly in the business and first-class market, making it a market leader.

Human resource issues
According to SIA (1997 : 27), 'SIA's strategy for future success is firmly entrenched
in attracting and developing staff who are highly qualified, trained and motivated
to do their best'. In 1997, SIA employed 27,516 employees, of whom 10,334 were
employed in the airline in Singapore and a further 2,973 by the airline overseas.
One notable innovation, Towards Optimal Productivity (TOP), launched in 1996,
encourages staff to contribute ideas on improving productivity and reducing waste.
It is similar to schemes operated by other airlines, and in 1996 it reduced annual
costs within the company by S$22.9 million. The company also spends significant
sums on staff training, retraining and updating. For example, one of its subsidiaries,

Singapore Airport Terminal Services (SATS), spent S$3.7 million on staff training and development in 1996/97, involving extensive training for 600 staff on its Krischeck personal computer-based check-in system.

Integration in SIA group activities
The global scale and distribution of airline services has now become increasingly dependent upon computer reservation systems (CRSs) for marketing (see Chapter 6). These systems have been developed by airlines and account for the majority of airline bookings in North America and Europe. SIA has participated in this recent expansion in CRS technology (Archdale 1991, 1992) to extend its products to retail travel agents whilst encouraging other airlines to join its system to market their services. By January 1998, the following partners had joined SIA's Abacus CRS:

- SIA
- Dragonair
- Cathay Pacific
- China Airlines
- Malaysia Airlines
- Philippine Airlines
- Royal Brunei Airlines
- Vietnam Airlines
- Garuda Indonesia
- Eva Airways Corporation
- SilkAir (an SIA subsidiary)
- All Nippon Airways
- WorldSpan (a CRS owned by Delta, Northwest and Transworld Airlines).

According to SIA, Abacus is the major CRS in Brunei, Hong Kong, Malaysia, the Philippines, Singapore and Taiwan, and is being extended to Korea and Australia to improve the distribution of airline services on a global basis.

The subsidiary and associated companies of the SIA group and their main activities are shown in Table 5.4, which highlights the involvement in the tourist transport system, particularly:

- tour wholesaling (package holidays)
- aviation insurance
- air transport (SilkAir)
- airport services (catering, airport ownership, security services)
- duty-free sales
- airport bus services
- aircraft leasing
- CRS (Abacus Travel Systems)
- airline software development
- aircraft engineering and maintenance
- hotel and property ownership (SIA Properties)
- quality service training
- Singapore Flying College
- cargo

Table 5.4   SIA subsidiary and associated companies at 31 March 1997

	Activities	Country of incorporation & place of business	% Equity held by groups on 31.3.1997
Subsidiary companies			
Singapore Airport Terminal Services (Private) Ltd	Investment holding company	Singapore	100
SATS Apron Services Pte Ltd	Airport apron services (in voluntary liquidation)	Singapore	100
SATS Airport Services Pte Ltd (previously known as SATS Cargo Services Pte Ltd)	Airport cargo, apron and passenger services	Singapore	100
SATS Catering Pte Ltd	Catering services	Singapore	100
SATS Security Services Pte Ltd	Security services	Singapore	100
SilkAir (Singapore) Private Limited (previously known as Tradewinds Private Limited)	Air transportation	Singapore	100
Tradewinds Tours & Travel Private Limited	Tour wholesaling	Singapore	100
Singapore Aviation and General Insurance Company (Pte) Ltd	Aviation insurance	Singapore	100
SIA Engineering Company Private Limited (previously known as Singapore Engine Overhaul Centre Private Limited)	Engine overhaul and related services	Singapore	100
SIA Properties (Pte) Ltd	Provision of building management	Singapore	100
Singapore Airport Duty-Free Emporium (Private) Limited	Dormant company	Singapore	100
Singapore Flying College Pte Ltd	Training of pilots	Singapore	100
Abacus Travel Systems Pte Ltd	Marketing of Abacus reservation systems	Singapore	61
Singapore Jamco Private Limited	Manufacture of aircraft cabin equipment	Singapore	51
Aero Laundry & Linen Services Private Limited	Laundry services	Singapore	100

Table 5.4   Cont'd

	Activities	Country of incorporation & place of business	% Equity held by groups on 31.3.1997
Cargo Community Network Pte Ltd	Provision and marketing of Cargo Community systems	Singapore	51
Star Kingdom Investment Limited	Real estate	Hong Kong	100
SATS (Curacao) N.V.	Catering services (in voluntary liquidation)	Netherlands Antilles	100
SH Tours Ltd	Tour wholesaling	United Kingdom	100
Auspice Limited	Investment company	Channel Islands	100
Singapore Airlines (Mauritius) Ltd	Aircraft leasing	Mauritius	100
Airline Software Developments Consultancy India (Pvt) Ltd	Airline software development	India	51

Associated companies

Island Cruises (S) Pte Ltd	Dormant company	Singapore	50
Service Quality (SQ) Centre Pte Ltd	Quality service training	Singapore	50
Asian Frequent Flyer Pte Ltd	Provision and marketing of frequent flyer programme	Singapore	33.3
Combustor Airmotive Services Pte Ltd	Repair of engine combustion chambers	Singapore	49
Asian Surface Technologies Pte Ltd	Fan blade repair and coatings	Singapore	29
Servair – SATS Holding Company Pte Ltd	Investment holding company	Singapore	49
Maldives Inflight Catering Private Ltd	Catering services	Maldives	40

Table 5.4   Cont'd

	Activities	Country of incorporation & place of business	% Equity held by groups on 31.3.1997
Beijing Airport Inflight Kitchen Ltd	Catering services	People's Republic of China	40
Beijing Aviation Ground Services Ltd	Ground handling	People's Republic of China	40
Asia Leasing Ltd	Aircraft leasing	Bermuda	21
PT Purosani Sri Lanka	Hotel ownership and management	Indonesia	20
PT Pantai Indah Tateli	Hotel ownership and management	Indonesia	20
Aviserv Ltd	Catering services	Pakistan	49
Pan Asia Pacific Aviation Services Ltd	Engineering services	Hong Kong	47.1
Asia Airfreight Terminal Company Ltd	Cargo handling services	Hong Kong	24.5
Tan Son Nhat Cargo Services Ltd	Cargo handling services	Vietnam	30
Taj Madras Flight Kitchen Private Ltd	Catering services	India	30
DSS World Sourcing Ltd	Sourcing of supplies	Switzerland	33.3
Asian Compressor Technology Services Company Ltd	Repair of aircraft engines and compressors	Taiwan	24.5

Source: SIA (1997 : 62–3).

The activities of SIA's subsidiary companies are reported separately in the annual report, although the majority of the document focuses on SIA. For example, the financial review of SIA identifies revenue generated, expenditure (Table 5.5), the capacity and break-even load factor (the point at which an aircraft makes a profit), as well as detailed accounting information for the group (e.g. taxation, dividends, the financial position, balance sheet, liabilities and assets). SIA also produces an individual breakdown of the profitability of its three principal activities:

● airline operations
● airport terminal services
● engineering services and others which are listed in Table 5.6

Table 5.5 Ten-year statistical review of SIA 1987/88 to1996/97

	1996/97	1995/96	1994/95	1993/94	1992/93	1991/92	1990/91	1989/90	1988/89	1987/88
Financial										
Total revenue ($ million)	6,519.3	6,252.4	5,940.7	5,560.5	5,134.6	5,012.7	4,601.7	4,730.7	4,271.8	3,778.2
Total expenditure ($ million)	5,866.0	5,496.5	5,124.2	5,026.4	4,479.6	41,149.0	3,760.2	3,601.8	3,406.7	3,301.6
Operating profit ($ million)	653.3	755.9	816.5	534.1	655.0	863.7	841.5	1,128.9	865.1	476.6
Profit before tax ($ million)	933.8	903.3	950.5	733.0	794.2	1,085.2	1,124.2	1,397.8	1,025.4	608.7
Profit after tax ($ million)	901.8	875.9	939.0	722.6	741.1	920.7	886.8	1,176.8	928.4	569.9
Internally generated cash flow[1] ($ million)	2,163.8	1,779.2	1,942.3	1,695.5	1,366.4	1,604.7	1,814.2	1,703.8	1,834.4	1,493.9
Capital disbursements ($ million)	2,365.9	1,395.1	1,790.7	1,835.4	1,619.0	1,620.2	1,077.3	1,140.0	912.6	723.7
Yield (c/ltk)	66.5	69.7	73.6	76.0	81.3	90.8	95.2	99.2	99.2	97.6
Unit cost (c/ctk)	43.8	43.6	46.0	49.8	50.6	54.9	58.4	57.8	58.3	61.8
Break-even load factor (%)	65.9	62.6	62.5	65.5	62.2	60.5	61.3	58.3	58.8	63.3
Fleet										
Aircraft (number)	80	71	66	64	57	48	43	40	37	34
Average age (months)	63	68	60	60	61	61	57	55	54	49
Production										
Destination cities (number)	77	77	75	73	70	67	63	57	57	55
Distance flown (killion km)	251.8	338.5	205.9	188.8	165.0	142.3	123.6	117.0	105.8	97.3
Time flown (hours)	325,085	294,880	264,096	241,346	211,435	180,744	157,039	149,355	136,632	124,175
Overall capacity (million tonne-km)	13,501.1	12,481.3	11,167.3	10,155.6	8,982.3	7,624.4	6,644.3	6,280.3	5,682.7	5,136.5
Passenger capacity (million seat-km)	73,507.3	68,529.4	64,074.0	59,290.4	53,077.6	47,454.3	41,701.2	39,236.4	36,461.6	34,438.0
Cargo capacity (million tonne-km)	6,203.9	5,585.1	4,773.6	4,231.3	3,630.0	2,898.3	2,494.5	2,380.7	2,080.8	1,736.1

Table 5.5   Cont'd

	1996/97	1995/96	1994/95	1993/94	1992/93	1991/92	1990/91	1989/90	1988/89	1987/88
Traffic										
Passengers carried (000)	12,022	11,057	10,082	9,468	8,640	8,131	7,065	6,793	6,182	5,618
Passengers carried (million pax-km)	54,692.5	50,045.4	45,414.2	42,328.3	37,860.6	34,893.5	31,332.2	30,737.0	28,785.1	25,757.0
Passenger seat factor (%)	74.4	73.0	70.9	71.4	71.3	73.5	75.1	78.3	78.9	74.8
Cargo carried (million kg)	674.2	603.8	550.5	483.4	399.1	342.8	295.7	278.1	241.3	207.6
Cargo carried (million tonne-km)	4,249.4	3,820.1	3,389.4	2,973.4	2,411.5	1,954.8	1,705.7	1,679.4	1,448.6	1,284.8
Cargo load factor (%)	68.5	68.4	71.0	70.3	66.4	67.4	68.4	70.5	69.6	74.0
Mail carried (million tonne-km)	99.2	89.4	72.7	64.3	55.2	55.0	41.8	50.3	42.1	43.5
Overall load carried (million tonne-km)	9,512.0	8,662.0	7,789.3	7,058.8	6,086.3	5,331.2	4,715.0	4,643.5	4,223.3	3,770.5
Overall load factor (%)	70.5	69.4	69.8	69.5	67.8	69.9	71.0	73.9	74.3	73.4
Staff[2]										
Average strength	13,258	12,966	12,557	12,363	11,990	11,418	10,818	10,052	9,246	8,653
Capacity per employee (tonne-km)	1,081,336	962,618	889,329	821,451	749,149	667,753	614,189	624,781	614,612	593,609
Load carried per employee (tonne-km)	717,454	668,055	620,315	570,962	507,615	466,912	435,848	461,948	456,770	435,745
Revenue per employee ($)	491,726	482,215	473,099	449,769	428,240	441,592	428,046	473,458	465,142	429,975
Value added per employee ($)	220,440	209,332	214,351	195,381	186,681	215,116	205,851	242,768	221,891	188,143

[1] Internally generated cash flow comprises cash generated from operations, dividends from subsidies and associated companies, and proceeds from sale of aircraft and other fixed assets.

[2] Figures for 1991/92 and prior years have been adjusted to exclude SIA engineering departments which became part of SIA Engineering Company from 1 April 1992.

Source: Singapore Airlines Annual Report (1997).

Table 5.6   Revenue, profit and employee strength of SIA business type

	Revenue ($ million)		Profit before tax ($ million)	
	1996/97	1995/96	1996/97	1995/96
Airline operations	6,627	6,337	827	783
Airport terminal services	308	283	120	139
Engineering services and others	287	270	129	160
Group	7,222	6,890	1,076	1,082

	Profit after tax ($ million)		Average number of employees	
	1996/97	1995/96	1996/97	1995/96
Airline operations	795	755	13,736	13,419
Airport terminal services	92	101	8,889	8,556
Engineering services and others	145	169	5,616	4,351
Group	1,032	1,025	27,241	26,326

	Total assets ($ million)		Capital expenditure ($ million)	
	31.3.1997	31.3.1996	1996/97	1995/96
Airline operations				
Aircraft, spares and spare engines	8,725	7,429	2,245	1,321
Others	4,594	3,964	95	77
Airport terminal services	918	860	77	152
Engineering services and others	284	942	33	23
Group	14,521	13,195	2,450	1,573

All inter-company balances and transactions have been eliminated upon consolidation.
Source: SIA (1997 : 45).

SIA also lists its current fleet and aircraft orders (Table 5.7) which highlight the company policy to retain one of the most modern international airline fleets. The airline has increased its fleet from 48 aircraft in operation in 1992 to 80 in 1997, which is impressive by any standards.

It is evident that company annual reports can be a useful data source from which to examine not only integration in the tourist transport system but also the performance of individual companies. Annual reports are an accessible data source which can be obtained direct from public companies' head offices. Although the amount of detailed information contained within annual reports may be somewhat daunting and complex, analysis of such sources will yield important insights into the commercial, operational and supply aspects of different tourist transport operators. For example, Table 5.8 outlines the prevailing shareholdings in the airline and yet more detailed knowledge of the airline, its shareholders and reasons for investment are not evident from an annual report.

Table **5.7**   SIA fleet and aircraft orders at 31 March 1997[1]

Aircraft	Engine[2]	In operation	On firm order	On lease to other operators	On option
B747-400 (Megatop)	PW 4056	36	8	1	10
B747-300 (Big Top)	PW JT9D-7R4G2	3	–	4	–
B747-300 Combi	PW JT9D-7R4G2	3	–	–	–
B747-400 Megaark	PW 4056	6	2	–	–
B747-200 Freighter	PW JT9D-7R4G2	1	–	–	–
B777	Rolls Royce Trent 800 series	–	36[3]	–	41[4]
A310-300	PW4152	17	–	–	–
A310-200	PWJT9D-7R4EI	6	–	–	–
A340-300 (Celestar)	CFM 56-5C4	8	9	–	20
Total		80	55	5	71

[1] Average age of fleet 5 years 3 months (as at 31 March 1997).
[2] PW = Pratt and Whitney Engines; CFM = GEC/CFM Engines.
[3] Includes 6 aircraft intended for Singapore Aircraft Leasing Enterprise.
[4] Includes 10 aircraft intended for Singapore Aircraft Leasing Enterprise.
Source: SIA (1997 : 73).

Summary

The analysis of tourist transport issues has attracted comparatively little research in contrast to the analysis of demand issues. The rise of transaction analysis is a useful way to view the supply chain in tourist transport systems and the contractual relationships which exist between consumers and suppliers. However, more recent theoretical syntheses of tourism supply issues by economists (Sinclair and Stabler 1997) highlight the importance of understanding the competitive conditions and markets in which tourist transport businesses operate. It is this more theoretically derived analysis that begins to advance the supply side beyond simple descriptive studies. The discussion highlights the dominant influence of the tour operator sector in the supply of package holidays and the purchase of transport services on behalf of customers at discounted prices. The ability of tour wholesalers to negotiate discounts with transport operators reflects the capital-intensive nature of the tourist transport business and the need to achieve high load factors to improve profitability. This reflects the indivisible nature of transport operations discussed in Chapter 2, where airline companies cannot operate half an aircraft if it is only 50 per cent full. The fixed costs of transport operations (e.g. repayments on loans to purchase capital equipment) mean that the incremental costs of selling existing capacity on a transport service are low once it has reached its break-even point. This is one explanation of the reduced price of airline tickets and stand-by fares as it

Table 5.8 Shareholders' investment in SIA in 1997[1]

Major shareholders	Number of shares	%
1. Temasek Holdings (Private) Limited	690,055,172	53.80
2. DBS Nominees Pte Ltd	80,373,957	6.27
3. HSBC (Singapore) Nominees Pte Ltd	53,708,412	4.19
4. Chase Manhattan (S) Nominees Pte Ltd	52,047,811	4.06
5. Delta Air Lines Holdings, Inc.	35,186,330	2.74
6. Overseas-Chinese Bank Nominees Pte Ltd	30,690,354	2.39
7. United Overseas Bank Nominees Pte Ltd	28,499,641	2.22
8. Citibank Nominees Singapore Pte Ltd	24,419,947	1.90
9. Post Office Savings Bank of Singapore	21,507,200	1.68
10. Raffles Nominees Pte Ltd	21,384,066	1.67
11. The Great Eastern Life Assurance Co Ltd	14,252,000	1.11
12. DB Nominees (S) Pte Ltd	13,809,329	1.08
13. NYUC Income Insurance Co-operative Limited	11,735,000	0.91
14. Swissair Swiss Air Transport Company Limited	8,000,000	0.62
15. Overseas Union Bank Nominees Pte Ltd	7,734,800	0.60
16. Prudential Assurance Company Singapore (Pte) Limited	7,334,000	0.57
17. Barclays Bank (Singapore Nominees) Pte Ltd	6,519,992	0.51
18. Chang Shyh Jin	4,256,000	0.33
19. ABN Amro Nominees Singapore Pte Ltd	3,871,830	0.31
20. Indosuez Singapore Nominees Pte Ltd	3,428,700	0.27
Total	1,118,814,541	87.23
Substantial shareholder (as shown in the Register of Substantial Shareholders)		
Temasek Holdings (Private) Limited	697,097,172[2]	54.35

[1] The limit on foreign shareholding is 27.5%.
[2] Includes shares in which the substantial shareholder is deemed to have an interest.
Source: SIA (1997 : 76).

is more efficient to sell a reduced-priced ticket if the carrier has capacity than to underutilise the capacity. The contractual relationships associated with the supply of tourist transport are often negotiated up to six months in advance and a great deal of market planning goes into the provision of a service.

The provision of services which meet certain quality standards is one of the reasons why labour costs are so high in the supply of tourist transport services. It is a labour-intensive activity which requires staff contact with travellers to ensure their needs are met at each stage of travel. Employees cannot easily be substituted where a service is dependent on face-to-face contact with customers. Employee training and corporate human resource management policies (Baum 1993) are assuming an important role in ensuring that the supply of tourist transport services is based on a sound understanding of service quality, maximising customer satisfaction and developing programmes with incentives to foster customer loyalty. Analysis of company reports provides a useful source to assess both the role of the transport operator's involvement in the supply chain and their development of more sophisticated ways of serving the customer's needs.

The example of SIA highlights how important these issues are for a market leader in the tourist transport business. The degree of vertical and horizontal integration within a company such as SIA highlights the significance of annual company reports. SIA has certainly ensured that its involvement in the supply chain enables it to develop more sophisticated ways to meet the needs of its customers.

Questions

1. How does the concept of a supply chain impact upon the analysis of the tourists' travel experience?
2. Can the market for tourist transport ever operate in conditions of perfect competition?
3. What is the value of using company annual reports to understand the operational and organisational structure of tourist transport providers ?
4. How can the case study approach be used for an analysis of tourist transport providers?

Further reading

Bywater, M. (1998) 'Who owns whom in the European travel industry', *Travel and Tourism Analyst*, 3 : 41–60.

Sinclair, M.T. and Stabler, M. (1997) *The Economics of Tourism*, London: Routledge. Although this is a difficult text for the uninitiated economist, it is well worth reading for detail on supply issues.

Witt, S., Buckley, P. and Brooke, M. (1991) *The Management of International Tourism*, London: Routledge.

This is a good general introduction on supply issues.

Internet sites

A good source on the prevailing conditions in financial reporting, especially the preparation of annual reports in the UK can be found at the website of the Chartered Institute of Management Accounting (CIMA):

http://www.cima.org.uk

The Singapore Airlines website can be found at:

http://www.singaporeair.com

See also Appendix 1 and 2 for other websites.

Chapter 6

Managing supply issues in tourist transport

Introduction

The provision of tourist transport services by public and private sector organisations is a complex process, requiring a wide range of human resource skills and managerial abilities and a sound grasp of the transport business and how it operates. In recent years, meeting the needs of consumers (travellers) has also assumed a higher priority in the supply of services. In Chapter 5, the conceptualisation and analysis of tourist transport supply issues highlighted the significance of understanding the broader strategic and contextual issues which affect the way different forms of transport supply perform in the marketplace. In this chapter, the emphasis is on a number of issues highlighted by the SIA case study, namely how successful transport providers can produce, manage and operate efficient supply systems to meet tourist needs. In particular, the chapter focuses on the mechanisms and tools used by operators to manage supply issues, together with the role of the public sector. This provides a comparison of the different objectives pursued by private sector operators, such as airlines, railway companies and cruise lines. Private sector operators are motivated by the financial performance of the business, to ensure that an efficient operation delivers products to its customers. In contrast, the public sector agencies directly and indirectly associated with the provision of tourist transport services are often not motivated by profit. They frequently have a more strategic view and are concerned with planning, coordination and liaison functions to ensure tourist transport provision meets public policy objectives at various spatial scales. The chapter commences with a discussion of one of the most important tools used by transport providers in the late 1990s – information technology (IT). This is followed by a review of one important strategy used by airlines to improve supply issues and competitiveness – alliances. Next, the role of the state in regulating tourist transport supply issues is discussed in relation to airline deregulation in the USA. The supply of transport services in destination areas is then reviewed.

Information technology and supply issues in tourist transport: a role for logistics and IT?

It is widely acknowledged that society has entered the 'information age' and that this has had implications for transport provision (Hepworth and Ducatel

1992). One of the immediate impacts for tourist transport providers is that up-to-date information flows are now vital when a supply chain exists, and the transport provider is just one component of the overall tourist product. As Christopher (1994 : 12) observes, the customer service explosion means there is a need for 'consistent provision of time and place utility. In other words, products do not have value until they are in the hands of the customer at the time and place required.' Christopher (1994) argues that logistics of service delivery are of paramount importance and enable organisations to add value and deliver a consistent product. Logistics is a vital concept to recognise, particularly when IT is also introduced, since IT and logistics enable transport providers to achieve their objectives in a competitive environment.

According to Quayle (1993 : 9),

> Logistics is the process which seeks to provide for the management and co-ordination of all activities within the supply chain from sourcing and acquisition, through production where appropriate, and through distribution channels to the customer.

Logistics provides a competitive advantage by offering a strategic view of operational issues and an understanding of the links in the supply system. It also assists in the coordination of the service delivery function, and transport in its own right is a vital element of logistics in moving the customer nearer to the product in a tourism context. Logistics is documented in detail by Quayle (1993) and Christopher (1994) and performs a vital role in providing the link between the marketplace and operating activity of the business. Figure 6.1 outlines the business functions which fall within the remit of logistics. What emerges from Figure 6.2 is that information flows are a critical component in logistics and the management of supply issues in transport.

Within the literature on IT in tourism, the seminal study by Sheldon (1997) is fast becoming the key reference source, replacing the earlier work by Poon (1993). The tourist industry generates large volumes of information that needs to be processed and used within a logistics context. For example, Sheldon (1997) notes that each airline booking generates 25 transactions that need processing. In Sheldon's (1997) model of tourism information flows, there are three main agents involved: travellers, suppliers and travel intermediaries. For the purpose of the discussion here, it is the supplier's use of IT to handle, utilise and manage these information flows which is of interest. From a transport supplier's perspective, information is essential to allow the organisation to function and for different departments to make decisions about corporate objectives, their consumers and competitors. It can also be harnessed in the marketing function (see Chapter 2). Sheldon (1997) cites the example of the airline industry which makes extensive use of IT in a wide range of contexts, including:

- Global Distribution Systems (see WTO 1994)
- frequent flyer databases
- yield management programmes (see Chapter 2)
- distribution and marketing of their products
- the design, operation and maintenance of aircraft and luggage handling
- check-in systems at airports (see SIA 1997)

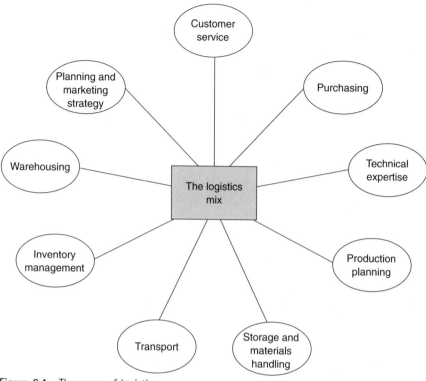

Figure 6.1   The scope of logistics

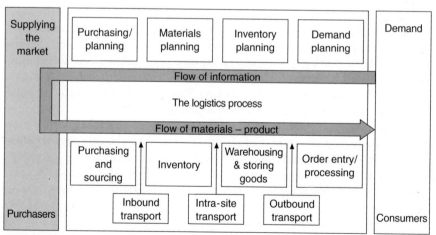

Figure 6.2   The logistics process

Although other transport sectors involved in tourism also make use of IT (including train operators, car rental agencies and coach and cruise ship operators), it is probably most highly developed in the airline sector, due to large investment in capital and the highly competitive nature of the business.

In that sense, IT is seen as integral to gaining a competitive edge, and in the case of SIA, in maintaining continuous product innovations. For example, on its Megatop-747s, all passengers have access to an in-flight telephone to make credit card calls from air to ground, and the developments listed in the SIA case study (page 171) all enlist state-of-the-art IT.

Sheldon (1997) traces the development of IT in the airline industry, where it is primarily seen as a means of improving the efficiency of operations and of management functions. The airline sector first developed computer techno-logy in the form of computer reservation systems (CRSs) in the 1950s. It is impossible in a book such as this to trace the rapid development of IT in the airline industry and all the other transport sectors, although it is pertinent to outline the current state of the art in IT and its organisation.

Following the rapid growth in CRSs (Knowles and Garland 1994) in the 1970s and 1980s (Archdale 1990, 1991), Sheldon (1997) outlines the typical configuration of an airline CRS; it comprises:

- a central site housing the computer systems driving the CRS (often up to 10 mainframe systems)
- the network hardware at the central site and computer staff to maintain it
- a series of front-end communication processors to process information and on-line storage devices at the central site.

This is complemented by satellite communications to remote communica-tion concentrators (RCCs) in key cities that relay data from the earth station. This is then relayed to reservation terminals and airports, providing rapid communications.

One of the major changes in the late 1980s and 1990s, has been the move from CRSs which contained only airline information for the proprietary air-line, to systems containing data for multiple airlines. Sheldon (1997) traces the development of CRSs into what have now been called global distribution systems (World Tourism Organisation 1994), like the Asian GDS, Abacus, mentioned in the SIA case study (Chapter 5). Table 6.1 outlines the principal developments contributing to the development of GDSs.

GDSs are CRSs which are affiliated with airlines. There has been a great deal of debate over the impact of airline affiliation on the competitiveness of air travel in North America. Sheldon (1997) argues that, following legislation, rules now exist to ensure all airlines are represented on GDS screens. Accord-ing to Sheldon (1997), the significance of major GDSs in 1996 was:

- Sabre   US$1,500–2,000 million
- Apollo   US$1,100+ million
- Abacus   US$650 million
- Amadeus   US$600 million
- System One   US$500 million
- Worldspan   US$500 million
- Galileo   US$400 million

  Source: Sheldon (1997 : 25).

Table 6.1   The development of GDSs

1976 Three North American airlines began to offer their systems – Apollo (United Airlines), Sabre (American Airlines) and PARS (TransWorld Airlines) as well as offering US travel agents terminals to access their systems.
1981 Eastern Airlines established System One Direct Access (SODA).
1982 Delta Airlines launched its DATAS II.
1987 In Europe, Galileo and Amadeus were formed and offered to travel agents. In Asia, Abacus was formed and also offered to travel agents.
1988 Japan Airlines formed Axess.
1990 System One was purchased by a non-airlines company – EDS. The merger of PARS and DATAS II resulted in the formation of Worldspan. In Japan, All Nippon Airways and Abacus formed Infini.
1993 Galileo and Apollo were merged to establish Galileo International.
1995 System One merged with Amadeus.

Source: Modified from Sheldon (1997 : 24).

Although a CRS will only show one airline's schedules, a GDS has the advantage of showing data on multiple carriers, including:

- flight schedules and availability
- passenger information
- fare quotes and rules for travel
- ticketing.

Since the advent of GDSs, airlines have also established 'a presence on the Internet and [are] using that as an important distribution channel especially to consumers' (Sheldon 1997 : 27). By accepting payment by credit card, airlines have harnessed a developing technology to complement GDSs and traditional distribution channels.

Sheldon (1997) also discusses other airline IT applications which include:

- baggage and cargo handling systems
- cabin automation (e.g. entertainment systems, visual route systems on in-flight screens using Geographical Positioning System equipment)
- safety systems
- decision support systems
- flight scheduling and planning
- crew scheduling and management
- gate management and control.

In the case of SIA, Sheldon (1997) discusses the current development of Krismax, a yield management system to assist with revenue and inventory control. The underlying principle behind Krismax is that the amount of space available on any flight segment can be controlled in relation to different classes of seats (as discussed in Chapter 2). The information is then fed into the CRS to control the availability of seats by segment. Sheldon (1997) observes that SIA's 'reservations from travel agents are taken care of with its major investment and involvement with Abacus, the Asian GDS, which has over 12,500 terminals in 5,000 travel agencies' (Sheldon 1997 : 40).

Sheldon (1997) also examines the developments in IT which have been introduced into land transport operations to help the logistics of fleet management in relation to car rental and other innovations which improve the tourists' travel experience.

Having examined the significance of logistics and IT in managing tourist transport in a supply context, attention now turns to the way in which one part of the transport sector – airlines – has pursued strategies to improve supply issues through strategic alliances.

Airline alliances

There is a very clear link between the use of IT in tourist transport sectors and the development and expansion of airline alliances, which were briefly discussed in Chapter 2. French (1997 : 99) argues that 'the development and exploitation of alliances on their current scale would not have been possible without the development from the early 1980s of immensely powerful global CRS systems, or without the parallel development of in-house computer programmes controlling the management of inventory, yield and revenue'. In fact, French (1997) implies that the rise of GDSs may prove critical to exploit the potential advantages global airline alliances may offer.

French (1997) argues that an airline alliance is 'any collaborative pooling of resources between two or more carriers, designed both to benefit all the partners in making their supply of capacity more efficient and in extending their market reach' (French 1997 : 81). Bennett (1997) views such developments in the context of a growing globalisation of the tourist transport business, indicating that 60 per cent of alliances have been formed since 1992. This recent phenomenon within the aviation industry is not as new as Bennett (1997) implies, since French (1997) traces the development back to the 1970s. However, it is the speed of this development that is critical to the aviation industry worldwide and that has implications for tourist supply issues. French (1997) points to the principal data source for analysing airline alliances and their current status. This is the survey published each year in the June issue of *Airline Business*. It is therefore an important reference source to trace the evolution, changing status and dynamics of alliances. Based on the 1997 data in *Airline Business*, Table 6.2 documents the changes in alliance structure. It highlights two basic issues:

- the continued growth of alliances
- a 'churn rate' (French 1997), indicating the formation and termination of agreements while the number has reached a critical mass.

Table 6.2 also highlights the need to explain the basic forms which alliances may take. Bennett (1997 : 214) distinguishes between two types of alliance:

- *Tactical partnerships*, comprising a loose form of collaboration designed to derive marketing benefits, characterised by code sharing and exemplified by the hub-and-spoke system of air travel in the US domestic airline market (discussed later in this chapter). This is reflected in the smaller feeder and regional airlines being aligned with key hub-based carriers.

Table 6.2 The development of airline alliances

	1994	1995	1996	1997
Number of airlines	136	153	159	177
Number of alliances	280	324	389	363
with equity states	58	58	62	54
non-equity alliances	222	266	327	309
New alliances	–	51	71	72

Source: *Airline Business*, June 1997; modified from French (1997) using data for 12 months to May of each year.

- *Strategic partnerships*, where an investment or pooling of resources by partners aims to achieve a range of common objectives focused on the partners' strategic ambitions. Bennett (1997 : 214–15) describes strategic alliances as incorporating

    Shared airport facilities (check-in lounges), improved connections (synchronised schedules), reciprocity on frequent flyer programmes, freight co-ordination and marketing agreements (code-sharing and block selling).

French (1997) by contrast, outlines ten principal features which may be included in strategic alliances developed by airlines:

- *Equity stake or equity exchange* may take two forms. First, a large carrier in an alliance may buy a share ownership in a smaller partner airline. If it is an overseas-based airline, this is likely to be a minority stake. Alternatively, if the airlines are located in the same country or region (e.g. the EU), it may lead to a majority shareholding. The second form in which the alliance may be developed is for two or more carriers to take part stakes in each other. A good example of this form is the Swissair, Delta Airlines and Singapore Airlines alliance established in 1989. Equity exchange or stakes are the strongest format an alliance may assume, given the investment and long-term commitment. However in 1997, only 54 of the 363 alliances were of this form.
- *Code sharing* is an agreement between two airlines, where one airline operates a single flight leg with its own flight code and that of its alliance partner. The advantage is that, aside from an airline expanding the flights it can offer, it can enhance its listing on CRSs. According to French (1997), two-thirds of airline pacts involve code sharing. There is considerable debate about the effects of alliances on the structure of the airline industry (Alamdari and Morrell 1997), particularly in the USA (Hannegan and Mulvey 1995) and Europe. However, as Hannegan and Mulvey (1995) show, the alliance between KLM and Northwest, formed in 1992, led to an increase of 350,000 passengers per annum on its transatlantic routes. The added revenue enabled Northwest to post an operating profit of US$830 million in 1994 compared to a loss of US$60.1 million in 1991. Thus, regulatory authorities are monitoring code sharing because of its potentially anti-competitive effects on consumers (it may reduce choice and improve airline profitability), together with the effects on the industry (Humphreys 1994; Burton and Hanlon 1994).
- *Joint services* are offered within a code-sharing agreement, and occur when one 'airline supplies and operates the crew and aircraft, the revenue and profits of flights are shared . . . between two partner airlines' (French 1997 : 85).

- *Block seat or block booking arrangements* occur where a partner airline reserves a block of seats on a connecting service for another airline so it is able to guarantee onward seats.
- *Joint marketing or a marketing agreement* is a collaboration on 'marketing, promotion and advertising of the flights of two or more partners. It generally recognises that some airlines have better local knowledge and experience in familiar markets, and can pass on the benefits of this to a partner airline through co-operation at the marketing level' (French 1997 : 85).
- *Joint fares* occur (if regulatory authorities allow such collaboration) and may result in fare matching on routes and revenue pooling.
- *Franchise agreements* are contractual agreements which allow an airline to let its partner assume its brand image. This has the advantage of allowing the partner to extend its network while assuming the livery and service standards of the franchiser. Code sharing and schedule coordination occur normally and the franchisee can also buy into the franchiser's extensive resources, marketing and organisational framework.
- *Wet-leasing* is where one airline company hires the aircraft and crew from another airline on a fixed-fee basis. The hiring airline normally markets the airline using its own designator code.
- *Frequent flyer benefits, cooperation and reciprocity* indicates that 'two or more alliance partners may choose to link their loyalty programmes, allowing passengers to accumulate frequent flyer mileage awards on all alliance partner airlines' (French 1997 : 86).

Bennett (1997) outlines the principal motivations which may explain why airlines enter into strategic alliances:

- to achieve economics of scale and learning, principally to improve profitability and perhaps also to benefit from economies of scope (Hanlon 1996)
- to gain access to the benefits of the other airline's assets
- to reduce risk by sharing it
- to help share the market, which may help reduce incapacity in mature markets and could reduce competition
- speed in reaching the market given the structural changes occurring in the airline industry (e.g. deregulation and privatisation – see Chapter 3).

According to French (1997 : 101), 'airline alliances are now such a central part of strategy' that a period of consolidation appears to have been reached after the initial high 'churn rate'. In contrast, Bennett (1997 : 217) maintains that 'the success is poor . . . because there are a whole host of challenges of both a practical and political nature that need to be faced' relating to how the alliance is organised, particularly when a number of partner airlines are involved. In fact, while alliances become more important to the operation and management of the airline industry, Bennett (1997 : 222) reiterates the concerns that 'less competition equates with less choice, while improved economic circumstances for airlines are tantamount to higher prices'. Inevitably alliances are part of the trend towards a number of mega-carriers dominating the airline industry as part of the globalisation process in tourism. However, attention now turns to another factor affecting the management and operation of tourist transport – deregulation and its impact on the domestic airline industry in the USA.

The state and the supply of tourist transport: airline deregulation in the USA

Air travel is a major form of tourist transport since 30 per cent of international travellers use air as the main form of transport (Mann et al 1992). International air travel provides an interesting example of how government policy (see Chapter 3) has led to different effects upon the supply of transport services for tourists. Sealy (1992) identifies two approaches:

- a regulated transport system where a country exercises sovereignty over its airspace
- a liberalised and unregulated system characterised by an open-skies policy.

In Chapter 3, it was clear that various historical and political factors may explain the aviation policy in a given country. The regulations governing airline operations by international bodies such as ICAO and IATA (see Holloway 1989; Mill 1992) and bilateral agreements were influential in shaping air travel in a regulated environment until the late 1970s (see Table 6.3). In addition to IATA (the United Nations body which facilitates the international regulation of air travel), national governments play an active role in regulating air travel (see Graham 1995).

The experience of domestic airline deregulation in the USA in 1978 led to a complete re-evaluation of the supply of air travel in terms of its organisation, operation and regulation by the state (see Milman and Pope 1997 for an up-to-date analysis of the US airline industry). Deregulation in North America was also a testbed for aviation strategies subsequently developed in Australia and those planned for the EC (now the EU). The case of the US domestic airline market is also interesting because 'it is more highly developed and more extensively used in the United States than in any other part of the world' (Graham 1992 : 188). For example, in 1988 air travel in the US exceeded 1.7 billion passenger kms and is set to increase to 2.2 billion by the year 2000. According to Button and Gillingwater (1991), the implications for the supply of tourist transport can be examined in relation to:

- the corporate response of airline companies to the new competitive environment for air travel
- the effect on the functioning of the transport system
- the effect on consumers, service provision and service quality
- the impact upon complementary infrastructure (e.g. airports).

To understand the impact of deregulation of the domestic market for air travel in the USA, it is pertinent to examine the regulatory framework prior to and following deregulation as a context in which to consider the changes on the supply of air services.

Airline regulation in the United States

In the USA, airline regulation can be dated to the passage of the 1938 Civil Aeronautics Act and the subsequent formation of the Civil Aeronautics Board (CAB) in 1946 which licensed routes and airline operations, regulated the

Table 6.3   International cooperation in air travel

International air travel requires countries to cooperate so that the movement of aircraft and people can occur in a reasonably flexible manner. To provide a degree of regulation and coherence in air travel, two important international agreements underpin present-day air travel: the *1944 Chicago Convention*, which established the principle of freedoms of the air and the *1946 Bermuda Agreement*, which provided a framework for bilateral agreements to implement freedoms of the air. A bilateral agreement is where two countries agree to provide air service on a reciprocal basis and it helps to facilitate and protect the rights of each country's airline irrespective of whether it is a profit or non-profit venture.

Consequently, the following 'Freedoms of the Air' can be observed:

Freedom 1: the right of an airline to fly over one country to get to another.

Freedom 2: the right of an airline to stop in another country for fuel/maintenance but not to pick up or drop off passengers.

Freedom 3: the right of an airline to drop off in a foreign country, traffic from the country in which it is registered, to a separate country.

Freedom 4: the right of an airline to carry back passengers from a foreign country to the country in which it is registered.

Freedom 5: the right of an airline to carry back passengers between two foreign countries as long as the flight originates or terminates in the country in which it is registered

Freedom 6: the right of an airline to carry passengers to a gateway in the country in which it is registered then on to a foreign country, where neither the origin nor the ultimate destination is the country in which it is registered.

Freedom 7: the right of an airline to operate entirely outside of the country in which it is registered in carrying passengers between two other countries.

Freedom 8: the right of an airline, registered in a foreign country, to carry passengers between two points in the same foreign country.

Note: According to Mill (1992) the first *two* freedoms are accepted internationally while freedoms 3–6 are the subject of bilateral agreements and the last two freedoms are rarely accepted.
Source: Based on Mill 1992 : 81.

pricing of fares and monitored safety issues. Federal regulation of civil aviation was firmly established within a government department. It also limited the number of domestic carriers until the 1970s to avoid excessive competition. Despite such measures, the postwar boom in domestic and international air travel in the USA was facilitated by a buoyant economy, innovation in air-craft design, reduced travel costs and stable fares maintained by the CAB. In addition, the CAB provided subsidies for local service carriers so that small communities could be connected to the emerging interurban trunk network of air routes, to achieve social equity in access to air travel. This also facilitated the development of major airlines as the CAB guaranteed loans for carriers who invested in new aircraft to serve such routes. To reduce subsidy payments, carriers were gradually awarded more lucrative longer-haul routes with a view to carriers cross-subsidising the shorter feeder routes from small communities to connect with trunk routes. These developments occurred against the back-ground of pressure from the airline industry to increase fares in the 1970s, which appeared to place the consumer at a disadvantage. In 1975, the CAB began to relax some of its restrictions on the operation and pricing of charter aircraft to compete with scheduled flights, permitted discounted fares and licensed new transatlantic carriers prepared to offer low fares. This provided the background for the 1978 Airline Deregulation Act which established greater flexibility in route licensing and abolished the CAB in 1984. As Goetz and Sutton (1997 : 239) argue, 'the Act stripped the CAB of its authority to control entry and exist, fares, subsidies and mergers'. As a result, some of the CAB functions were transferred to the Department of Transportation, including responsibility for:

- the negotiation of international air transport rights and licensing of US carriers to serve the airline market
- the monitoring of international fares
- the maintenance of air services to small communities
- consumer affairs and complaints
- airline mergers.

In contrast, the Federal Aviation Administration powers included:

- the promotion of air safety and use of navigable air space
- regulations on the competence of pilots and airworthiness of aircraft
- the operation of air traffic control (ATC) systems.

From these regulatory responsibilities it is evident that the structure of the US airline business comprises:

- airlines
- airports and ATC providers
- aircraft manufacturers (e.g. Boeing and its recent acquisition, McDonnell Douglas)
- consumers
- third parties, such as government agencies (e.g. FAA and Department of Transportation).

In the case of the airline business in the USA, Shaw (1982 : 74) identifies the following structure for airline companies:

- the majors – earning in excess of $1 billion per annum
- the nationals – based on a regional network
- new entrants
- small regional and commuter airlines which provide the short-haul link-ups with the majors to feed into the networks of their partner airlines.

The effects of deregulation on the supply of tourist transport

Within the literature on the supply of tourist transport, one area which has been well researched is airline deregulation (Goetz and Sutton 1997). This has focused on the controversy over the effect of such measures on the commercial environment for airline operations, and there is not space within this chapter to review the specialised range of papers generated by researchers on this topic. One approach is to consider a limited number of issues which are constantly referred to by researchers. According to the US Department of Transportation, following deregulation the number of carriers serving the USA increased from 36 in 1978 to 72 in 1980 and 86 in 1985, dropping to 60 by 1990 (including air cargo carriers); see Table 6.4. However, a range

Table 6.4 Airlines providing interstate jet services during the deregulation era

Origin and name	Began service[1]	Date	Status
Trunk carriers (11)			
American	pre-1978	1989	1st ranking carrier[2]
Braniff	pre-1978	1982	Ceased operation due to bankruptcy
		1984	Resumed limited service
		1989	Ceased operation due to bankruptcy
Continental	pre-1978	1989	6th ranking carrier (under Texas Air Corp)
Delta	pre-1978	1989	2nd ranking carrier
Eastern	pre-1978	1989	Declared bankruptcy; conducting limited operations (under Texas Air Corp)
National	pre-1978	1980	Acquired by Pan Am
Northwest	pre-1978	1989	5th ranking carrier
Pan Am			
TWA	pre-1978	1989	8th ranking carrier
United	pre-1978	1989	3rd ranking carrier
Western	pre-1978	1986	Acquired by Delta
Local service carriers (8)			
Frontier	pre-1978	1985	Acquired by People Express
Hughes Airwest	pre-1978	1980	Acquired by Republic
North Central	pre-1978	1979	Merged within Southern to form Republic
		1986	Republic acquired by Northwest
Ozark	pre-1978	1986	Acquired by TWA
Piedmont	pre-1978	1987	Acquired by USAir
Southern	pre-1978	1979	Merged with North Central of form Republic
		1986	Republic acquired by Northwest
Texas International	pre-1978	1982	Acquired by Continental
USAir	pre-1978	1989	4th ranking carrier
Intrastate carriers (5)			
Alaska	pre-1978	1989	15th ranking carrier
AirCal	1979	1987	Acquired by American
Air Florida	1979	1984	Ceased operation due to bankruptcy
		1985	Acquired by Midway
PSA	1979	1987	Acquired by USAir
Southwest	1979	1989	9th ranking carrier
Charter carriers (2)			
Capitol	1979	1984	Ceased operation due to bankruptcy
World	1979	1985	Ceased operation due to bankruptcy
Commuter carriers (3)			
Air Wisconsin	1982	1989	18th ranking carrier
Empire	1980	1986	Acquired by Piedmont
Horizon	1983	1986	Acquired by Alaska

Table 6.4   Cont'd

Origin and name	Began service*	Date	Status
New carriers (17)			
Air Atlanta	1984	1986	Ceased operation due to bankruptcy
Air One	1983	1984	Ceased operation due to bankruptcy
American International	1982	1984	Ceased operation due to bankruptcy
America West	1983	1989	11th ranking carrier
Florida Express	1984	1988	Acquired by Braniff
Frontier Horizon	1984	1985	Ceased operation due to bankruptcy
Hawaii Express	1982	1983	Ceased operation due to bankruptcy
Jet America	1982	1986	Acquired by Alaska
Midway	1979	1989	16th ranking carrier
Muse (Transtar)	1981	1985	Acquired by Southwest
New York Air	1980	1985	Acquired by Continental
Northeastern	1982	1984	Ceased operation due to bankruptcy
Pacific East	1982	1984	Ceased operation due to bankruptcy
Pacific Express	1982	1984	Ceased operation due to bankruptcy
People Express	1981	1986	Acquired by Continental
Presidential	1985	1987	Became feeder carrier for United
Sunworld	1983	1988	Ceased operation due to bankruptcy

[1] Date carrier began interstate service with jet aircraft.
[2] Size ranking based on passengers carried during 12 months ended September 1989.
Source: Button (1991).

of factors such as financial insolvency, mergers and acquisitions reduced the number of operators to 10 carriers of regional or national scale by 1988. This was followed by a period of consolidation, and by 1991 the situation had worsened, with both Pan Am and Eastern Airlines having faced bankruptcy: the existing carriers had either prospered or lost market share to competitors (see Table 6.4).

Thus a 62 per cent increase in the number of domestic travellers carried on US airlines during 1978–90 has been followed by a greater degree of concentration and integration in the airline business Table 6.5 outlines the situation in the mid-1990s, with the performance and passenger volumes for the main US airlines. Milman and Pope (1997 : 4) argue that in the late 1990s 'the US airline operates in an oligopolistic market structure . . . [and] . . . in recent years, access for airlines new to the industry has become more difficult due to the limited availability of terminal space and gates, a lack of departure and landing slots at major airports, and price competition from the dominant carriers' (see Table 6.5). In terms of route structure, Table 6.6 outlines the ten most important domestic markets in the USA, highlighting the popularity of short-distance markets such as New York–Boston, Dallas–Houston and New York–Washington.

Table 6.5 Leading US airlines' statistics and performance, 1995/96

	Passengers (million)			Revenue passenger miles (billion)			Employees		Number of aircraft		Aircraft departure (000s)	
	1995	1996	% change[1]	1995	1996	% change[1]	1995	1996	1995	1996	1995	1996
Delta	86.9	97.2	11.8	85.1	93.9	10.3	58,621	58,935	539	544	950	943
American	79.5	79.3	-1.2	102.7	104.5	1.8	84,915	82,571	635	642	823	787
United	78.7	81.9	4.1	111.5	116.6	4.5	75,634	79,205	558	564	780	785
US Airways (US Air)	56.7	56.6	-0.1	37.6	38.9	3.5	40,780	39,417	394	390	794	738
Southwest	50.0	55.4	10.7	23.3	27.1	16.1	18,816	21,863	224	243	685	748
Northwest	49.3	52.7	6.8	62.5	68.6	9.8	43,604	45,320	380	399	566	586
Continental	35.0	35.7	2.1	35.5	37.4	5.2	30,443	29,199	317	317	458	443
Trans World (TWA)	21.6	29.3	35.9	24.9	27.1	8.9	23,121	24,731	186	192	277	284
America West	16.8	18.1	7.9	13.3	15.3	15.1	9,402	9,357	93	99	195	207
Alaska	10.1	11.8	16.6	8.5	9.8	14.6	6,732	7,440	74	74	143	153
Total	**484.6**	**518.0**	**6.9**	**504.9**	**539.2**	**6.8**	**392,068**	**398,038**	**3,400**	**3,464**	**5,671**	**5,674**

[1] Calculated on unrounded data.

Source: Air Transport Association of America cited in Milman and Pope (1997 : 7).

Table 6.6   Top ten US domestic markets, 1995/96

Markets	-	Passengers[1] 1995 (000s)	1996[2] (000s)
1. New York	Los Angeles	2,991.1	3,149.0
2. New York	Chicago	2,981.6	2,996.5
3. New York	Miami	2,675.6	2,777.6
4. Honolulu	Kahului, Maui	2,761.5	2,750.0
5. New York	Boston	2,491.4	2,400.9
6. New York	San Francisco	2,185.8	2,282.5
7. New York	Orlando	2,005.1	2,234.9
8. Dallas/Fort Worth	Houston	2,208.4	2,205.1
9. Los Angeles	Las Vegas	1,956.4	2,102.9
10. New York	Washington	2,118.2	2,087.4

[1] Outbound plus inbound passengers.
[2] 12 months to end-September 1996.
Source: Milman and Pope (1997 : 8).

Goetz and Sutton (1997 : 239) examine the effects of concentration in the airline industry post-deregulation in detail, noting that 200 carriers were absorbed or went bankrupt in 1983–88, with the nine largest airlines (American, United, Delta, Northwest, Continental, USAir, TWA, Pan Am and Eastern) responsible for 92 per cent of domestic revenue passenger miles. By 1995, 'the industry has been transformed from a regulated oligopoly of ten trunk carriers controlling 87 per cent of the market [in 1978] to an unregulated oligopoly of eight major carriers controlling 93 per cent in 1995' (Goetz and Sutton 1997 : 239). This is an interesting situation given the rationale for deregulation: that it would lead to a situation of perfect competition with no major economies of scale or barriers to entry.

Goetz and Sutton's (1997) excellent synthesis of deregulation observes that while average air fares (in constant dollars) have declined since 1978, discount pricing and fare wars have also reduced fares. The result is that higher fares have been levelled for short-notice business travel; but under conditions of severe discounting, some fares have dropped below cost levels. However, airlines servicing more peripheral routes where one airline dominates have set fares 18–27 per cent higher on average than on trunk routes. Largely as a result of discount fares, domestic passenger volumes have increased dramatically under deregulation. In the period 1978–93, an 87 per cent growth occurred, with passengers increasing from 256 million to 478 million. Flight departures in the same period also grew from 5 million to 7.2 million. But between 1990 and 1993, airlines in the USA made record losses of US$13 billion – the largest losses ever in history (Goetz and Sutton 1997). What has this meant for the structure and provision of services through American airline networks?

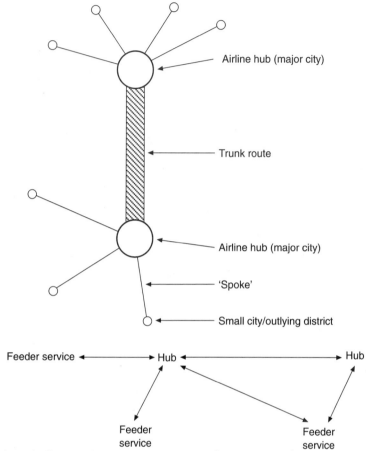

Figure 6.3 Schematic diagram of a hub-and-spoke system (after Page 1994)

The spatial effects of deregulation

From the transport geographer's perspective, a distinctive spatial structure in air travel has emerged in the USA (see Chou 1993 and Shaw 1993), whereby the major US airlines have developed a hub-and-spoke structure as spatial and commercial strategies for organising airlines' operations in a deregulated environment. This contrasts with the CAB regulation era where interurban routes were often 805 km or more in length and little attention was given to integrating the route networks among operators. However, in a deregulated environment where cost reductions are a central element of the commercial strategy, least-cost solutions and network maximisation are a priority to achieve efficient operations. Airline services need to be responsive to demand and there has been a greater emphasis on airlines connecting all the nodes in their network. In this context, a hub-and-spoke system of provision (Figure 6.3) may enable airlines to serve a large number of people over a wide area, the hub acting as

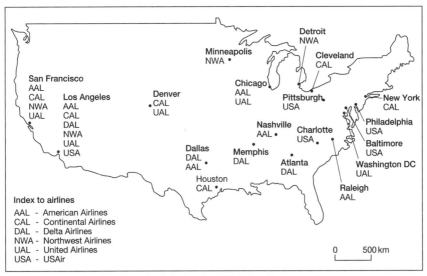

Figure 6.4 Distribution of major hubs in the USA (after Page 1994)

a switching point for passengers travelling on feeder routes along the spokes which cannot support a trunk route.

According to O'Kelly (1986) hubs are least-cost solutions for airlines and may combine a range of airports in a region, assisting the airline in running a high-frequency service along trunk routes between hubs. Along the spokes, regional carriers, often code sharing under the major's identification, provide the feeder services. The result has been a geographical concentration of airline hubs in major US cities, based on historical ties with certain airports, airline mergers, the servicing of niche tourist markets and responses to competitors, so that major operators provide 100–200 departures each day from some of the key hubs. Thus, a spatial concentration has occurred among the six largest airlines, leading to the development of four major hub cities – Atlanta, Chicago, Dallas and Denver (see Figure 6.4). This was illustrated in Shaw's (1993) recent analysis where the following national hubs emerged (with the regional hubs in parentheses):

- American Airlines   Dallas/Fort Worth, Chicago (Nashville, Raleigh/Durham)
- Delta Airlines   Atlanta, Cincinnati, Dallas/Fort Worth (Salt Lake City)
- USAir   Pittsburgh, Charlotte (Baltimore and Philadelphia)
- United Airlines   Chicago (Washington DC, Denver)
- Northwest Airlines   Detroit, Memphis, Minneapolis/St Paul
- Continental Airlines   Houston, New York, Denver, Cleveland

Figure 6.4 also incorporates two other hubs (Los Angeles and San Francisco) of lesser significance in the overall pattern of airline routes to illustrate the principal airports serving the western seaboard.

In spatial terms, Continental Airlines' network covers a wide area within the USA, while Delta Airlines' network is heavily concentrated in the southern states, with regional hubs serving the western seaboard. In contrast, Northwest Airlines serves the Mid West, while United Airlines has hubs located on an east–west axis across the USA, offering the best connections to cities on the western seaboard. USAir, however, has four hubs in close proximity, due to its interest in the East Coast market. Shaw (1993) observes that USAir is unique as it operates many other nodes as smaller hubs to serve other areas. This is one factor which has made the airline attractive to foreign airlines such as British Airways, which took a $200 million stake in the airline in April 1993, providing it with a foothold in the lucrative North American and transatlantic market.

Goetz and Sutton (1997) explain the spatial patterns of deregulation in terms of core–periphery concepts (see Chapter 2). They distinguish between two types of hubs

- domestic hubs
- international gateways.

They also expand upon Shaw's (1993) analysis as Table 6.7 shows. Goetz and Sutton (1997) identify the international gateways as funnels for international services and connections to domestic destinations but they do not function as domestic hubs. The gateways are mainly located in coastal cities while the domestic hubs are the focal point for the domestic system, and Dresner and Windle (1995) raise a range of issues related to such developments. In terms of the geographic concentration in the airline industry which occurred post-regulation, Goetz and Sutton (1997 : 244) argue that

> Between 1978 and 1993, every domestic hub core except Cleveland experienced an increase in single carrier concentration. In 12 of the 22 hub cores as of 1993, one carrier accounted for more than 60 per cent of their traffic, and 9 hubs reported more than 70 per cent concentration. These high levels of concentration at hub cores reflect the increases in both hub and spoke operations and industrial consolidation. Once entry and exit regulations were removed and carriers adopted hub-based networks, carriers concentrated traffic, personnel, and infrastructure at key points in their systems. The development of 'fortress hubs' – cities where no other carriers were able to establish beachheads of operation – emerged as a key strategy for major airlines facing competitive threats from new entrants into the industry. (Goetz and Sutton 1997 : 244)

One key factor in hub dominance was CRSs, with CRS-owning airlines having a 13–18 per cent greater chance of selling airline tickets through travel agents. Williams (1993) also observed that travel agent use of CRSs became concentrated in hub cities, with Sabre processing 87 per cent of travel agent revenues in Dallas, and United's Apollo system handling over two-thirds of bookings in Denver.

Following deregulation, there has been a considerable debate amongst researchers over the effect on consumers (tourists and non-tourists). For example, Kihl (1988) argues that deregulation led to a decline in service quality as

Table 6.7   Domestic and international airline hubs in the USA

Domestic hubs	Principal airline
Atlanta	Delta
Baltimore	USAir
Charlotte	USAir
Chicago	United, American Airlines
Cincinnati	Delta
Cleveland	Continental
Dallas Fort Worth	American Airlines, Delta, Southwest
Denver	United
Detroit	Northwest
Houston	Continental, Southwest
Las Vegas	American West
Memphis	Northwest
Minneapolis	Northwest
Nashville	American Airlines
Newark	Continental
Philadelphia	USAir
Phoenix	American West
Pittsburgh	USAir
Raleigh-Durham	American Airlines
St Louis	TWA
Salt Lake City	Delta
Washington	United, USAir
International gateways	
Boston	Delta, USAir, Northwest
Los Angeles	United, Delta, American Airlines, Northwest
Miami	American Airlines, Delta, USAir, TWA
San Francisco	United
Seattle	United, Alaska, Northwest

Source: Goetz and Sutton (1997).

smaller communities not directly connected to trunk routes faced fare increases and less frequent services. Goetz and Sutton (1997) observe that 'smaller turbo-prop carriers' have replaced jet services and service quality has declined. It is clear that airline mergers may have led to a decline in the number of carriers serving some communities, but Jemiolo and Oster (1981) maintained that any changes in service provision to less-accessible communities were a result of recession and greater fuel costs rather than deregulation. Yet Goetz and Sutton (1997 : 252) argue that 'small communities overall have experienced frequent interruptions in service and many have been dropped from the air network altogether', a feature emphasised by Kihl (1988). Of 514 non-hub communities with an air service in 1978, 167 had lost their service by 1995, with only 26 communities gaining a new service. The Airline Deregulation Act did include provisions for Essential Air Service subsidies so that small communities were still served. In 1995, some 77 communities received such a subsidy.

Debate has also centred on the effect of deregulation on passenger safety (Moses and Savage 1990). Golich (1988) asserts that the development of hubbing operations led to a decline in safety as more services and takeoffs/landings were concentrated into specific areas, increasing the potential for accidents. However, the US civil aviation statistics for the period 1975–90 show that the number of fatalities actually dropped from 663 to 424, while the rate per million aircraft miles flown has remained constant at 0.001 for scheduled services (see Abeyratne 1998; Braithwaite et al 1998 for the situation elsewhere). However, Golich's (1988) assessment may have a great deal of validity in view of the congested nature of many US airports which is discussed later.

Following deregulation and the growth in new entrants to the domestic airline market, consumer complaints increased to a peak of 40,985 in 1987 but dropped to 6,106 for 1991 and 4,629 in 1995 (US Bureau of the Census 1996). The most commonly reported sources of dissatisfaction in 1995 were:

- flight problems, such as cancellations and delays (1,133)
- baggage problems (628)
- problems associated with refunds (576)
- customer service issues such as unhelpful employees, inadequate meals and poor cabin service (667)
- errors in ticketing and reservations (666)
- incorrect or incomplete information about fares (185)
- the process of 'bumping' passengers off overbooked flights (by offering incentives not to travel on a specific flight) (263)
  Source: US Bureau of the Census (1996).

Airlines have exercised greater control over their workforces since deregulation, in pursuit of continued increases in productivity and greater economies of scale (Humphries 1992). Critics have also argued that deregulation has led to the larger carriers developing sophisticated marketing campaigns to foster customer loyalty to influence the choice of carrier through 'frequent flyer programmes' (Laws 1991) and the use of airline-controlled CRSs (Archdale 1991) to promote their services via travel retailers, as noted by Goetz and Sutton (1997). Investment in technology and marketing tools to shape the pattern of supply to potential customers has also been reinforced through the hub and spoking operations of the 'majors'. For new entrants to gain a foothold in the US domestic airline market, not only would they need to secure a slot at an airport but they would need to be able to offer services to compete with the existing frequencies and service available. This would require major capital investment and commercial nerve at a time of considerable flux in the aviation market. Furthermore, Graham (1992) identifies the influence which major carriers exercise on hub airports, particularly their ability to limit new competition. Ironically, this has meant that hub airports have become less financially secure in the US as restrictive practices such as 'majority-in-interest' (Graham 1992) may limit airport revenue generation through the airlines' power of veto on capacity improvements to encourage competition.

The changes brought about by deregulation have also had a pronounced effect on the US airport system as the demand for air travel has continued to grow despite constraints on the supply of airport capacity (Sealy 1992). Deregulation and the development of trunk routes and hubs have intensified congestion at major US airports. Sealy (1992) notes that 16 of the world's 25 leading airports handling in excess of 15 million passengers a year were in the USA.

Having examined the effect of deregulation on the supply of air transport, attention now turns to the interface of supply and demand issues in relation to land-based tourist transport systems. The example of Bermuda has been chosen to reflect some of the problems and management issues associated with the supply of tourist transport services in a small island where the available space for tourism and transport competes and conflicts with other land uses.

The supply of tourist transport in destination areas

The supply of tourist transport services in the destination area is one area neglected in Buckley's (1987) transaction analysis. Two studies which deal with this issue are Teye's (1992) study of Bermuda and Heraty's (1989) examination of supply-related problems which characterise less developed countries (LDCs). For this reason, the case of Bermuda is now considered, followed by an overview of the supply-related issues raised by Heraty (1989).

Land-based tourist transport systems: Bermuda

Tourism research on the Caribbean has generated a great deal of literature in learned journals such as *Tourism Management, Annals of Tourism Research* and more recently in publications by Gayle and Goodrich (1993) and Todd and Mather (1993). The Caribbean is noteworthy as a major destination for over 10 million visitors from developed countries each year and is the focus of the world's cruise ship business (Lawton and Butler 1987; Peisley 1992a, 1989; Dickinson 1993). Therefore, it is an appropriate area on which to focus, particularly as tourist arrivals increased by 40.3 per cent in the period 1985–90, while world tourist arrivals only increased by 25 per cent. Tourist arrivals in the Caribbean generated $96 billion in gross expenditure for the region in 1990 (Gayle and Goodrich 1993). Research by Archer (1989, 1995) considers the economic impact and volume of tourism in the Caribbean basin and how international arrivals and cruise ship arrivals affect the economy of small islands (Table 6.8). The literature on small island tourism is still developing (Pearce 1987), but Wilkinson's (1989) analysis of such small island microstates provides a useful review into:

- ways of classifying small islands
- why tourism has assumed such a prominent role
- the problems related to scale and size on small islands
- the impact of external influences seeking to control the supply chain (e.g. multi-national corporations and overseas investment companies).

Table 6.8 Visitor arrivals and expenditure in Bermuda 1980–93

	Visitor arrivals (000s)			Visitor expenditure[1] ($ million)		
	Air	Cruise	Total[2]	Air	Cruise	Total[2]
1980	491.6	117.9	609.5	277.6	12.4	290.0
1981	429.8	104.7	534.5	275.0	13.3	288.3
1982	420.3	124.2	544.5	286.2	15.4	301.6
1983	446.9	120.8	567.7	319.3	16.2	335.5
1984	417.5	111.4	528.9	324.1	14.9	339.0
1985	406.7	142.0	549.6	335.6	21.9	357.5
1986	459.7	132.2	591.9	402.2	20.1	422.3
1987	477.9	153.4	631.3	439.6	25.0	464.6
1988	426.9	158.4	585.2	414.1	26.4	440.5
1989	418.3	131.3	549.7	426.5	24.3	450.8
1990	434.9	112.6	547.5	467.9	22.2	490.1
1991	386.2	128.2	514.3	423.9	31.7	455.6
1992	375.2	131.0	506.2	410.7	30.8	441.5
1993	413.1	153.9	567.1	465.2	39.3	504.5
1994	417.0	172.9	589.9	483.3	41.9	525.3

[1] Within Bermuda, the Bermudan dollar exchanges almost at par with the US dollar.
[2] Not all of the row totals equal the sum of the figures included in the row because of rounding.
Source: Archer (1995 : 27); processed from (1) Bermuda Government E/D cards and (2) Department of Tourism/Statistics Office Exit Survey data.

The effect of multinational corporations on microstates was reviewed by Britton (1982) in relation to tour operators, airlines and hotel corporations, which seize development and marketing opportunities to control the tourism economy through foreign ownership. Bermuda is an interesting example in this context. Wilkinson (1989) notes that its relatively strong economic and political will to restrict the large-scale expansion of tourism led to a moratorium on new hotels and limits on the number of cruise ships entering the harbour at any one time.

The development of tourism in microstates such as Bermuda has generated a requirement for a land-based tourist transport system to facilitate development and the movement of visitors around the island (Conlin 1995). Although infrastructure provision is not without environmental impacts (Holder 1988), Bermuda is a good example where research into the role of tourist transport has led to a better understanding of the interface of transport and tourism and the conflicts they can cause.

CASE STUDY   The tourist transport system in Bermuda

Bermuda comprises almost 150 small islands in the Caribbean, although only 20 are actually inhabited. A series of bridges and causeways connect the main islands where approximately 62,000 people live in an area of 55 km². Bermuda is unique within the Caribbean due to its high per capita income of $22,540 (1987) which is greater than both the USA and Switzerland (Teye 1992). Such prosperity derives from:

- international businesses in the islands
- military bases
- tourism, which is the largest contributor to the economy (see Archer 1995 for a detailed review of international tourism in Bermuda).

The high-spending markets (e.g. the North Americans) attracted to the islands require good quality air and sea transport links (Bertrand 1997) and a transport network which is coordinated and efficient. The reliance upon land transport for tourists within the destination to provide airport/port transfers to resorts, transport to tourist attractions and sightseeing has a major impact on the transport system on the islands.

One of the main concerns which Teye (1992) observes was the decline of the islands' railway system and the growth in car ownership which stimulated the development of the country's road network. As in many microstates, cars have become the symbol of economic prosperity, yet space constraints and the continued development of roads have contributed to high densities of car ownership (e.g. 220 vehicles per km² in Bermuda) with associated environmental impacts. Bermuda's government examined the problem posed by private car ownership in the 1970s and 1980s and the complex nature of transport planning for residents and tourism (see Teye 1992 : 399–400 for a detailed analysis of government research on the issue). The principal measures within Bermuda's transport policy have been to regulate ownership through licensing, speed controls and limits on vehicle replacement. The consequences for tourist transport are restrictions on car-hire, which have led to an expansion in motorcycle and moped use. This has led to conflicts between the use of mopeds by tourists (less than 50 cc engine size) for sightseeing and touring, and local traffic. For example, in 1970 there were 3,296 accidents and 24 fatalities, although a crash helmet law in 1977 and improved traffic education for tourists by moped hire companies led to a significant reduction in the number of accidents.

In common with many other small islands which have developed tourism, Bermuda has traffic problems related to non-tourist (commuting) and tourist travel to the capital city (Hamilton) and tourist travel within tourist zones and corridors along the south shore, where much of the tourist accommodation is located (Figure 6.5). Given the primary role of Hamilton as a commercial and entertainment centre and the hub of the island's bus services, congestion generated by resident and tourist transport use is virtually inevitable, unless these activities are decentralised to other areas.

Teye (1992) also notes the effect of seasonality in tourist movements on the islands (May–October) and the effect of cruise ships. The cruise ship business in the Caribbean in the 1980s is well documented by Lawton and Butler (1987) and there is widespread agreement that the arrival of such ships can swamp small islands when the visitors disembark. Table 6.8 (on page 204) outlines the growth of air and cruise ship arrivals to 1994. Unlike other cruise ship ports, Bermuda has attracted long-stay visits (up to four days) where the ships dock and remain rather than arriving in the morning and leaving that evening. As a result, Teye (1992) argues that 'cruise ships in Bermuda literally become docked resorts'. Yet prior to 1984, up to seven cruise

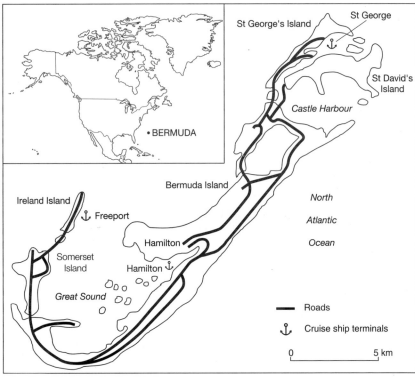

Figure 6.5   Bermuda: location and infrastructure (after Page 1994)

ships a week were docking in Bermuda. The impact on the tourist transport system can be seen in traffic congestion due to visitors renting taxis for sightseeing.

One solution to the management of this problem was the introduction in 1988 of limitations on cruise ship passengers by restricting the number of ships to four in port at any one time, and to attempt to restrict cruise ship visits (see Table 6.8, page 204). This deliberate policy by Bermuda's government eliminated weekend cruise ships from docking, thereby reducing overcrowding and temporary congestion on the island's transport network. This serves as a useful example of how the management of the supply of tourist transport infrastructure can induce a reduction in the demand. According to Peisley (1996 : 137) 'Bermuda remains the most tightly controlled of cruise ship destinations' in the Caribbean. In the 1990s, the number of ships which can operate in the area is limited to five of 4-star ranking or higher: two from the Celebrity Company, one from each of RCCL, NCL and Majesty. Cunard can make occasional visits with its three ships *QE2*, *Vistafjord* and *Royal Viking Sun*. The aim is not only to preserve the island's up-market image, but also to control visitor numbers and to ensure a relatively high onshore visitor expenditure (Peisley 1996).

One additional measure to manage tourist travel patterns in Bermuda, which was adopted as government policy in 1973, was the development of tourist-specific routes (see Lew 1991 for a discussion of this concept). Although little progress has been made in this direction, Teye (1992) argues that the severity of the traffic problems facing the islands may require a clear strategy to designate a road hierarchy to reduce the conflict between tourist and non-tourist use of roads. This would need to encourage the rational and managed touring of the islands' natural and man-made attractions and is preferable to further road building in a country already short of space. It is ironic that while tourism has generated prosperity for Bermudans, the symbol of success – the private car – is now posing major transport problems for the country, in common with many other microstates which have followed tourism as a route to economic development. As Teye (1992 : 405) readily acknowledges, 'most microstates have very few options if they allow poor transportation planning to destroy their tourist industry'. Visitors expect good destination-specific transport systems and they form an important part of the tourist's overall experience of travel. For this reason, it is pertinent to focus on the issues raised by Heraty (1989) in the context of the LDCs in terms of how the supply of tourist transport needs to meet both the expectations and needs of visitors if the service encounter in the destination is to reach minimum standards. Although some of the problems Heraty (1989) examined are unique to LDCs, the general issues raised also have implications for developed countries. The issues which highlight tourists' expectations in the developed world and the maintenance of service quality, include:

- *Airport transfers* Tourists arriving in a destination after a tiring flight, require convenient and comfortable transfer vehicles. The tendency to use a limited number of transfer vehicles to shuttle a large number of different tour groups to a dispersed range of hotels often adds to the inconvenience associated with long-haul travel in LDCs. The trend towards the use of baggage trucks to transport tourists' baggage to hotels due to the lack of space on old vehicles is disconcerting for tourists and can cause delays due to misdelivery.
- *Sightseeing tours* For LDCs receiving high-spending tourists, one lucrative tourist transport service which offers potentially high profits to private operators is sightseeing tours (see Plate 6.1). However, visitors from developed countries expect coaches with air-conditioning and a public address system, vehicles which are safe and give good all-round visibility. In many cases, the capital cost of such vehicles is prohibitively high for private operators in LDCs who are forced to import them. Yet where they are available, they significantly enhance the tourists' experience. Well-trained tour guides, able to provide commentaries and answer questions in a variety of languages, are also an important asset. Guides who are able to convey the local culture, history, customs and lifestyle to visitors will be able to contribute to the tourists' pleasant memories of their holiday and adequate stopping places at cafes, restaurants or clearings with toilets and refreshment facilities also have a part to play.

Independent travel by tourists

The more adventurous tourist often wishes to travel on local public transport systems (see Plate 6.2) and adequate information needs to be made available (e.g. timetables). The use of hire cars is also a major tourist service in LDCs which need to be supplied according to a code of good practice. According to Heraty (1989 : 289), good practice should include:

- high-quality vehicle standards, insurance and contract conditions
- provision of tourist-oriented maps and leaflets
- signposting of routes to tourist attractions and sights
- training for the police in dealing with tourist drivers

Plate 6.1   Tourist sightseeing in a less developed country – Fiji in the Pacific Islands (Source: S.J. Page)

Plate 6.2   On small islands tourists tend to use local bus services, as the main bus terminus at Valletta, Malta, shows. All services operate from this hub on the island. The main problem for the island is in financing the replacement of its ageing fleet of buses (Source: S.J. Page)

- action to address road-based and car-related crime affecting tourist hire cars
- safety standards for mopeds where governments permit tourists to hire such transport
- incentives to encourage short-stay visitors (e.g. cruise ship passengers) to venture away from the port to visit other locations
- licensing and regulation of taxi companies and drivers to prevent tourists being exploited.

From the transport policy makers' perspective, Heraty (1989 : 290) notes that in accommodating tourist transport needs, a range of problems need to be addressed at government level to facilitate tourist travel in destination areas, including:

- Import duties may need to be relaxed to facilitate the acquisition of new vehicles and to make them more affordable and able to absorb 'standing costs' (when the vehicle is off the road) as tourist use declines outside of the peak season.
- Roads and tourist transport infrastructure need to be maintained to reduce wear and tear on vehicles.
- Skilled mechanics and vehicle drivers need to be trained to ensure an adequate supply of labour to meet demands.
- Traffic congestion related to peaked seasonal demand by tourists needs to be managed.
- Sufficient transport operators need to be licensed to prevent limited competition and cartel-type operations resulting under prohibitively expensive situations.

A recent review by Grant et al (1997) also outlines the transport policies and strategies which local authorities need to develop to facilitate a more efficient supply of tourist transport services. They highlight the practical and organisational issues involved in the development of specific forms of tourist transport and in meeting the needs of users (e.g. car-based, coach-borne, pedestrians, people travelling on public transport) and needs of sightseers and the importance of integration (May and Roberts 1995). In a recent study of New Zealand's premier South Island destination, Milford Sound, congestion and inadequate road provision emerge as a potential threat to the area's tourism potential (Anon 1998). The region's single-lane bridges and tight corners make it a difficult terrain for drivers. Since about 300,000 people travel to Milford Sound per annum, capacity problems may inhibit future growth. Research by Page and Meyer (1996) also illustrates the problem of foreign driver accidents on New Zealand's road system, where tight corners and driving conditions can raise safety issues for foreign visitors as Figure 6.6 shows.

Summary

This chapter has examined the various issues facing transport providers in the 1990s in managing the supply of services and associated operational issues. In particular, it has emphasised how various transport providers, such as airlines, have used the two major tools, IT and logistics to improve the flow of information to assist in the management and delivery of services to customers. The growing importance of IT is demonstrated by the competition between international airlines (Oum and Yu 1998) and the greater use of IT to gain competitive advantage. This also applies to the role of infrastructure providers discussed in the next chapter, in optimising the use of the facilities (see Haghani and Chen 1998) to gain cost savings and optimal use of major sunk costs in

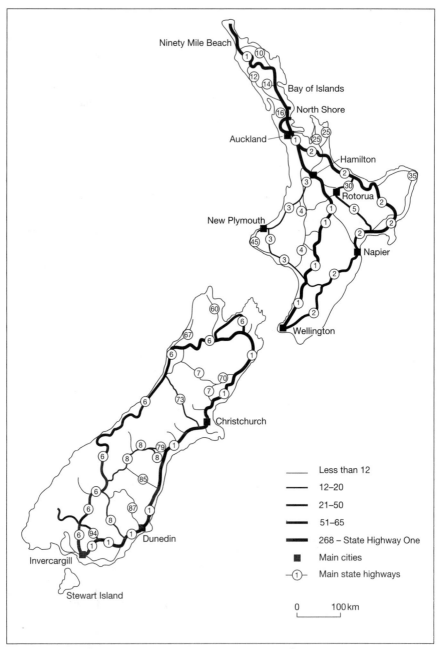

Figure 6.6 Tourist road accidents in New Zealand in 1994 (redrawn from Page and Meyer 1996)

infrastructure. The other tools now being harnessed in terms of strategic management by transport providers such as airlines, are the growing use of alliances and strategic partnerships. This is a function of enhanced competition in the marketplace (Charlton and Gibb 1998), a feature discussed in considerable detail in the context of air transport in the EU (Graham 1998) and Australia (Hooper 1998), rail travel in the UK (Knowles 1998), coach travel in the UK (White and Farrington 1998) and the US railway industry (Jahanshahi 1998). Competition is a theme running throughout this book and is not simply confined to any one chapter. In fact competition is now assuming a greater importance as many governments seek to deregulate former state-planned transport systems (e.g. Cline et al 1998), placing a greater emphasis on the individual transport businesses to meet the needs of tourists in a competitive environment where all the tools available are needed to operate in a cost-efficient and safe manner (Braithwaite et al 1998).

Despite the emphasis on aviation, due to its dominant effect on the tourist sector, the example of land-based supplies issues on small islands and in the less developed world is frequently overlooked as peripheral to the main concern of transport planners and researchers. Yet, as this chapter has shown, the tourist experience does not stop when the tourist disembarks from an aircraft and leaves an airport terminal. Thus, as chapters five and six show, the integration and linkages between air transport and land transport need to be viewed as part of the seamless travel experience, particularly in cases where package holidays are produced and sold by organisations with a horizontally integrated tourism and transport operation (e.g. Thomson Holidays as discussed in chapter five). In fact there is growing evidence from research by Page and Meyer (1996) that the role of safety is also assuming a growing importance for the transport sector in individual countries. Recent research on the wider issue of tourist safety (Abeyratne 1998) and accidents and injuries on transport as a form of recreational activity (e.g. jetboating and white water rafting) also needs to be set in the wider context of tourists' experience (Page 1997). As a result, this chapter does illustrate that the supply issues facing the transport operator cover a diverse range of fields from business operations, including IT and logistics, through to safety and management of the services provided to ensure the company operates in a profitable and sustainable manner.

The lessons of deregulation in the USA show that in a business environment free of state regulation, where profit is the motive for service provision, transport operators behave in the most efficient and restrictive manner in order to protect market share. The emergence of a hub-and-spoke structure may indicate how the European airline business will respond to a greater degree of deregulation in the 1990s. But as the discussion by Heraty (1989) suggests, the supply of tourist transport services in destination areas needs to be carefully targeted at the market to ensure both customer satisfaction and to maximise revenue generation for private sector transport operators. As the case of Bermuda shows, where there is a significant demand for a destination, government action may be taken to restrict the activities of tour operators and tourists without adversely affecting the tourism industry. Understanding the expectations and motives of visitors confirms the importance of setting the demand and supply for tourist transport services in a systematic framework where the wider issues and interrelationships can be understood. In this context, the next chapter considers the specific implications of supply and demand for infrastructure development.

Questions

1. What is the role of logistics in transport service provision?
2. How do transport companies utilise IT in the sales of transport?
3. Has airline deregulation in the USA affected the safety of domestic air travel?
4. What problems do tourist transport operators face in less developed countries?

Further reading

Abeyratne, R. (1998) 'The regulatory management in air transport', *Journal of Air Transport Management*, 4 (1): 25–37.

Braithwaite, G., Caves, R. and Faulkner, J. (1998) 'Australian aviation safety – observations from the lucky country', *Journal of Air Transport Management*, 4 (1): 55–62.

French, T. (1998) 'The future of global distribution systems', *Travel and Tourism Analyst*, 3: 1–18.

Goetz, A. and Sutton, C. (1997) 'The geography of deregulation in the US airline industry', *Annals of the Association of American Geographers*, 87 (2): 238–63.

Hilling, D. (1996) *Transport and Developing Countries*, London: Routledge.

Milman, A. and Pope, D. (1997) 'The US airline industry', *Travel and Tourism Analyst* 3: 4–21.

Sheldon, P. (1997) *Tourism Information Technology*, Oxford: CAB International.

Internet sites

See Appendix 1 for sites relating to air travel in the USA.

Chapter 7

Managing tourist transport infrastructure: the role of the airport

Introduction

The development of tourism requires a transport infrastructure to facilitate the free movement of tourist traffic, and much of the research in this context has focused on modal forms of travel (e.g. rail travel, air travel and car-based trips). One of the fundamental links that has been overlooked in the tourist transport system is the way in which demand and supply are brought together and managed, and the infrastructure used to ensure the system functions efficiently. In both the transport and tourism literature, terminal facilities, which provide the context in which the tourist embarks on the mode of transport, to ensure the interaction of supply and demand takes place smoothly, have been largely overlooked. Yet it is widely recognised that the 'holiday experience' begins when the tourist arrives at a terminal, ready to embark on a journey. In fact some commentators even suggest that the experience effectively begins when the traveller leaves their home environment.

This chapter examines the challenge of managing the interaction of supply and demand for tourist travel with reference to one type of terminal facility – the airport. Airports are probably the most complex environments and systems in which this interaction occurs and yet they often remain poorly understood in a tourism context. As Ashford et al (1991 : 1) argue, 'the airport forms an essential part of the air transport system, because it is the physical site at which a modal transfer of transport is made from the air modes to land modes'. The chapter commences with a discussion of the management challenge posed by terminal facilities, emphasising the organisations involved in the management process. This is followed by a discussion of the locational and planning issues associated with airport development, particularly access to markets. Future development plans for world regions experiencing a rapid growth in tourist travel are then discussed, as arrivals are likely to outstrip existing airport capacity. But why focus on air travel rather than other types of terminal? What literature does exist is airport-specific and ports, bus/coach terminals and railway terminals have not attracted much attention, with a notable exception being Bertolini and Spit (1998). However, in Chapter 3 the importance of the terminal environment for InterCity rail travel was discussed, albeit briefly, in relation to service quality and the travel experience. In addition, ATAG (1993) provided a convincing argument for a focus on air travel:

in 1993, the volume of air travel worldwide was equivalent to one-fifth of the world's population travelling once a year. The air transport industry also generated 24 million jobs and US$1,140 billion in gross output (see ATAG 1993; http://www.atag.org/AH/Index.htm). Therefore, while some of the principles and issues discussed in this chapter relate solely to air travel, some of the broader issues (i.e. developing a customer-focused approach) have a wide application to the management of the tourist–transport interface which occurs in transport terminal facilities.

The management challenge of tourist transport terminals: airports

To the uninitiated, occasional traveller, terminal facilities can be a bewildering, seemingly chaotic and unnerving experience (a theme explored in more detail in Chapter 8). The semblance of chaos is conveyed by Barlay (1995):

> The airport cavalcade can baffle or startle the inexperienced passenger . . . Laden with suitcases and packages, calm and rational people grow uptight, defensive with aggression, fail to allow themselves time to familiarise themselves with the layout or study the free guides to terminals. (Barlay 1995 : 48)

The entire psychology of travel and the change in the behaviour of the traveller in airports (e.g. tunnel vision or airport syndrome) embraces a whole series of emotions among the diversity of passengers: 'joy, grief, anxiety, expectations, aggravation, yearning and fulfilment'. From a tourists' perspective, numerous factors affect their experience, according to Barlay (1995 : 49):

- speed of check-in (see Plate 7.1)
- efficiency of passport control and customs clearance (the UK government is now placing the onus on airlines to check outbound travellers' passports in order to devote more resources to inbound travellers)
- luggage retrieval
- availability of shops, duty-free goods and associated services
- a spacious and relaxed environment in which to wait prior to boarding the aircraft (see Plate 7.2).

From the airport's perspective, operational issues have dominated its work. The management challenge is to recognise the tourists' needs and to minimise likely problems while ensuring the terminal operates as a smooth series of complex systems. Doganis (1992) defines airports as complex industrial organisations that

> . . . act as a forum in which disparate elements and activities are brought together to facilitate, for both passengers and freight, the interchange between air and surface transport. (Doganis 1992 : 7)

In some countries, for a range of historical, legal and other reasons, the scope of airport activities can vary from the highly complex and all-embracing to the very limited. In physical terms, Doganis defines an airport as

Plate 7.1 Check-in desks at London Stansted airport operated by BAA (reproduced courtesy of BAA London Stansted)

Plate 7.2 The newly opened London Stansted airport won awards for the Sir Norman Foster design and the incorporation of an airport on one level with spacious waiting areas to reduce the stress of travel (reproduced courtesy of BAA London Stansted)

Essentially one or more runways for aircraft together with associated buildings or terminals where passengers . . . are processed . . . the majority of airport authorities own and operate their runways, terminals and associated facilities, such as taxiways or aprons. (Doganis 1992 : 7)

Doganis (1992 : 7) distinguishes between the three principal activities of airports:

- essential operational services and facilities
- traffic-handling services
- commercial activities

although 'at most, if not all, airports, the major consideration must be passenger flows' (Ashford et al 1991 : 27) and this in itself requires management measures to ensure smooth operation (e.g. pricing and flow management – see Matthews 1995). (See also Appendix 2, which outlines a range of Internet sites for world airports which provide more in-depth access to data on many of the principal activities of airports.) This comprises a broad range of activities within any airport environment, whether in an international gateway airport, a regional or local airport:

- ground handling (Soames 1997; Caves 1994)
- baggage handling (Neufville 1994)
- passenger terminal operations
- airport security
- cargo operations
- airport technical services
- air traffic control (where applicable – see Majumdar 1994, 1995)
- aircraft scheduling (takeoff/landing slot allocation)
- airport and aircraft emergency services
- airport access

which are described in detail by Ashford et al (1991) in what remains the principal study on airport operations. Figure 7.1 highlights some of the relationships which need to be managed in the airport system so that the airport–airline–traveller interactions occur in a professional and smooth manner. Figures 7.2 and 7.3 also illustrate the spatial interactions in the airport system, indicating how the airport enables an aircraft to take off/land and to unload and load passengers. As part of these functions, it enables travellers to change their mode of transport and be processed efficiently (e.g. ticketing and documentation).

Each of these functions impinges upon the management of the tourists' experience of the terminal facilities. Table 7.1 outlines the weekly pattern of scheduled departures from the world's top 30 airports, which are only a fraction of the ICAO list of 1,012 airports worldwide. Of the top 30 airports, 19 are located in the USA, confirming a pattern similar to that observed by Sealy (1992). This is followed by Europe and Asia-Pacific.

At a global scale, the management of airports is affected by government policy which, in part, determines the pattern of ownership. As Doganis (1992) observes, there are four main types of ownership. These are:

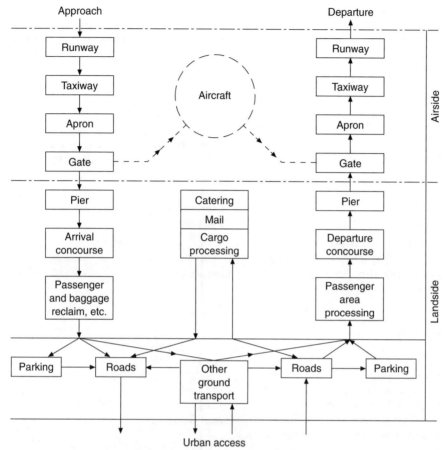

Figure 7.1   An airport system (redrawn from Ashford et al 1991)

- *State ownership with direct government control*, characterised by a single government department (i.e. a Civil Aviation Department) which operates the country's airports. The alternative to a centralised government pattern of control and management is localised ownership such as municipal ownership.
- *Public ownership through an airport authority*, usually as a limited liability or private company. For example, the British Airports Authority (BAA) was one of the early examples of a national airport authority. Aer Rianta in the Republic of Ireland is another example of a national airport authority. There are also cases of regional airport authorities in the USA.
- *Mixed public and private ownership* is an organisational model adopted at larger Italian airports, where a company manages the airport, with public and private shareholders.
- *Private ownership* was a model of limited appeal prior to the wave of privatisation in the 1980s. One of the early examples in the UK was London City Airport in London Docklands, opened in 1987 (Page 1987, 1989b, 1991; Page and Fidgeon 1989). However, Doganis (1992) points out that the major impetus to private ownership was the privatisation of BAA in 1987.

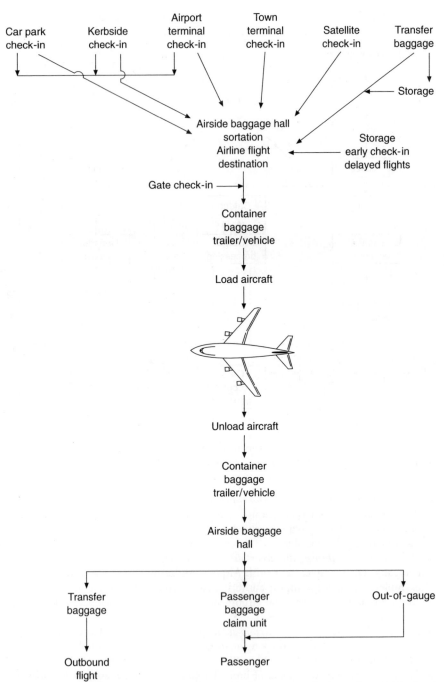

Figure 7.2 Luggage loading and unloading in the airport system (redrawn from Ashford and Moore 1992)

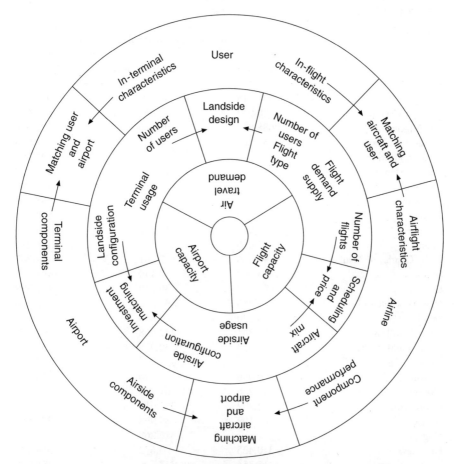

Figure 7.3 Airport relationships (redrawn from Ashford et al 1991)

In generic terms, Doganis (1992) argues that a prime function of airport management is to determine the objective of any economic and management strategy for an airport by addressing four questions:

- Should airports be run as commercially oriented profitable concerns?
- How should one improve airport economic efficiency?
- Should profits from larger airports be used to cross-subsidise loss-making smaller airports?
- Should airports be privatised?

These four questions essentially raise the issue of what form of airport ownership a government deems important and the prime objectives which will affect the economic strategy each airport pursues. This in turn will require very different management strategies. For example, airlines and bodies such as IATA still argue that airports are public utilities when arguing against increased airport charges.

Table 7.1   The world's top airports, January 1998

	Aircraft seats (scheduled departures)[1]		
	Domestic	**International**	**Total**
Atlanta	937,414	68,725	1,006,139
Chicago/O'Hare	870,957	114,656	985,613
Los Angeles	688,819	206,210	895,029
Dallas–Fort Worth	788,265	70,723	858,988
London Heathrow	95,150	697,641	792,791
Tokyo (Maneda)	548,003	10,626	658,829
Frankfurt	113,771	444,882	559,593
San Francisco	461,281	87,065	548,346
Denver	509,794	6,594	516,388
Phoenix	486,953	17,323	504,276
Miami	243,010	256,202	498,212
Paris (Charles de Gaulle)	47,566	447,201	494,767
St Louis	483,087	8,440	491,527
Detroit	441,469	36,129	477,598
New York/Newark	372,380	90,890	463,270
Seoul	243,230	209,848	453,078
New York/JFK	208,331	242,458	450,787
Minneapolis–St Paul	401,973	30,976	432,948
Hong Kong	–	432,389	432,389
Boston	370,036	48,140	418,176
Houston	343,266	61,734	405,000
Amsterdam	4,292	391,610	395,902
Bangkok	101,192	289,998	388,188
Singapore	–	376,268	376,268
Las Vegas	370,425	3,672	374,097
Tokyo (Narita)	15,854	347,696	363,550
New York/La Guardia	334,881	28,089	362,970
Philadelphia	324,531	36,293	359,824
Rome	162,781	185,112	347,893
Seattle	322,408	25,430	347,838

[1] The statistics are based on services ending week of 12 January 1998 and code-shared flights have been removed where possible.
Source: Modified from *Airline Business*, January 1998.

It is useful to examine a number of specific issues associated with airport management which will then be exemplified by a case study of BAA. As Doganis (1992 : 45) rightly argues, 'matching the provision of airport capacity with the demand while achieving and maintaining airport profitability and an adequate level of customer satisfaction is a difficult task'. One of the principal problems facing any airport manager is that of planning. The time lag between the decision to build and the opening of a new airport terminal is five to ten years, as the case of Heathrow Airport's planned Terminal 5 illustrates. In addition,

Table 7.2 Operating costs at BAA airports in 1996 and 1997

	1997 £ million	1997 £ million	% change 1996/97
Staff costs	238	246	3.4
Retail expenditure	166	176	6.0
Depreciation	100	110	10.0
Maintenance	64	72	12.5
Rent and rates	64	68	6.2
Cost of development property sales	–	20	100
Utility costs	77	77	0
Police costs	39	38	−2.6
General expenses	62	75	21.0
Total	810	882	8.9

planners also need to ensure forecasts of future growth are within realistic bounds so that terminal facilities can accommodate demand for at least another decade. However, for any airport development, probably the most fundamental issues to understand are

- costs
- the economic features of airports
- sources of revenue
- methods of charging and pricing airport aeronautical services (see Forsyth 1997b; Reynolds-Feighan and Feighan 1997)
- the type of commercial strategy to adopt
- potential sources of commercial revenue (see Zhang and Zhang 1997)
- the most appropriate management structure for an airport as a commercial/non-commercial organisation
- financial performance indicators (see Hooper and Hensher 1997; Gillen and Waters 1997)
 Source: Based on Doganis (1992) and Ashford and Moore (1992).

Costs and economic characteristics of airports

Research undertaken in the 1980s and reported by Doganis (1992) identified the following costs of airports based on an analysis of European airports:

- *Staffing* accounts for 42 per cent of costs; it is normally the major operational cost for most airports.
- *Capital* charges comprise 22 per cent of costs; they include interest payments on commercial loans and the cost of depreciation of capital assets.
- *Other operational items* (e.g. electricity, water and supplies) comprise 11 per cent of costs.
- *Maintenance* accounted for 9 per cent of costs while *administration* resulted in 4 per cent of costs.

These costs are listed in Table 7.2 for the British Airports Authority for 1996/97. Doganis (1992) also draws attention to the differences between US and

Table 7.3   Employment at UK airports 1994/95

Airport	Number of employees
BAA owned airports	
Heathrow	3,562
Gatwick	1,857
Stansted	574
Glasgow	552
Edinburgh	315
Aberdeen	269
Southampton	133
Non-BAA airports	
Manchester	2,200
Birmingham	681
Newcastle	512
Belfast International	288
Luton	500
East Midlands	396
Bristol	308
Highlands and Islands	194
Cardiff	133
Leeds/Bradford	181
London City	166
Liverpool	54
Teesside	143
Humberside	80
Norwich	152
Exeter	121
Bournemouth	153
Total	13,590
Others	336

European airports; airport costs are reduced by airline rental or lease agreements on terminal facilities in the USA. Likewise, many US airports are not directly involved in baggage handling, which is left to airlines. But financing costs for US airports tend to be a major element of expenditure, often 44 per cent of total costs if depreciation is included, although staff costs tend to remain at approximately 22 per cent.

Table 7.3 shows the number of employees at the main airports in the UK in 1994/95 (CIPFA 1996); a total of 13,590 people were employed at national and regional airports, which indicates the significance of the direct employment-generating potential of such infrastructure. Three airports (London Heathrow, London Gatwick and Manchester) employed over half of the airport workers in 1994/95. In addition, Langley (1997) provides detailed data on UK airports from an operational and financial perspective.

Two of the principal economic characteristics of airports are that:

- Economies of scale exist as the volume of traffic increases, though congestion can lead to increases in unit costs.
- Development programmes for airports increase unit costs, particularly when new terminals are opened and are operating below their design capacity.

Airport revenue

According to Ashford and Moore (1992), airport revenue can be divided into two categories:

- *operating revenues*, which are generated by directly running and operating the airport (e.g. the terminal area, leased areas and grounds)
- *non-operating revenues*, which include income from activities not associated with the airport core business

while Doganis (1992) divides it into aeronautic or traffic revenues and non-aeronautical or commercial revenues.

In terms of *aeronautical revenues*, a range of possible revenue sources exist (though not all airports necessarily collect or use the revenue in a set way):

- landing fees
- airport air traffic control charges
- aircraft parking
- passenger charges
- freight charges
- apron services and aircraft handling (where the airport provides such services).

In terms of *non-aeronautical revenue*, Doganis (1992) outlines the following sources:

- rents or lease income from airport tenants
- recharges to tenants for utilities and services provided
- concession income (e.g. from duty-free and tax-free shops)
- direct sales in shops operated by the airport authority
- revenue from car parking where it is airport-operated
- miscellaneous items
- non-airport-related income (e.g. through land development or hotel development).

Within a European context, Doganis (1992) noted that aeronautical revenue accounted for 56 per cent of revenue and non-aeronautical revenue for 44 per cent of income. In the USA, the situation was slightly different, with airports generating more revenue from commercial sources (e.g. concessions 33 per cent; rents 23 per cent; car parking 4 per cent; other non-aeronautical sources 17 per cent and aeronautical fees 23 per cent). CIPFA (1996) provides a range of detailed data for all UK airports.

Pricing aeronautical services

According to Doganis (1992), aeronautical fees (i.e. those costs commercial aircraft have to pay to land at an airport) will continue to remain a crucial element of airport finances since they can be adjusted to offset revenue losses. The impact of such fees on airlines varies from nearly 20 per cent of airline operating costs to less than 2 per cent. This directly affects the price the consumer pays for their ticket. Landing fees continue to be a key source of revenue (Ashford and Moore 1992). Doganis (1992) and Ashford and Moore (1992) identify a range of approaches to charging fees, including:

- a fixed rate per tonne multiplied by a unit charge
- a rate per tonne with weight-break points, with fees increasing in steps
- single fixed charges irrespective of size of aircraft, which are used to fund the following types of services:

  - air traffic control facilities
  - landing facilities such as the runway and taxiways
  - parking of the aircraft on a stand or apron for a specified time, after which a separate fee is levied
  - use of aircraft gates, airbridges and terminal facilities (though increasingly airports are levying departure taxes to pay for the use of such facilities by travellers)
  - takeoff facilities.

CIPFA (1996) compared the landing charges for an international inbound flight using a Boeing 737-400 at various UK airports in 1994/95. A comparison of Cardiff and Gatwick indicates that the combined cost of the landing fee and passenger facility charge levied to land at Cardiff at 9.30 am in June was £2,589.00. This equates to a revenue per passenger of £17.26. The same flight would cost £1,814.35 to land at Gatwick at the same time, with a revenue per passenger of £12.10. Such variations are marked within the airport system in the UK. However, as Doganis (1992) notes, each airport has in place its own levels of rebates/surcharges for flights and airlines, so these may not in fact be the true costs once such factors are taken into account.

In Europe, additional surcharges apply for the use of aircraft which do not meet noise requirements (see Chapter 8 and the British Airways case study and research by Perl et al (1997) on pricing such impacts). Doganis (1992 : 69–111) outlines many of the issues associated with pricing airport charges and this should be read in conjunction with Toms (1994).

Commercial strategies for airports

With the move towards privatisation in the UK and other countries, airport managers have had to adopt more commercially driven business principles when looking at the best strategy to adopt in managing airport infrastructures and services. Carter (1996), for example, points to the growing trend towards airport privatisation so airports can seek private sector finance to invest in the development strategy needed to compete in the new millennium. In fact Carter (1996) describes the move from a simple public utility view of airports towards a management-led approach which is more commercially driven. For example, O'Hare airport in the USA invested US$618 million in a new terminal facility in 1993 as well as a customs and baggage hall able to process 4,500 passengers an hour. Atlanta airport undertook a US$200 million refurbishment in the 1990s, resulting in an airport with 158 domestic departure gates. It has a daily average of 2,066 takeoffs and landings and 25,000 parking spaces (yielding US$50 million in revenue). From such an example, one can see the investment needs and importance of a commercial strategy to ensure airports maximise their revenue-generating potential and invest in the most judicious manner. As Skapinker (1996 : I) argues, 'most of the world's airports . . . have barely begun to exploit their full commercial potential'. In many cases, seeking to generate additional revenue from non-aeronautical sources is far more attractive to many airport managers than resorting to increases in landing fees.

Doganis (1992 : 112–13) argues that, in establishing a commercial strategy for future development:

> Airport owners and operators have to make a choice between two alternative strategies. They can follow the *traditional airport model* . . . where airports see their primary task as being to meet the basic and essential needs of passengers, airlines, freight forwarders and other direct airport customers or users . . . The alternative strategic option is that of the *commercial airport model.*

In the commercial model, revenue maximisation from a wide range of customers offers the airport manager a greater flexibility in the finance available to invest in a demand-led approach to airport development. In many ways, the traditional model epitomises the government-managed approach, characterised by Athens airport in the late 1980s (Doganis 1992 : 113). In contrast, Frankfurt Airport exemplifies the commercial model, which requires airport managers to recognise a number of discrete markets amongst airport users:

- passengers (departing, arriving and those transferring between flights)
- the airlines, which are major consumers of space for storage, maintenance, staff and catering
- airport employees
- airline crews
- meeters and greeters
- visitors to airports
- local residents
- the local business community
  Source: Based on Doganis (1992 : 115).

As a result of recognising the business opportunities afforded by these groups, there is potential to derive income from activities and services which these users require. Income can be derived from rents from service providers and concessionaires which offer services on the airport premises as well as from involvement in those activities identified above (e.g. retailing, car parking and aircraft-related services). Developing the most appropriate organisational structure to manage the airport also needs to take into account the relationship with government bodies (i.e. should a commercial approach be developed?),

... the pressures for change, not least of which are those that relate to the personalities and abilities of individuals with directional responsibilities within the organisation ... [and whether] ... the airport authority assumes ... a brokering function
with minimal operational involvement in many on-airport activities (the US model)
to direct operational involvement in many of the airports' functions (the European
model). (Ashford et al 1991 : 13)

For example, if a commercial model is pursued, it will require an organisational
structure which is responsive to change. Ashford et al (1991) review the various
management structures adopted by airports. Doganis (1992) and Ashford and
Moore (1992) point to the preoccupation of airport managers with operational
issues prior to the onset of the commercial model. As a result, airport management was traditionally organised into functional areas (e.g. security, finance,
administration, human resource management, engineering and so forth). Each
area was headed by a specialist manager reporting to the airport manager. The
problem with such a structure was that commercial activities did not assume
a clear focus, with different aspects of business dispersed across each of the
airport's functional areas. The result was that no one assumed responsibility
for pursing a commercial strategy. The commercial model therefore requires
a greater focus on business and revenue-generation as a core activity of the
airport and Doganis (1992) points to the BAA model at Gatwick Airport Ltd.
Here the emphasis is on providing a service for external users and internal users
within the organisation. Each department is headed by a director and the necessary business skills are drawn from within or from outside BAA. In 1989, the
commercial director was responsible for the management of activities that
yielded 59 per cent of the airport's revenue.

The development of commercial opportunities, most notably retailing and
concessions, has been a major result of the more commercially driven model
of development. Yet how can an airport evaluate the success of its management and business strategy? Ashford and Moore (1992) cite the earlier work
of Doganis and Graham (1987) *Airport Management: The Role of Performance
Indicators* which evaluated the performance of 24 leading European airports
based in 11 countries. They examined five key areas in airport management:

- costs
- revenue
- labour productivity
- capital productivity
- financial profitability

The basic principle is to assess whether the resources being used at a given airport are effective, efficiently employed and able to achieve the desired outcome.
Where measures are less satisfactory, it allows airport managers to address the
issue (see Doganis 1992 : 159–87 for more detail on operationalising performance indicators).

Having examined the scope and issues confronting airport management,
attention now turns to a case study of the British Airports Authority, which

will provide more detail on the current operational issues and achievement of this airport operator. BAA also illustrates the progress one leading airport organisation has made through privatisation.

CASE STUDY   Airport privatisation in the UK: the role of BAA

In 1985 the British government produced its White Paper *Airport Policy* (HMSO 1985) which Doganis (1992 : 27) interpreted as stating:

(a) that airports should operate as commercial undertakings
(b) that airports' policy should be directed towards encouraging enterprise and efficiency in the operation of major airports by providing for the introduction of private capital.

The ensuing 1986 Airports Act then turned BAA into a limited company, BAA Plc, which was floated on the Stock Exchange in 1987. This was then organised into a single company operating seven airports: London Heathrow, London Gatwick, London Stansted, Southampton, Glasgow, Edinburgh and Aberdeen. The 1986 Act also led to the UK's largest regional airports (with a turnover exceeding £1 million) becoming public limited companies and being detached from their former local authority owners. In New Zealand, the three largest airports were also privatised in 1988 as individual corporations (see Doganis 1992 : 28–33 for an analysis of the costs, benefits and issues of airport privatisation), while in South Africa a semi-privatised model has been pursued (Prinz and Lombard 1995). Yet it is the BAA model which remains widely cited for its commercial success in the 1980s and 1990s.

BAA's activities in the 1990s
One of the most accessible sources of current data on BAA is its annual report (see Chapter 5 for a review of the value of company reports as data sources). In the BAA *Annual Report 1996/97*, the company mission statement identified its commercial focus:

Our Mission is to make BAA the most successful airport company in the world. This means always focusing on our customers' needs and safety. Achieving continuous improvements in the costs and quality of our processes and services. Enabling our employees to give of their best. (BAA 1997 : 1)

The scale of operations at BAA is reflected in the following key statistics for 1996/97:

- a profit of £444 million (before tax and exceptional items)
- 98 million passengers passed through BAA airports
- the total revenue from BAA airports was £1,373 million, comprising:

  - airport traffic charges of £467 million (34 per cent)
  - retailing revenue of £606 million (44.1 per cent)
  - property services revenue of £252 million (18.4 per cent)
  - other sources of £48 million (3.5 per cent)

- capital expenditure at BAA airports was as follows:

  - London Heathrow £321 million
  - London Gatwick £60 million
  - London Stansted £14 million
  - Scottish airports £21 million
  - Capitalised interest £51 million
  - Other investment £29 million

- BAA airports handle 71 per cent of UK air passenger traffic.

The organisational abilities of BAA are also focused on a range of non-operational activities, including:

- airport management at Melbourne Airport, Australia, and Indianapolis in the USA
- retail management
- property management
- project management

In terms of airport management, BAA discusses various aspects of its activities, including:

- *Airport charges*, including the BAA's review of airport charges in which it ranked 16th (Heathrow) and 18th (Gatwick) among European airports in terms of landing and takeoff charges.
- *Passenger traffic growth*, which recorded a 4.6 per cent increase in 1995/96 traffic volumes, with the strongest growth occurring in the North Atlantic (UK/USA) markets, which grew by 7 per cent. London Stansted represented an initial £400 million investment by BAA in 1991, and is now reported to be operating profitably. Following the lifting of restrictions on passenger volumes in 1996, London Stansted can now expand to 8 million passengers a year.
- *Meeting customer needs*, where the UK Monopolies and Mergers Commission (MMC) provided the means for customers to express their views of BAA's activities. The MMC reported that 'BAA's quality of service is generally rated high by customers'. This is supported by BAA's ongoing Quality of Service Monitor (QSM) which has interviewed over 1 million travellers since its inception in 1990/91. The QSM interviews 150,000 passengers a year in relation to the airport experience and customer needs. The results for the 1996/97 QSM are reported in summary form in Table 7.4.
- *Safety and security*, where BAA reports that it has developed 'the world's most advanced and effective hold baggage screening technology' (BAA 1997 : 13), to be introduced at all its airports at a cost of over £175 million.
- *Maximising employees' contribution to business success*, where employee empowerment measures have been introduced as part of BAA's commitment to quality (see Chapter 9 on Total Quality Management). BAA also introduced a 'Freedom to Manage' initiative, where a greater emphasis on teamwork, simpler management structures and a greater commitment to customer service has been developed through centres of excellence. The improvements are reported in terms of airport employee productivity, where 1996/97 saw a 4.4 per cent increase in the number of passengers per employee handled at each UK airport.

In the field of non-aeronautical services, it is BAA's retail activities that have attracted a great deal of interest from analysts.

Table 7.4 Quality of Service Monitor scores 1996/97[1]

Measures	Heathrow	Gatwick	Stansted	Southampton	Glasgow	Edinburgh	Aberdeen
Cleanliness	3.9	4.0	4.4	4.5	4.3	4.0	4.2
Mechanical assistance[2]	3.9	4.0	4.1	3.9	4.1	3.8	3.9
Procedures[3]	4.0	4.2	4.3	4.5	4.4	4.2	4.3
Comfort	3.9	4.0	4.3	4.5	4.2	4.1	4.1
Congestion	3.5	3.7	4.3	4.7	4.2	4.1	3.9
BAA staff	4.0	4.2	4.3	4.5	4.3	4.1	4.3
Value for money	3.5	3.7	3.6	3.7	3.6	3.7	3.5

[1] Scale: 1 = Extremely Poor; 2 = Poor; 3 = Average; 4 = Good; 5 = Excellent

[2] This includes services such as luggage trolleys, aircraft access and flight information.

[3] This covers check-in procedures, security issues, baggage reclaim facilities and immigration procedures.

Source: Modified from BAA (1997 : 12).

Airport retailing

According to Skapinker (1998a : VI), 'an airport is, in many respects, the ideal environment for retailers. The potential customers are affluent, or reasonably so, and are often on the lookout for presents for family or friends. Above all they are trapped.' This is the retailer's dream: a captive audience with disposable income and time to shop. BAA is a good example of an organisation which carefully assesses customer demand through a product market focus and this is reflected in its contract to manage retailing at Pittsburgh Airport. However, one of the criticisms which Skapinker (1998a : VI) reports is that 'some airline executives are unhappy about the way shops are taking over large areas of airport terminals... [which] ... means there is less space for airline facilities, such as check-in desks... passengers become absorbed in their shopping, do not hear departure announcements, and arrive late at gates, delaying flights'. For airport operators such as BAA, such trends seem set to continue because of the returns on investment. In fact BAA and Amsterdam's Schiphol Airport have been shortlisted by the South African government to compete for a 20 per cent stake in the Airports Company of South Africa to provide new investment and private sector expertise, particularly in retailing.

In 1996/97, BAA generated a total of £425.2 million of net retail income, comprising:

- duty-free and tax-free shops, £248.8 million (59 per cent)
- bookshops, £21.2 million (5 per cent)
- car parking, £64.1 million (15 per cent)
- car rental, £17.8 million (4 per cent)
- catering, £21.2 million (5 per cent)
- bureaux de change, £27.6 million (6 per cent)
- advertising, £11.7 million (3 per cent)
- other, £12.8 million (3 per cent).

As BAA succinctly describe its position, 'a successful airport business without a successful retail operation is unthinkable' (BAA 1997 : 19). The BAA strategy for 1997/98 is to increase revenue by 9 per cent, double the rate of traffic growth. This does not seem unrealistic, given the fact that net retail income rose by 11 per cent in 1996/97. This reflects a focus on growing product markets as opposed to selling manufacturers' products. In 1996, BAA also introduced a bonus points customer loyalty scheme, yielding 170,000 members and a marketing database on regular customers using the airport. In physical terms, the £33 million redevelopment of Heathrow's Terminal 2 building created an additional 2,320 m^2 of retail space. In Pittsburgh, BAA has increased the number of retail outlets in its Airmall from 46 to 102, increasing sales per passenger from US$2.40 to US$7.10. Overall sales have risen from US$20 million to US$72 million.

In Europe, the possible loss in 1999 of duty-free sales for intra-European EU travel could affect BAA's retailing by £120 million, which is why Skapinker (1998 : VI) argues that those airports 'with more advanced retailing skills will attempt to compensate by dreaming up more imaginative ways of parting passengers and their money'. In 1998, BAA was mounting a public awareness campaign at its major airports to highlight the effect of losing intra-European duty-free sales for passengers in terms of increased costs of travel. To maintain the cutting edge of its commercial acumen and management skills, BAA also needs to invest in future projects to improve the airport environment. In 1996/97, BAA invested £496 million in its 10-year strategy which aims to invest a total of £4.4 billion in addition to the previous £3.3 billion in the first nine years of BAA's operation. The driving force is to meet the continued forecast growth in air travel and provide facilities to accommodate such growth. One of the largest investments is in the £440 million Heathrow Express rail link from Paddington in Central London to the airport. The rail link, operational in 1998, runs as a wholly owned subsidiary of BAA. Such investment also raises the broader issues associated with airport planning and development, since planning is a vital element of management.

Airport planning and development

According to O'Connor (1995 : 269), 'airports are among the most important elements of modern cities . . . in many places, the traffic at the airport reflects the vitality of the tourist industry. For these reasons, airports are critical to the vitality of metropolitan areas' (see Figure 7.4), a feature observed in the seminal study by Hoare (1974) of London Heathrow as a growth pole. The significance of airport planning and development has been a feature often overlooked in many analyses of tourist transport. However, a number of UK (e.g. Sealy 1976) and North American texts have examined the theme of airport development indirectly (e.g. Wells 1989; Taneja 1988) and in a more focused manner (e.g. Ashford and Moore 1992; Smith et al 1984; Ashford et al 1991; Horonjeft and McKelvey 1983), although probably the most up-to-date and accessible source is French (1997a). O'Connor (1996) outlines the importance of locational factors in the development of airports in Asia-Pacific, notably:

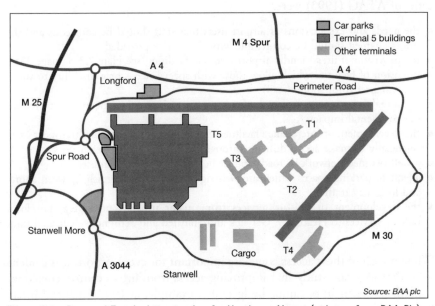

Figure 7.4   Proposed Terminal 5 extension for Heathrow Airport (redrawn from BAA Plc)

- physical and technical considerations (i.e. appropriate sites)
- economic forces, where air traffic now is a vital element in the national economy of most countries
- the role of government in forward planning (e.g. Changi Airport in Singapore developed two parallel runways and has well-organised ground services and facilities to ensure the smooth passage of travellers, backed by high levels of investment).

In physical terms, locational factors are important (e.g. the availability of a large expanse of flat land, access to urban centres which are the source of travellers and an environment able to accommodate the noise and physical pollution from airport operations). As the example of Kansai International Airport in Japan shows (see Chapter 3), airport developers are having to seek new sites away from major population centres. Speak (1997) examines the example of Kai Tak, Hong Kong, which handled 21,370,000 passenger movements and 150,118 aircraft movements in 1995. The development and planning of the new Chek Lap Kok site was originally advocated in 1989, 25 km west of Hong Kong's central business district (CBD) (Figure 7.5). In 1996 an airport development authority was established to develop the project, and the airport opened in 1998. The physical development of the 1248 ha site involved flattening a small island and land reclamation. The airport will be capable of accommodating 35 million passenger movements a year using the first runway; and it could expand to a capacity of 87 million once the second runway is completed. The need for access to Hong Kong's CBD required two bridges and a dedicated airport railway to be built (Figure 7.5) and has necessitated 10 associated infrastructure projects listed in Figure 7.5.

ATAG (1993) also acknowledged the importance of surface transport access to airports (http://www.atag.org/AH/Index.htm). Among the principal concerns of ATAG (1993) were:

- the needs and expectations of airport users that staff should be courteous and the prices charged should be compatible with the service provided
- the provision of user-friendly airport terminal facilities (see Plate 7.3 on page 235).
- provision of facilities which allow those with special needs or a disability to travel (see Chapter 8 for more detail)
- encouraging passengers to use public transport to journey to the airport to reduce environmental impacts
- the development of interchange facilities that do not penalise public transport users financially or make access difficult through poorly designed interchanges
- a well-planned network of roads and dedicated airport access routes
- adequate parking spaces which are well signposted to distinguish between short- and long-stay traffic
- the development of rail–air services from major urban centres (e.g. London's rail–air services to Heathrow, Gatwick and Stansted and Schiphol's dedicated train service).

While not all of these access issues are important for every airport, it is evident from the BAA case study that improving access from urban centres consumes a significant element of capital budgets. For example, in 1997 Auckland International Airport in New Zealand was reported as requiring NZ$1 billion in

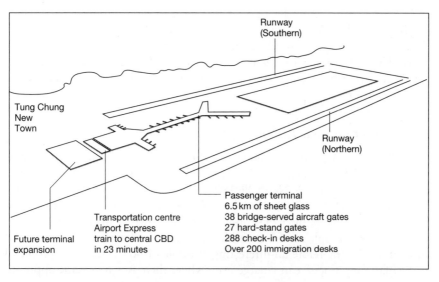

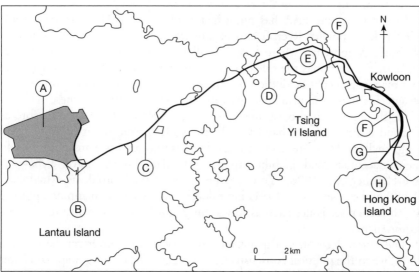

A Chek Lap Kok Airport
   1,248 ha site built on a
   man-made platform from
   bedrock of 2 islands and
   938 ha of land reclaimed
   from the sea; cost $12.1 bn

B Tung Chung New Town;
   cost $1.38 bn

C North Lantau Expressway; cost $307 m

D Tsing Ma Bridge; cost $1.77 bn

E Tsing Yi Island; cost $5.6 bn

F Airport Railway; cost $8 bn

G Western Harbour Crossing Tunnel;
   cost $1.9 bn

H Hong Kong Island land reclamation:
   cost $8.07 bn

   Airport Cost: $40 bn

Figure 7.5   Chek Lap Kok airport. *Source*: Hong Kong Airport Authority data

capital expenditure over the next 15–20 years, a large proportion of which will be devoted to access issues and terminal design.

It is the process of physical planning of airports to which attention now turns. Through a focus on Denver International Airport (see Dempsey et al 1997), it is possible to understand the procedures and processes followed to build an airport.

Denver International Airport

According to Goetz and Szyliowicz (1997), Denver International Airport (DIA) opened five years late in 1995, mainly due to problems with its auto-mated baggage system (de Neufville 1994). The cost overrun rose from US$1.7 billion to over US$5 billion, and though the airport was forecast to handle 56 million passengers by 1995, only 30 million used it in that year. It also received noise complaints from residents, despite seeking to minimise noise impacts. DIA is therefore an interesting example of how short-term problems in the planning process can emerge even where there is a systematic approach to development.

In the USA, the FAA has published guidelines on the approach towards airport development. Rycroft and Szyliowicz (1980) examine the limitations of the FAA approach where access to all forms of data is not possible, and argue that it overlooks the role of power and political variables. It also assumes that planning occurs in a rational manner, whereas there is all too often a lack of agreement on the principal goals of large projects such as these. Further-more, this approach also assumes that it is possible to derive accurate forecast demand estimates (see Chapter 4). In the wider literature on transport plan-ning, such problems remain intractable as traffic forecasts are notorious for their inaccuracy and inability to accommodate human travel behaviours (see Hensher et al 1996). To overcome some of these problems, Maldonado (1990) and de Neufville (1991) have developed a strategic planning approach which recognises issues such as uncertainty and flexibility in future planning requirements.

The initial decision to expand airport capacity in Denver can be dated to 1974, when the political consensus was to expand the existing Stapleton Inter-national Airport. By the time deregulation occurred in the 1980s, expansion was not considered feasible, given the FAA's and Denver's consultants' fore-casts for future traffic growth. Forecasts suggested that the existing traffic of 25 million passengers in 1983 would grow to 56 million by 1995. Runway capacity at Stapleton had already caused a bottleneck in the national air trans-port system during adverse weather conditions due to runway design. Goetz and Szyliowicz (1997) provide an interesting commentary on the politics of the decision to select a site for DIA. They also examine how the airline users exacted major concessions from the developer – the city of Denver. The most serious concession was exacted by United Airlines, which sought and gained an automated baggage system to service its concourse. This concession also contributed to major delays in the airport becoming operational. Delays to the

Plate 7.3   Airport retailing provides concessionaires with a captive audience and offers travellers an activity prior to travel (reproduced courtesy of BAA London Stansted)

airport's opening due to technological problems associated with the baggage system added to cost overruns. The airport traffic forecasts also assumed that Denver would emerge as a major hub. But in March 1994, Continental Airlines abandoned its hubbing operation at Denver, leaving United as the main hubbing operator while the FAA assumed that American Airlines might consider establishing a hub operation by 1995 (see Bania et al 1997 on hubbing operations). The DIA project was downsized by reducing the number of runways from six to five and gates from 120 to 88 – this was made possible by the modular design of the airport.

The airport also aimed to reduce noise impacts, being located 24 miles from Denver's CBD. But according to Goetz and Szyliowicz (1997 : 272–73):

> In all of 1994, Stapleton received 431 noise complaints, but in the first twelve months since DIA opened the city received over 57,000 complaints. Much of this can be attributed to previously non-impacted populations now receiving some amount of aircraft noise. But it remained unclear whether these complaints indicate that the original noise contours developed in the Environmental Impact Statement are still accurate.

(The complex issues associated environmental assessment techniques and noise pollution are discussed in more detail in Chapter 8.)

Goetz and Szyliowicz (1997) view the DIA example as highlighting the airlines' role as something which could not have been foreseen in the rational comprehensive planning model advocated by the FAA. The goals of each airline were not in sympathy with those of the city of Denver and the FAA, and the political system affected both the decision-making and implementation phases of planning. As Szyliowicz and Goetz (1995) argued, a range of actors with different strategies, agendas and purposes ensured that no one unitary actor drove the project forward. As a result, political and cognitive factors were powerful in shaping the airport development process. The ability to accommodate future risk and uncertainty also assumes a much greater role given the 'critical nature of aviation forecasts for airport planning and design and of demand forecasts' (Goetz and Szyliowicz 1997 : 274). This means that the planning process needs to be able to accommodate likely errors in forecasts. Goetz and Szyliowicz (1997) also point to the need for decision makers and planners to adapt to new business situations with flexibility rather than remaining locked into a project that may not be suitable on completion. Vested interests, political careers and an unwillingness to admit that earlier decisions may have been made on inaccurate information still affect megaprojects such as airports. A similar problem can be seen in the CAA's Swanwick air traffic system being developed for the UK. This is a mega-IT project which has so far revealed 15,000 errors in the computer software and is expected to be £100 million over budget and several years late. It was envisaged that it would replace the ageing West Drayton air traffic control centre, although some critics believe it may even be abandoned if it cannot be made operational (see CAA 1997 for detail on air traffic control and Swanwick). Both DIA and the Swanwick example highlight overoptimistic expectations that technology which is not tried and tested can be put into practice immediately. What emerges is a need for flexibility and constant monitoring once the decision has been taken to implement a project and a willingness to halt the project if technical problems cannot be solved.

Global airport development

French (1996a) reviews the environmental constraints which now affect airport expansion, notably community objections to aircraft noise and emissions. These have resulted in a range of state-of-the-art airport development projects which can meet future demand on new sites while mitigating environmental concerns. A significant number of these new projects are located in Asia-Pacific. Nevertheless, French (1996a) argues that planners are not keeping pace with demand. In late 1995, there were only 11 new airports under construction:

- in Europe: Oslo, Athens and Berlin
- in Asia-Pacific: Sydney, Hong Kong, Seoul, Bangkok, Kuala Lumpur and three sites in China (Shanghai, Guangzhou and Shuhai)
- in the Middle East: three new airports.

In Japan, government agencies are currently identifying sites for Tokyo's third airport, likely to cost US$32 billion, with construction starting in the year

2000. Japan also has a US$500 million plan for a domestic airport at Kobe, to be operational by 2003 and a five-year US$40 billion programme for other airports. In India, a feasibility study is considering a new US$3.2 billion airport for Bombay, to open in 2005. A large number of airport enhancement projects are also under way in Asia-Pacific to upgrade terminal buildings and to construct new runways. French (1996a) points to concerns that by 1995 half the airports in Asia-Pacific are capacity-constrained. A number of efficiency gains at Asia-Pacific airports may reduce pressure at some airports (e.g. improvements to surface transport access and customs procedures).

French (1996a) points to the improvements achieved in European air travel by new air traffic control systems. The cost of congestion was estimated at US$6 billion for European airports to the year 2000, based on estimates by ATAG. Among the most pressing problems are airport congestion at peak times and runway capacity constraints. The Association of European Airlines argued that 26 of Europe's 29 largest airports would need additional terminal capacity in 1995, and 25 will need extra runway capacity by the year 2005. These airports are forecast to handle 1.2 billion passenger movements and 16 million aircraft movements. Those airports which the Association of European Airlines believe will be worst affected are the following hubs:

- London Heathrow and Gatwick
- Frankfurt
- Paris (Charles de Gaulle)
- Schiphol Airport, Amsterdam
- Rome
- Madrid
- Zurich

As French (1996a : 16) observes, 'by around 2000 most of the world's major city airports will be faced with critical decisions on how to ensure that their status as leading traffic hubs is maintained, both for their own economic benefit and for the . . . the air transport industry itself'.

Conclusion

There is no doubt that 'airports are complex businesses with functions that extend substantially beyond the airfield or traffic side of operations' (Ashford et al 1991 : 11). Airports provide the vital infrastructure which links:

- the airline
- the traveller
- air and surface transport

and they are complex systems which require clear management guidelines and structures to ensure they operate efficiently.

The late 1980s and 1990s have seen the process of privatisation sweep across many of the world's airport organisations, as the public sector is unable to provide access to large sources of capital to invest in upgrading schemes. The full or partial privatisation of airport companies has also changed the airport as a travel terminal. Under privatisation, a greater focus on non-traffic revenue sources and a broad revenue strategy have transformed the airport from a waiting area and functional environment. The transformation, which has been inspired by business principles to generate revenue from retailing, has generated the airport mall as a cultural icon. The airport is often viewed as a post-modernist expression of the conspicuous consumption now pervading many developed societies, epitomised by leisure travel. In this new environment where marketers have seized the initiative to develop products and services for a wide range of target groups (including the traveller) the airport is now firmly part of the broad travel experience. The example of BAA and its QSM embodies the current concern with customer needs in a market characterised by high-volume business and a captive clientele. The consumer's interests are also represented by the UK Consumers' Association publication, *Holiday Which*. Its Winter 1998 survey, *Which Airport*, undertakes a qualitative review of UK airports and compares value for money, access, areas in need of improvement (e.g. catering) and the best departure lounges as well as the range of shops available. Such surveys illustrate how important the airport is now becoming in the wider travel experience.

The case study of BAA also illustrated the diverse range of activities associated with airport management, the sheer scale and volume of their operations, and the fact that they now have a customer-driven focus to improve the traveller's experience. Even so, large organisations such as airports cannot necessarily respond to issues as quickly as the customer would like. For example, on a flight from Scotland to London in March 1998, the author experienced an incident which highlights this issue: the ground staff held the departure up to allow an inebriated passenger on board, who was served additional alcohol during the flight and who then created difficulties during and after the flight. At the destination, ground crew were faced with this situation and it took an inordinate time for the airport security staff to be called. While this reveals total ineptitude by the airline staff for allowing a passenger on board who was unfit for travel, it also illustrates the unexpected events with which airport staff have to deal and the time delay in putting the appropriate system and action in place. It was also the airport staff who had to deal with the outcome of the airline staff's inexperience and lack of understanding of the situation. Thus, airport management embraces a wide range of operational and human interactions which affect the way travellers perceive the airport environment.

In terms of development strategies for airport growth, the example of DIA illustrates the complex interaction of factors that can affect the final outcome. Airport development involves extremely large funding, increasingly from the private sector, to meet the complex needs of airlines and travellers as well as the airport operator. To describe such entities as mega-projects is a good description, given their scale and vastness. Managing such entities once they are fully operational and meeting a broad range of customers' needs in the fast-changing world of domestic and international travel is a key challenge.

Questions

1. Why are transport terminal facilities an integral part of the tourist travel experience?
2. What are the primary functions of an airport?
3. How has the introduction of airport privatisation affected the management and operation of airports globally?
4. Is the airport of the future likely to look considerably different in terms of design, internal organisation, layout and function?

Further reading

The literature of airports and tourism is very limited. Aside from the standard texts on the subject, the following special issues of journals offer a number of insights: the special issue of *Built Environment* in 1996 edited by K. Button and the special issue of *Transportation Research E: Logistics and Transportation Review* 33(4), which examines recent research on airport performance and pricing. The *Journal of Air Transport Management* also remains an important source of research articles on the activities associated with airports. In addition, Anon (1998) 'Asian airport development', *Travel and Tourism Analyst*, 2 : 1–22 offers an excellent overview of the issue in the context of Asia-Pacific.

Internet site

Appendix 2 lists a good range of world airport websites. ATAG also lists a range of airport websites in North America which can be accessed at:

http://www.atag.org

The site for the British Airports Authority (BAA) is at:

http://www.baa.co.uk

Schipol, Amsterdam, is at:

http://www.schipol.nl

Chapter 8

The human and environmental impact of tourist transport: towards sustainability?

Introduction

The international expansion of tourism and the development of transport systems to meet this demand have had a range of direct and indirect social, cultural, economic and physical impacts on both host populations affected by the operation of tourist transport, and destination areas. In the 1970s and 1980s this led to a growing concern about the impact of tourist travel on the environment, but little attention has been given to the experiences for tourists whilst in transit. This chapter examines the effect and impact of tourist transport from two perspectives: the effect of travelling on the tourist's experience and the impact of transport systems on the environment. In Chapters 1 and 2, the concept of the service encounter was discussed (by using Bitner et al 1990), which recognises that dissatisfaction with tourism services is associated with three types of incident: employee failure to respond adequately to customer needs; unprompted and unsolicited employee actions and service delivery failure (Ryan 1991 : 42–43). The first part of this chapter focuses on service delivery failure, particularly how travel delays and service interruptions may contribute to the stress of tourist travel, and some of the measures taken by transport operators to address such problems. This is followed by a discussion of environmental issues from the transport provider's perspective, including the role of Environmental Auditing and Environmental Assessment to address the long-term implications of new tourist transport projects for the environment. The chapter concludes by considering the extent to which sustainable tourist travel may assist in identifying ways of reducing transport's impact on the environment. A landmark study in this area is Banister (1998).

The human consequences of modern tourist travel

Previous chapters have shown that the tourist travel experience is a complex phenomenon to understand. Social psychologists (Pearce 1982) and marketers are continually trying to understand the relationship between consumer behaviour (Qaiters and Bergiel 1989; Schiffman and Kauk 1991) and tourist travel (Goodall 1991), tourists' degree of satisfaction with travel services and their propensity to revisit destinations in the future. One area which has really lacked serious academic research is the tourists' feelings and the trauma sometimes associated with international travel; a more detailed discussion of tourist health can be found in Clift and Page (1996). In terms of foreign travel, stress is a feature often overlooked since tour operators and travel agents often extol the virtues of taking a holiday to fulfil a deep psychological need (see Chapter 4). Ryan (1991) notes that tourist travel experiences offer many potential avenues of research.

The stress associated with international, and to a lesser degree, domestic travel is the result of various psychological factors which are compounded by the effect of congestion on transport systems. For example, in March 1998, a British Airways jet in Florida experienced technical problems and after two days and two attempts to repair the aircraft, a replacement was sent to Florida to ferry passengers back to the UK. BA offered full refunds as compensation but the stress and anxiety such events can cause for certain passengers on holiday cannot be underestimated (see Barlay 1995 for a range of interesting insights into the delights of air travel). McIntosh (1990a) argues that the stress of travel could be attributed to:

- preflight anxieties
- airside problems
- transmeridian disturbance
- fears and phobias
- psychological concerns

while in-flight health problems can also be added to the stress involved in modern-day long-haul travel (Harding 1994).

Preflight anxieties emerge when tourists commence their journey by travelling to the place of departure, often to meet schedules imposed by airlines. McIntosh (1990a : 118) suggests that the marketing of travel insurance to cover eventualities such as missed departures can also heighten the inexperienced traveller's sense of anxiety. Once at the departure point, the preflight check-in and the complex array of security checks associated with luggage can subject the traveller to a significant amount of stress in an unfamiliar environment. In addition to this is 'the apprehension . . . initially generated by preflight security . . . searches . . . [which are] . . . a reminder of the risk of hijack and in-flight explosion' (McIntosh 1990a : 118). Overcrowding in terminal buildings associated with the throughput of passengers at peak times can overwhelm and disorientate travellers, whereas seasoned travellers (e.g. business travellers) often have access to executive lounges and a more relaxed and welcoming environment free from some of these stressors. In fact Air New Zealand's Koru Club membership scheme is an excellent example of one such marketing tool to help relieve preflight stress.

Airside problems, including the design and layout of holding areas for passengers travelling economy class, may contribute to an impersonal and dehumanising process prior to departure, which is exacerbated by an absence of information about the nature and duration of delays. As Ryan (1991 : 43) argues '. . . passengers delayed in air terminals might be observed as passing through a process of arousal to anxiety, to worry, to apathy, as they become initially frustrated by delays [and] eventually reach apathy because of an inability to control events'.

Transmeridian disturbance associated with time zone changes during long-haul travel is a major problem for some travellers (Petrie and Dawson 1994). The condition is often associated with a lack of sleep on long-haul flights and a sleep–wake cycle which can cause exhaustion, commonly referred to as 'jet lag'. Travel agents may need to be sensitive in their advice to some clients as to the effect of transmeridian disturbance on those suffering from depression. Taking a long-haul holiday to forget their problems may heighten their sense of depression on east–west travel across the world's main time zones. One solution which Barlay (1995) identifies is the use of mild sleeping pills or use of melatonin. Melatonin is a 'naturally occurring neuro-hormone secreted by the pineal gland, a small pea-like organ at the back of the brain. Its rate of secretion is increased by darkness, causing the individual to feel sleepy' (Barlay 1995 : 166). However, there is a medical debate on the possible side-effects of such a drug, and in New Zealand it is now only available on prescription.

Fears and phobias associated with the likelihood of political insurrections, how hospitable the host population will be and potential language difficulties in the destination region all contribute to the traveller's apprehension in transit. This stress can be alleviated by in-flight entertainment and public relations campaigns by national tourism organisations to reduce travellers' fears. The threat of terrorism or hijack is also an underlying worry for some travellers. Travellers' anxiety appears to follow a cyclical pattern, being heightened after an incident and then subsiding in response to the ensuing public relations exercises by airlines to reassure passengers of the increased security measures which are in place. Yet in extreme cases, terrorism may pose a major threat to travel. For example, terrorist activity and threats in Europe actually deterred North American visitors from travelling to popular destinations such as London in the 1980s.

Psychological concerns, such as loneliness and a sense of isolation can also contribute to the traveller's feelings of anonymity during their journey, particularly if travelling alone (McIntosh 1995). The experience is often heightened on a busy jumbo jet carrying approximately 450 passengers, where an individual feels a sense of anonymity and of being confined in a strange environment 10,000 m up in the sky. Safety issues also induce a sense of unease amongst travellers following an incident such as an air crash. Although air crashes are rare occurrences (Steward 1986) in terms of the volume of passengers carried and the number of takeoffs and landings undertaken, they do assume a prominent role in the psychology of tourist travel. According to Barlay (1995 : 7) 'the worst pterophobia sufferers confess their debilitating fear' but as many as 80 per cent of regular fliers have apprehension when boarding an aircraft. Although few people experience outright panic, mild pterophobia can affect many passengers. On a boat, travellers reason that they could potentially swim to safety in the event of an accident. In a car or train, they feel a greater degree of control. However, travellers can choose which airline they fly with, based on their safety record, which may help overcome some of the worst effects of pterophobia. While claustrophobia in the aircraft cabin may also exacerbate the anxiety passengers feel, airlines such as SIA have recognised that gadgets such as Krisworld and high-quality catering (Frapin-Beange et al 1995) can assist in passing the time and breaking the monotony on long-haul flights. At the destination, tourists may need reassurance when using local transport systems where operators give the impression of being blasé or unconcerned about safety issues and passenger welfare.

In-flight health-related problems may also affect passengers on long-haul flights where immobility, reduced air pressure within the flight cabin and dehydration may occur due to the recirculation of dry air within the aircraft. Barlay (1995) shows that most airlines use a mix of recycled and fresh air, leading to a build-up of CO_2 that can exacerbate jet lag, nausea and the onset of migraine. There is also growing concern over the effect of smoking on long-haul flights as recirculated air may lead to passive smoking risks when air filters are not adequately maintained.

Barlay (1995 : 164–68) also lists a number of measures for passengers to consider to ensure a comfortable flight:

- *air pressure* – mild flu can cause extreme pressure in the ears which may be relieved by pinching the nose and swallowing, sucking a sweet or in extreme situations, some airlines carry a decongestant which may be inhaled.
- *shoes* – comfortable footwear and regular exercise on the aircraft are essential to help prevent swollen ankles.
- *clothing* – layers of loose, roomy clothing are ideal for flying, and can be adjusted depending on the changing cabin temperatures.
- *skin dehydration* – frequent application of moisturising creams is highly recommended.
- *liquids and alcohol* – dehydration is a major problem on long-haul flights as eyes can become dry and sore. Consumption of alcohol exacerbates dehydration, compounded by high-altitude flights. Mixing drinks (tea, coffee, non-carbonated water and fruit juice) are recommended by the British Airline Pilots Association to keep the body topped up with fluids (see Leggat and Nowak 1997 for more detail).
- *food* – eating with moderation, especially in business class and first class, is strongly recommended to avoid indigestion and feelings of being bloated.
- *exercise* – gentle body movements can assist in avoiding the effects of tiredness and aching limbs on long-haul flights, together with walking up and down the aisles.

There are also a range of other more persistent physical problems that affect the tourist's experience in transit, most notably *motion sickness*.

McIntosh (1990b : 80) provides a useful overview of motion sickness as a 'debilitating but relatively short-lived illness which indiscriminately affects air, land and sea travellers'. Yardley (1992) examines the literature associated with the concept of motion sickness, casting doubt over previous explanations of its causes and the tourist's susceptibility. It is clear from the existing literature on travellers' health that this affliction is not fully understood (Oosterveld 1995). Some researchers believe it is associated with the way in which different modes of transport stimulate an alteration in the perceived stability of the travel environment (i.e. motion changes such as swaying from side to side or violent changes in altitude due to turbulence in air travel) and this affects one's sensory system. This tends to overstimulate the sensory perceptors. It may affect the traveller's perception of the environment and causes various symptoms such as drowsiness, vomiting, increased pulse rate, yawning, cold sweats, nausea and impaired digestion. Although some drug therapy may attempt to block the effects of motion sickness, no comprehensive cure exists and McIntosh (1990b : 82–83) reviews measures to assist the traveller in overcoming sea, car and air-sickness. However, Barlay (1995) argues that motion sickness is now a rare event in modern passenger transport, though Barlay has obviously not experienced some of the challenging weather conditions in which airlines fly.

The experience of travel stressors and health-related problems may be severe among certain groups such as the elderly. McIntosh (1989) reviews the range of problems which the elderly may experience on tourist transport systems such as immobility and confusion when a number of time zones are crossed during a journey. In view of the increasingly aged structure of tourism markets in developed countries (see Viant 1993; Smith and Jenner 1997), the welfare of elderly travellers and their service experience in transit is assuming a greater significance (McIntosh 1998) among the more innovative transport carriers. According to Viant (1993), the 'senior travel market' (those over 55 years of age) in Europe accounts for 20 per cent of domestic and international tourist trips and is forecast to increase from 142.1 million trips in 1990 to 255.2 million by the year 2000, an increase of 79.6 per cent. As growing numbers of senior travellers experience excellent health, it is evident that this niche market will present many opportunities for the tourist transport system.

There is a growing concern for disabled travellers in the transport literature (Oxley and Richards 1995). A recent study by Abeyratne (1995) considers a range of international and national regulatory measures for the carriage of elderly and disabled persons by air. Abeyratne (1995) examines the scope of the issues for airlines and airports in terms of:

- contacts with airline reservations and ticket sales agents who can advise travellers.
- specific fares, charges and related travel conditions, since some airlines require some elderly and disabled passengers to be escorted
- accessibility of aircraft, via wheelchair or airbridges for incapacitated travellers
- movement, facilities and services on board aircraft to ensure that the passengers' carriage can be undertaken in a way that provides a safe and comfortable environment.

In the context of the USA, Abeyratne (1995) produced a set of detailed guidelines (see Appendix 1) to facilitate further the 'passage of elderly and disabled persons'. These guidelines highlight a range of policy issues for airports and airlines and the practical measures needed to facilitate the further growth in travel by disabled and elderly tourists.

With these guidelines in mind, it is pertinent to consider the measures transport operators can take to reduce the stressful experience associated with different aspects of tourist travel:

- Provision of special assistance at airports for senior travellers and disabled tourists, building on Thomas Cook's innovative Airport Travel Services for group travel to reduce the stress for group organisers taking large parties of tourists abroad

- Development of 'fear of travelling' programmes for different modes of transport, especially air travel (e.g. Thomson Holidays in the UK offers such a scheme through its own airline Britannia and British Airways also offer a one-day fear-of-flying programme – see Barlay (1995) for a humorous review).
- Planners and designers can improve the structure and appearance of terminal buildings so that they are built with the customer in mind, reducing the stress of being in an unfamiliar environment. The award-winning design of the Stansted Airport terminal building (London) is one example of how to incorporate these principles into new terminal buildings. This contrasts markedly with smaller '1930s-style' regional airports, e.g. Jersey in the Channel Islands (Perks 1993), which has inadequate space to accommodate departing tourists in its check-in area, an absence of air-conditioning in the departure area, and too few seats at the departure gates, requiring passengers to sit on the floor in the height of the summer season. This was one result of bunching charter flights on a Saturday in the 1993 season rather than distributing them across the week. The overcrowding and unnecessary queuing induced through poor planning and inadequate staffing levels undoubtedly contribute to the stress of tourist travel. It is apparent that many of the first-generation '1930s airports' are now too small and not designed with the 1990s tourist-traveller in mind
- Provision of accurate and up-to-date information when travel delays occur
- Airline staff should inform travellers prior to takeoff about the aircraft sounds they will hear (e.g. as wheels are retracted and the change in engine sound at the cruising altitude) to allay any fears
- Provision of accurate in-flight advice for travellers, such as KLM Royal Dutch Airlines' *Comfort in Flight* brochure (see Leggat 1997 for more detail on the 15 airlines surveyed)
- Replacement of 'anxiety-provoking intensive security screening' (McIntosh 1990a : 120) with low-profile security checks at ports of departure to reduce the potential for passenger stress
- In extreme cases, general practitioners may prescribe mild medication (e.g. diazepam) to relax the traveller in-flight, but this is often a last resort

To date, research on tourists' experience of travel has focused on travellers' health, health precautions prior to departure and problems encountered at the destination (see the journal *Travel Medicine International* and the *Journal of Travel Medicine*). This discussion, however, has shown that throughout the transport system, greater attention needs to be paid to the tourist's experience in transit, due to the range of problems that travel may engender. As Gunn (1988 : 163) suggests

> tourists seek several personal travel factors and will opt for the best combination . . . [of] comfort (freedom from fatigue, discomfort, poor reliability), convenience (absence of delays, cumbersome systems, roundabout routines), safety (freedom from risk, either from the equipment or other people), dependability (reliable schedules and conditions of travel), price (reasonable, competitive) and speed'.

Thus tourist 'transportation is more than movement – it is an experience' (Gunn 1988 : 167). The operator needs to recognise this so that the total travel experience is as free from inconvenience and stress as possible.

Tourist transport not only has an impact on the traveller, but also affects the environment. For this reason, the second part of the chapter considers some of the issues associated with the environmental effects of tourist transport.

The environmental impact of tourist transport ─────────────

During the 1980s, there was increasing concern with environmental issues and the impact of different forms of economic development, particularly tourism. This international growth in *environmentalism* has meant that there is a greater emphasis on the protection, conservation and management of the environment as a natural and finite resource. Within the tourism and transport business, this concern has emerged in the form of the concept of *sustainable* tourism which highlights the vulnerability of the environment to human impacts from tourism and the need to consider its long-term maintenance (see Hall and Lew 1998). Much of the work on sustainability can be dated to the influential 1987 World Commission on Environment and Development report *Our Common Future* (Brundtland 1987) which asserts that 'we have not inherited the earth from our parents but borrowed it from our children'. In other words, sustainable development is based on the principle of 'meeting the needs of the present without compromising the ability of future generations to meet their own needs' (Brundtland 1987). This requires some understanding of the natural environment's ability to sustain certain types of economic activities such as transport and tourism. However, research on transport and tourism has often been considered in isolation, as the following discussion will show, although the use of research techniques such as Environmental Auditing and Environmental Assessment may help to bridge this gap and, to recognise the specific impacts of tourist transport systems.

Transport and the environment

Tourist transport is one component of a much wider concern for more sustainable forms of development as problems relating to the impact of transport on the environment are symptomatic of the need for more environmentally sensitive forms of development (Banister and Button 1992). Within the context of transport planning, there has also been a greater understanding of the complex and sometimes detrimental impact of certain forms of transport on the environment (Carpenter 1994 offers a detailed insight in relation to railways). The emerging environmental research on the impact of different modes of transport (TEST 1991) has focused on the implications for transport and policy making (Department of the Environment 1991; HMSO 1997b) in relation to controversial new tourist and non-tourist infrastructure projects. This interest in the impact of infrastructure projects has led to measures for environmental mitigation. The emphasis on the environment has also led to detailed research on specific components of environmental problems induced by transport such as:

- health and safety
- air pollution
- noise pollution
- ecological impacts
- the environmental effects of different modes of transport.

One consequence of such research is that policy makers have focused on the direct costs and problems associated with the development of new transport infrastructure, which is now subject to more rigorous environmental safeguards to minimise the detrimental impacts. This concern for the environmental dimension has also been mirrored in tourism research.

Tourism and the environment

As discussed in Chapter 2, the increasing sophistication among tourists has been reflected in the development of a 'new tourism' (Poon 1989), accompanied by a greater emphasis on the consumer requirements of tourists in terms of their search for more authentic holiday experiences and individualised tourism services. One consequence of this 'new tourism' phenomenon is a greater concern for the natural and built environment in which tourism activities are undertaken and their impact in different localities. This greater awareness of environmental issues related to tourism is reflected in the rapid expansion and diversity of research on 'sustainable tourism' (Smith and Eadington 1992), which emphasises the need for a more holistic assessment of how tourist-related activities (e.g. tourist transport) affect the environment.

Recent reviews of research on the environmental dimension in tourism have identified the scope and nature of this growing body of knowledge as well as the existing weaknesses in the structure and form of such studies (Pearce 1985). The recognition of the symbiotic relationship between conservation and tourism (Romeril 1985) has led to the need for a greater integration of interdisciplinary and multidisciplinary approaches to research on tourism and the environment in order to achieve sustainable tourism development, of which transport is an integral component. Central to sustainable tourism development is the need to overcome tourism's tendency on occasions to destroy the very resources on which it depends. This is the focus of the *Tourism and the Environment* report (Department of Employment/English Tourist Board 1991) aimed at encouraging the UK tourism industry to recognise that the environment is its life-blood and that it needs to consider the long-term consequences of tourism activity and development. Although Romeril (1989) argues that appropriate strategies and methodologies are required to understand the complex inter-relationships between tourism and the environment, no universal environmental methodology appears to have been adopted by researchers in their assessment of tourism and the role of transport in affecting the environment.

It is evident from the discussion so far that tourism and tourist transport systems are consumers of the environment (Goodall 1992) since the provision of tourist infrastructure has a direct impact on the environment, particularly in destination areas. Selman (1992) and Newson (1992) discuss the concept of Environmental Auditing as one way of examining the extent to which tourist transport systems and their activities are environmentally acceptable. Does tourist transport cause unnecessary pollution? Can measures be taken to mitigate and reduce the harmful effects on the environment without compromising the commercial objectives of the tourist transport operator? According to Goodall

(1992 : 62) one needs to distinguish between the existing and the future impacts of tourist transport. Two types of research methodology can be used here:

- Environmental Auditing of existing transport systems and their performance and effect on the natural and built environment
- Environmental Assessment to consider the impact of proposed developments in the tourist transport system.

Each methodology has been developed as a multidisciplinary technique requiring an input from disciplines such as economics, atmospheric science, environmental science, geography, management studies and planning. Within these techniques, a systems approach is often used as a method of examining how different tourist transport inputs affect the environment, and how to mitigate the effects of outputs which contribute to environmental degradation (Wathern 1990).

Environmental Auditing and tourist transport

Within tourism and transport studies the two most notable studies published on Environmental Auditing are Goodall (1992), Goodall and Stabler (1997) and Sommerville (1992), which illustrate how this research technique is used as a response to the growing interest in sustainable development (Banister and Button 1992). Environmental Auditing is a voluntary exercise which tourist transport operators and tour operators, who contract transport services for clients, may undertake to assess how their activities affect the environment and how they can reduce this impact by making modifications to existing business practices. Newson (1992 : 100) notes that 'the term auditing, borrowed from finance, implies a thoroughness and openness which is essential in a meaningful desire to reform commercial practices' but few Environmental Audits have been publicised. Some examples from the field of consumer products (e.g. the Body Shop and Proctor and Gamble in the UK) have followed the lead of North America in terms of consumer demand for more 'green products' (Selman 1992). Critics argue that such companies have harnessed new-found environmental awareness to gain competitive advantage and increase market share by offering environmentally friendly services and products as part of the move towards total quality management within their organisation.

Research by Forsyth (1997) raises a wider debate on tourism and environmental regulation, since auditing is a voluntary process. Forsyth surveyed 69 UK-based companies involved in tourism (e.g. tour operators, travel agents, hotels, passenger carriers, tourism associations, national tourist offices and consultancies advising companies sending tourists overseas). Forsyth argues that self-regulation is viewed as preferable in the tourism–environment debate because environmentally responsible practices can be harnessed to increase competitive advantage. Forsyth's survey was sponsored by the World Wide Fund for Nature (UK) and Tourism Concern and prior to any interviews, respondents were sent a copy of *Beyond the Green Horizon: A Discussion Paper on Principles for Sustainable Tourism* (Tourism Concern and World Wildlife Fund 1992). Forsyth's (1997) results identify four main types of practice:

- cost-cutting measures (e.g. paper recycling)
- adding value to the product (e.g. information on destinations as sympathy booklets)
- long-term investment (e.g. staff training)
- legislation (e.g. tourist taxes – see Abeyratne 1993 for a fuller discussion of air transport taxes).

Businesses, however, saw themselves as powerless to effect change, given the threat of competitors who did not adopt similar measures. The main obstacles businesses perceived to developing practices compatible with sustainable tourism principles were:

- it is the responsibility of government to initiate sustainable tourism
- it may leave businesses at a disadvantage if competitors do not embrace similar practices
- operators are powerless to produce change
- a potential lack of interest among customers in sustainable tourism

The potential for businesses was in 'labelling green or sustainable tourism as quality tourism, and by acknowledging that populist market demand may lead to stereotypical approaches to minorities or ecotourism not helpful to equitable development' (Forsyth 1997 : 270).

For those businesses which view environmental self-regulation as offering benefits to their image and product, there is evidence that if committed transport operators undertake an Environmental Audit, it may prompt other companies to follow suit, thereby improving the awareness of environmental issues within their sector of the tourism business. The establishment of the British Standards Institution's (BSI) new Environmental Management System, mirroring the BS 5750 quality system for service providers is evidence of the significance of Environmental Auditing as a potent force in the 1990s which will encourage companies to establish a benchmark of acceptable standards of environmental management in commercial activities. Tourist transport systems are no exception to this environmental awareness and it is likely to increase in the 1990s and in the new millenium. For example, P&O European Ferries have undertaken a comprehensive environmental review and are implementing environmental policies to reduce the company's impact on the atmosphere, marine environment and on shore.

Goodall (1992) identifies the role of Environmental Auditing in corporate policy making among tourism enterprises (e.g. transport providers) which includes:

- *a consideration stage,* where the legislation and scope of environmental issues are considered
- *a formulation stage,* where an environmental policy is developed
- *an implementation stage* where both existing and proposed activities can be considered
- *a decision stage* where transport operations are either modified or left unchanged in pursuit of a corporate environmental policy.

More specifically, Goodall (1992 : 64) recognises that policy statements and action to minimise environmental impacts need to consider:

Table 8.1   Types of Environmental Audit

Audit	Description
Activity	An overview of an activity or process which crosses business boundaries in a company, e.g. staff travel by employees of a hotel chain.
Associate	Auditing of firms which act as agents, subcontractors or suppliers of inputs, e.g. tour operators using only hotels which have adequate waste water and sewage treatment or disposal facilities and which are in keeping with the character of a destination.
Compliance	Relatively simple, regular checks to ensure the firm complies with any current environmental regulations affecting its operations, e.g. airline checking on noise levels of its aircraft at takeoff.
Corporate	Typically an audit of an entire company, especially a transnational one, to ensure that agreed environmental policy is understood and followed throughout the firm.
Issues	Concentration upon a key issue, e.g. ozone depletion, and evaluation of company operations in relation to that issue, e.g. hotel chain checks aerosols used are CFC-free, uses only alternatives to CFC-blown plastic foams for insulation and retrieves any CFCs used in air-conditioning plant for controlled disposal.
Product	Ensuring that existing products and proposed product developments meet the firm's environmental policy criteria, e.g. tour operator designs holiday based on walking once destination reached, using locally owned vernacular accommodation and services.
Site	Audit directed at spot checks of buildings, plant and processes known to have actual or potential problems, e.g. hotel checking energy efficiency of its heating and lighting systems, airport authority checking aircraft noise levels near to landing and takeoff flight paths.

Source: Goodall (1992 : 68).
© John Wiley and Sons.

- the extent to which transport operations and associated activities comply with environmental legislation through company regulations
- ways of reducing negative environmental impacts such as polluting emissions and use of energy-efficient modes of transport and equipment based on state-of-the-art technology
- the development of environmentally friendly products
- how to encourage a greater understanding of environmental issues among staff, customers and people affected by tourist transport.

Translating these principles into commercial practice is a complex process even though organisations such as the World Travel and Tourism Council recommend that such audits should be undertaken annually to foster more responsible forms of development (Goodall 1992). As Table 8.1 shows, the nature and scope of Environmental Auditing in the tourist transport system may be determined by the objectives, commitment of senior management, and the size of their organisation to resource such an exercise. One tourist transport operator which has developed a corporate audit is British Airways.

CASE STUDY   British Airways and environmental management

Consumer interest in environmental issues in the late 1980s prompted tourist transport operators in the UK (e.g. P&O European Ferries and British Airways) to undertake Environmental Audits to provide a public image of 'environmentally conscious' companies. Purchasers of tourist transport services (e.g. tour operators such as Thomson Holidays) also undertook Environmental Audits to respond to this trend as previously mentioned with reference to Forsyth (1997). This case study focuses on the extent of environmental management by one of the world's largest tourist transport operators – British Airways. The significance of British Airways (hereafter BA) as a tourist transport company has been documented elsewhere (see Laws 1991) and need not be reiterated here. Commercial aviation dominates the world's communications infrastructure since over '8,800 subsonic jet aircraft . . . flew over 1.7 billion passenger kilometres in 1990' (Sommerville 1992 : 161). BA's role in the world airline industry is reflected in its £8.37 billion turnover in 1996/97, having carried 38 million passengers on its fleet of 258 aircraft (British Airways 1997). Air travel is a useful example to focus on as airline operation has a variety of impacts on the environment. These can be dealt with under the following headings.

Noise
Early tourist travel on turbo-prop and jet-propelled aircraft generated a significant noise impact during takeoff, in flight and on landing (see Farrington 1992 for a discussion of the technical issues). Modern aircraft technology has reduced the level of noise impact in response to international conventions and legal requirements at specific airports which aim to reduce noise impacts for local communities. For example, Air New Zealand's ageing fleet of Boeing 737s had to be hush-kitted to operate at certain New Zealand airports. The sheer volume of air travel creates a persistent problem for those affected by airlines' flight paths. As Sommerville (1992) notes, since the 1970s, the number of people affected by noise nuisance at Heathrow within a 35 Noise Index Number contour has dropped by almost 75 per cent, but this was accompanied by a dramatic increase in the volume of air travel. Increasingly airports are monitoring individual aircraft (e.g. their noise footprint) to ensure they meet noise regulations (see M. Smith 1989 for further information). The phasing out of older aircraft is one way of reducing the noise impact in line with recent guidelines issued by countries abiding by ICAO recommendations. From April 1995, aircraft over 25 years old are to be replaced in countries observing ICAO guidelines, while additional regulations will apply in European airspace.

BA's (1997) Annual Environmental Report lists its fleet composition and it plans to replace its older Chapter Two aircraft in advance of the ICAO guidelines and introduction of Chapter Three aircraft by 1 April 2002 in Europe and 31 December 1999 in the USA. As Table 8.2 shows, BA's aircraft are progressively being updated to meet the Chapter Three requirements for European and transatlantic operation. As at 31 March 1996, 18.6 per cent of BA's fleet were Chapter Two aircraft. In terms of the impact of noise on residents at Heathrow, the number affected by the 57 Leq contour (noise threshold) had dropped by 6.5 per cent in 1994. In the UK, London

Table 8.2   British Airways fleet composition as at 31 March 1997

Aircraft type	Engine type	ICAO Annex 16 Chapter/FAR Part 36 stage	Number of aircraft	Average age of aircraft (years)
Concorde	Olympus 593	Exempt	7	20.3
Boeing 747-100	JT9D 7/7A	2	15	24.9
Boeing 747-200	RB211-524D4X	3	16	16.4
Boeing 747-400	RB211-524G/H2	3	22	5.1
Boeing 777-200	GE90-76B	3	5	1.4
Boeing 777-200/GW 777-200	GE90-85B	3	4	0.2
McDonnell Douglas DC 10-30	CF6-50C2	3	8	17.8
Boeing 767-300	RB211-524H	3	16	5.4
Boeing 757-200	RB211-535C/E4	3	37	10.2
Airbus A320 100/200	CFM56-5A1	3	10	8.1
Boeing 737-200	JT8D-15A	2	41	14.9
Boeing 737-400	CFM56-3C1	3	13	5.1
BAe ATP	PW126[1]	5/Stage 3	13	7.4
DHC7-100	PT6A-50	5/Stage 3	2	13.0
DHC8-300	PW123	5/Stage 3	8	5.2
Total			230	10.0

[1] Brymon Airways aircraft.
Source: British Airways (1997 : 18).
Reproduced courtesy of British Airways.

Heathrow and London Gatwick and Manchester International Airport use aircraft noise monitoring systems; more detailed measurements and data can be found in BA's *Report of Additional Environmental Data* (BA Report No. 6/97) which accompanies its 1997 environmental report (BA 1997). BA also lists its day and night noise infringements at a range of airports, together with the noise cost of BA's Chapter Two aircraft at airports around the world. BA (1997) also discusses how to reduce the noise impacts and records its progress towards targets it established to improve its environmental impact. Table 8.3 highlights the noise targets.

Emissions, fuel efficiency and energy

The growing concern over global warming and 'greenhouse gases' (e.g. CFCs, CO_2, NO_x and methane) has meant that atmospheric pollution from aircraft has come under increasing scrutiny. BA (1992) identified the range of emissions from aviation (Table 8.4). At a global level, BA (1997) estimated that civil aviation accounts for 400–500 million tonnes of carbon dioxide from 20,000 million tonnes of fossil fuels. This recognises that almost 50 per cent of global warming may be a consequence of

Table 8.3   British Airways noise objectives 1996/97

Last year's objectives/targets	Status
To work with airports and other bodies to reduce the operational noise impact of aircraft.	Representation on Noise and Track Keeping Groups and representing the industry on appropriate consultative groups. No scheduling of aircraft in the half-hour prior to night quota period. Simulator trials to test steeper angles of approach. (Ongoing)
To have no Chapter 2 aircraft at Heathrow by the summer of 2000 with an overall phase out ahead of the compliance deadline.	Increase of 0.9% in British Airways aircraft to Chapter 3. Over 92% of the Heathrow fleet (at 31/3/97) were Chapter 3. (Restated)
At Heathrow and Gatwick, the number of flights and their noise impact to stay within the current movement and noise quota limits.	Night movement quotas not exceeded. British Airways has a voluntary restriction of only using QC2 aircraft in the night quota period. (Ongoing)
Aircraft noise infringements at Heathrow and Gatwick not to exceed the base year of 1995 (assuming no change in the regulatory standard).	Reduction of 50% in noise infringements at Heathrow. Noise infringements at Gatwick increased threefold as a result of additional operations. Overall total increased 2.4%. (Remains target but regulatory regime likely to change)
To use fixed electrical ground power supply for aircraft on all stands where it is available, subject to requirements for start-up and shutdown of aircraft systems and for short turnarounds.	Operation of GPUs in the night period monitored by HAL. British Airways charged £4,500 for operating GPUs at night on stands with serviceable FEGP (Jan–Mar 97). (Restated)

Table 8.3   Cont'd

Last year's objectives/targets	Status
To have modified or replaced all ground power units at Heathrow and Gatwick to meet the improved noise standard by the end of 1999.	Programme of replacement proceeding according to plan. (Ongoing)
Testing at night of aircraft engines in the noise enclosure on No.1 maintenance base at Heathrow not to exceed an average of three runs per night at high thrust settings with a maximum of 30 minutes duration for any single night at high thrust.	New procedure developed with HAL to reflect reduced need for engine ground running. For example, total permitted running time reduced from 240 minutes to 150 minutes per night. Survey of engine ground running carried out. (Met)
To develop a long-term strategy for the central location of engine ground run facilities at Heathrow, in order to reduce further noise impact on our neighbours.	Strategy now settled. It depends on planning approval of new centrally located engine ground run pens on British Airways West Base and at Terminal 5 Forward Maintenance Unit. (Ongoing)
To improve the operational noise performance of the Heathrow East Maintenance Base engine ground run facility and the impact of night testing of engines at Heathrow over the next 12 months.	Project manager appointed to carry out research on potential improvements. (Ongoing)
It is an objective that the Forward Maintenance Unit ground run pen, planned as part of the proposed Terminal 5 at Heathrow, will meet recognised standards for noise impact.	Engine runs at the Forward Maintenance Unit ground pen will not exceed 65 dB(a) at the nearest residential property, meeting criteria which avoids sleep disturbance. British Airways will accept planning conditions on this basis if permission is granted. (Ongoing)

New or restated objectives/targets

To have no Chapter 2 aircraft at Heathrow and no Chapter 2 Boeing 747 aircraft by the summer of 2000 with an overall phase-out ahead of the compliance deadline.

To review by August 1997 opportunities for the earlier scheduling of the service to Nairobi/Entebbe/Dar es Salaam in order to reduce noise infringements at Gatwick.

To reduce to zero the number of GPUs used on stands with serviceable FEGP and to prevent the unnecessary use of APUs on stand.

To establish a record keeping system for engine ground running at Gatwick by the end of 1997.

Source: BA (1997 : 17).
Reproduced courtesy of British Airways.

Table 8.4   Airline emissions[1]

Emission	Environmental effects	Approximate emissions (millions of tonnes)	
		Commercial aviation	Worldwide (fossil fuels)
Oxides of nitrogen	Acid rain. Ozone formation at cruise altitudes and smog and ozone at low levels	1.6	69[2]
Hydrocarbons	Ozone and smog formation at low levels	0.4	57[2]
Carbon monoxide	Toxic	0.9	193[2]
Carbon dioxide	Stable – contributes to greenhouse effect by absorption and reflection of infrared radiation	500–600	20,000[2]
Sulphur dioxide	Acid rain	1.1	110[2]
Water vapour	Greenhouse effect by absorption and reflection of infrared radiation	200–300	7,900[3]
Smoke	Nuisance – effects depend on composition	negligible	N/A

[1] Other emissions, mainly from paints and cleaning solvents are associated with aircraft maintenance and also from ground transport supporting the airline's operation.
[2] OECD Secretariat estimates (for 1980), from OECD Environmental Data 1989.
[3] Derived from BP Statistical Review of Energy 1991.
Aviation figures from AEA estimates apart from NO_x (Egli, Chimia 44, 369–371, 1990).
Source: British Airways (1992 : 8).
Reproduced courtesy of British Airways.

man's activities, with civil aviation's contribution accounting for 3 per cent of the total effect of global warming. Table 8.5 shows that BA's fuel consumption increased in 1995/96 and 1996/97, due to increased loads and a greater amount of operational activity. Table 8.6 is interesting since it shows the approximate fuel consumption of specific types of aircraft in kilograms of fuel per 100 available tonne kilometres (ATKs) flown, which is the number of tonnes of capacity available for the carriage of revenue load multiplied by the distance flown (BA 1997 : 52). Table 8.6 illustrates the relative efficiency of long-haul aircraft, while the use of Boeing 757s on short-hop 'shuttle' services explains why their consumption is relatively high.

Wastewater, energy and materials
In 1990, BA's expenditure on waste disposal was £1.5 million and it has pursued a corporate 'reduce, reuse and recycle' philosophy since then. Recycling of aircraft materials (e.g. waste oil, tyres, batteries and metals) has been in place since the 1950s and aluminium and paper recycling occurs, while water and effluent management schemes have been reviewed to ensure the quality of waste management is improved. In

Table 8.5   British Airways emissions from worldwide flying operations (mainline)

	Tonnes 1995/96	Tonnes 1996/97	% change
Fuel consumption	4,468,051	4,880,000	9.2
Carbon dioxide	14,063,701	15,392,000	9.4
Water	5,524,582	6,046,000	9.4
Unburned hydrocarbons	5,175	5,400	4.6
Carbon monoxide	14,511	16,000	10.1
Nitrogen oxides	54,399	59,400	9.2
Sulphur dioxide	4,113	4,500	9.2

Source: BA (1997 : 26).
Reproduced courtesy of British Airways.

Table 8.6   British Airways approximate fuel consumption by aircraft type 1996/97

Aircraft type	Consumption (kg/100 ATK)
A320-111/211	286
BAe ATP	327
B737-200/400	357
B747-100	265
B747-200	255
B747-400	239
B757-200	331
B757-200 (E4)	358
B767-300	221
B777-200 (GE90/76B)	203
B777-200 (GE90/85B)	206
DC-10-30	281

Source: BA (1997 : 29) based on data calculated from aircraft flight recorders.
Reproduced courtesy of British Airways.

1996/97, BA generated £97,860 in recycling revenue, a 54 per cent increase on 1995/96 and £25,000 of the revenue was donated to 81 charities (compared to £18,000 given to 60 charities in 1995/96). Complex energy efficiency monitoring is also undertaken to identify energy savings. BA's use of CFCs and chlorocarbon (CC) in its engineering operations for cleaning purposes has been reviewed and alternatives are being sought, with aerosol use replaced wherever possible by trigger sprays. Deicing fluid used in BA's airport operations is biodegradable and there is evidence of a decline in its use between 1989 and 1997 (except in 1995/96). The annual environmental report also provides a range of detailed tables of data associated with the chemicals BA uses in its airline operations and documents a gradual reduction since 1993/94.

Table 8.7 Estimates of additional fuel burned due to congestion at London Heathrow and London Gatwick airports 1995/96 and 1996/97

Quantities and costs	Location	1995/96	1996/97
Fuel burned due to carrying extra fuel (tonnes)	Heathrow	13,800	15,800
	Gatwick	3,900	5,900
	Total	17,700	21,700
Cost of fuel burned due to carrying extra fuel (£)	Heathrow	1,792,000	2,391,000
	Gatwick	503,600	922,000
	Total	2,295,600	3,313,000
Total additional fuel burned due to arrival congestion (tonnes)	Heathrow	22,700	24,600
	Gatwick	2,800	3,400
	Total	25,500	28,000
Total cost of fuel burned due to arrival congestion (£)	Heathrow	2,951,000	3,721,700
	Gatwick	366,300	530,900
	Total	3,317,300	4,252,600

Note: Heathrow fuel cost £151 per tonne and Gatwick £155 per tonne for the year 1996/97; for 1995/96 a figure of £130 per tonne was used for both airports. For 1995/96, fuel burned due to carrying extra fuel; figures are estimated based on the last six months.
Source: BA (1997 : 38).
Reproduced courtesy of British Airways.

Congestion

Congestion is viewed by BA as 'the most immediate problem facing the aviation industry in Europe. It is estimated that delays in the air and on the ground cost the industry some $5 billion per year ... and could rise to $10 billion per year by 2000' (BA 1992 : 22) and in 1996/97 BA spent £4,252,600 on fuel due to congestion at Heathrow and Gatwick (BA 1997 : 38). Congestion increases fuel consumption, resulting in additional emissions, particularly where a lack of airspace, air traffic control problems and inadequate runway capacity delay flights. BA has estimated that in 1996/97 aircraft burnt an extra 28,000 tonnes of fuel due to holding (stacking) in the air at Heathrow and Gatwick (Table 8.7). An additional 15,800 tonnes of fuel were also burnt due to carrying excess fuel in case delays occurred at Heathrow, while a further 5,900 tonnes were burnt for the same reason on flights to Gatwick. This is accentuated by taxiing delays prior to takeoff and after landing. Measures to improve airport capacity at Heathrow are currently being evaluated in terms of the proposed Terminal 5. The public inquiry for Terminal 5 entered its third year in 1997 and is expected to conclude in 1998. This is expected to delay the government's response until the year 2000, with the new terminal opening no earlier than the year 2004 if permission is granted. The brief discussion of the Terminal 5 public inquiry outlined by BA (1997) is interesting in the light of the discussion in Chapter 7. It highlights the lobbying by the British Airports Authority (BAA) and BA, given BAA's announcement in May 1997 that BA would be the main occupant of Terminal 5. BA has intimated that if approval for Terminal 5 is granted, it would work with BAA to reduce vehicle pollution and congestion on the ramp by using automated baggage and cargo handling systems and other stand services. In BA's view, the consolidation of all its services in one terminal would also reduce congestion by alleviating

Table 8.8   Tourism and conservation objectives 1996/97 in British Airways environmental report

Last year's objectives/targets	Status
To assess the British Airways Assisting Conservation programme in terms of benefits to conservation and the costs and benefits to British Airways.	Review carried out by an external consultant, programme largely endorsed. (Met)
To develop the contribution of British Airways Tourism for Tomorrow Awards to responsible tourism.	An additional award was given in 1996 to the British Airways Holidays Hotel displaying the best environmental performance. (Ongoing)
To work within British Airways and the industry to improve understanding and awareness of the interaction of tourism with the environment.	Seminar held at the Royal Geographical Society to discuss the relationship between tourism and conservation. Some 150 individuals from industry, academia, and other parties attended (Restated)
To undertake an additional destination audit of St Lucia with British Airways Holidays in 1996–97 and communicate recommendations for improvement.	Audit commenced in March 1997. Results will be published later in the year. (Ongoing)

New or restated objectives/targets

To lead improvement in the environmental performance of the tourism industry.

To improve communication on the airline's assistance given to conservation efforts worldwide.

Source: BA (1997 : 43).
Reproduced courtesy of British Airways.

inter-terminal transfers of passengers and baggage, though congestion would not be entirely eliminated as BA passengers also travel on non-BA/BA-alliance airlines. BA also summarises the infrastructure implications of providing new transport links to Terminal 5 to reduce potential congestion. For example, this includes extending the Heathrow Express rail service from Paddington and involves various feasibility studies associated with the London Airports Surface Access Study, published in 1996. But it is inevitable that Terminal 5 and an additional runway may lead to increased environmental impacts in terms of the landtake and effect on the local community.

Tourism

BA (1992) acknowledges that the environmental impact created by tourists travelling to destinations by air is an issue which falls under the remit of its environmental management programme. BA (1992 : 25) perceives its role as one of environmental education by:

● raising awareness of the [environmental] issues within the industry and with customers
● persuading governments and tourist authorities to impose discipline and appropriate planning regulations and management procedures to ensure future tourist development is managed in an environmentally responsible way.

Table 8.8 outlines BA's objectives and achievements in relation to tourism and conservation. Its aim to lead in the environmental performance of tourism is evidence of Forsyth's (1997) argument on the benefits of self-regulation. Figure 8.1 illustrates

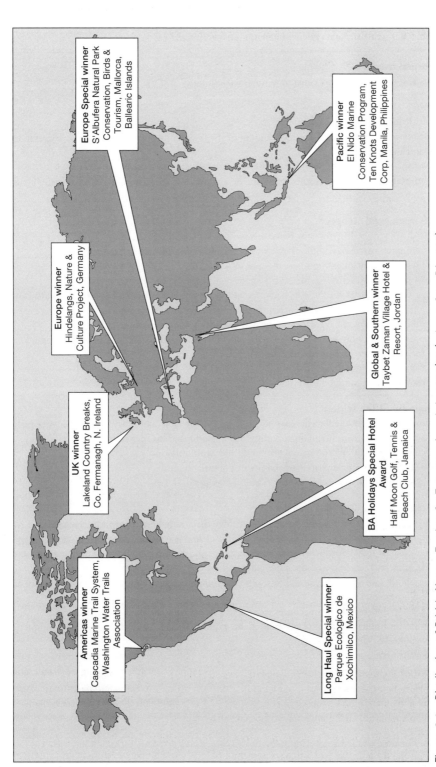

Europe Special winner
S'Albufera Natural Park
Conservation, Birds &
Tourism, Mallorca,
Ballearic Islands

Pacific winner
El Nido Marine
Conservation Program,
Ten Knots Development
Corp, Manila, Philippines

Europe winner
Hindelangs, Nature &
Culture Project, Germany

Global & Southern winner
Taybet Zaman Village Hotel &
Resort, Jordan

UK winner
Lakeland Country Breaks,
Co. Fermanagh, N. Ireland

**BA Holidays Special Hotel
Award**
Half Moon Golf, Tennis &
Beach Club, Jamaica

Americas winner
Cascadia Marine Trail System,
Washington Water Trails
Association

Long Haul Special winner
Parque Ecologico de
Xochimilco, Mexico

Figure 8.1   Distribution of British Airways Tourism for Tomorrow Awards in 1996/97 (redrawn from BA 1997)

Table 8.9   British Airways environmental policy 1996/97

Corporate goal
To be a good neighbour, concerned for the community and the environment.

Policy
British Airways will seek:
- to develop awareness and understanding of the interactions between the airline's operations and the environment
- to maintain a healthy working environment for all employees
- to consider and respect the environment and to seek to protect the environment in the course of its activities.

Environmental strategy
British Airways will strive to achieve this by:
- setting clearly defined objectives and targets addressing our environmental issues
- taking account of environmental issues in our commercial decision making
- working constructively with organisations concerned for the environment
- promoting our environmental activities with our staff, customers and other stakeholders and letting them know of our concern for the environment
- observing rules and regulations aimed at protecting the environment
- providing support and advice to staff, suppliers and other stakeholders on environmental matters relating to our operations
- using natural resources efficiently
- monitoring, auditing and reviewing our performance.

Responsibilities: staff
All staff are responsible for safeguarding, as far as they are able, both their working environment and the greater environment surrounding our operations. This includes:
- complying with environmental standards and procedures
- notifying management and supervisors of potential hazards
- avoiding needless wastage of energy and materials.

Responsibilities: line management
All line managers, in relation to activities under their individual control, are responsible for identifying and ensuring compliance with environmental regulations affecting our environment. Each director shall address environmental matters regularly, identify items requiring action and make sure they are followed up. Line managers must:
- establish individual responsibilities, objectives and accountabilities for subordinate staff in environmental matters
- develop and maintain procedures to protect the working and external environment
- monitor implementation of procedures and working practices and take swift and appropriate steps to put deficiencies right
- ensure that a statement of environmental impact, tailored to specific requirements, is prepared as part of the planning of facilities and operations, and for modifying or abandoning them
- provide channels for employees and contractors to be consulted on environmental matters
- investigate and report all environmental incidents and near misses and take necessary follow up action
- set quality standards covering relevant discharges and disposals including any that are not covered by statutory requirements

Table 8.9   Cont'd

- review regularly the use of materials and energy in order to reduce waste, optimise recycling and select materials compatible with environmental objectives
- maintain accurate and comprehensive records of discharges and other waste disposals to the environment, including breaches of compliance limits
- report any breaches to the relevant regulatory bodies or internally as appropriate and take action to bring operations within compliance.

Source: BA (1997 : 51).
Reproduced courtesy of British Airways.

the scope and geographical distribution of the Tourism for Tomorrow awards which have been in operation for seven years. In 1996/97 there were 100 entries from 44 different countries, with 30 tourism experts giving their time to judge the entries. In 1997/97, BA also sponsored a seminar following the awards at the Royal Geographical Society on the theme 'tourism is essential to environmental conservation'. Research conducted by BA's Tourism Council Australia Environment Scholarship in 1996 found that of the 402 passengers surveyed, 24 per cent stated they were aware of British Airways' environmental programme. In 1996/97, as part of its British Airways Assisting Conservation scheme, BA also offered 3,055 individuals travel awards, valued at £453,000, which cost £22,000 to administer.

So how can one evaluate BA's performance in the environmental management of tourist transport? One initial issue to consider is BA's stated environmental policy, which is reproduced in Table 8.9. This illustrates an integrated approach to environmental management so that the company complies with all existing environmental regulations. It also highlights the corporate ethos – to ensure 'all staff are responsible for safeguarding, as far as they are able, both their working environment and the greater environment surrounding our operations' (BA 1997 : 51). Table 8.10 also provides a useful summary of BA's overall environmental costs, investments and savings in 1996/97.

Figure 8.2 outlines the way in which BA manages its environmental performance, with the Environment Branch playing the lead role to 'advise, support, stimulate and monitor environmental performance' (BA 1997 : 10). To assist with communications between each department in the organisation and the Environment Branch, there are 'environmental focal points' in each department. Within BA, the network of 'environmental champions' within the organisation enables staff to take part in environmental initiatives. In 1996/97, there were 305 such champions within BA. The champions and the Environment Branch communicate by means of an 'environment letter' which is produced three to four times a year.

In evaluating the scope of BA's environmental measures, it is evident that the organisation is establishing a benchmark by 'taking the lead [but] it is up to other sectors of the industry to extend the initiative' (Sommerville 1992 : 173). Environmental researchers have also developed a greater interest in the future impact of tourist transport systems in terms of the requirement for additional infrastructure and its impact on the environment, which is now considered in relation to Environmental Assessment.

Table 8.10 A selection of environmental costs, investments and savings made by British Airways 1996/97

Project/Initiative	£000s
Environment Branch including consultancy work, sponsorship and publications	587
Acquisition of aircraft over next four years (Boeing 747-400, 777, 757-200 and 737–300.	2,800,000
A significant part of the expenditure will relate to noise, emission and fuel efficiency performance	
Extra fuel burned as a result of congestion delays at Heathrow and Gatwick	4,250
Replacement of chillers utilising CFCs at Technical Block A, Heathrow with non-ozone-depleting alternative	1,000
Investment during the next two years on sub-metering (should be almost 1,000 meters).	200
Halon replacement programme	250
Recycling of halons from ground systems for essential use in aircraft fire protection systems	50
Noise infringements at Heathrow and Gatwick (1995/96)	75
Charges for the three months January to March 1997 for operating GPUs on stands at Heathrow with serviceable fixed electrical ground power	5
Specific noise charges at airports from operating Chapter 2 aircraft (1995/96)	7,230
With HAL, British Airways is sharing support of £30,000 for research on active noise at Cranfield University	15
Investment over the next two years on sub-metering	200
Heathrow maintenance base gas heating project – conversion of oil-based district heating system to a more efficient local gas system (investment over next two years)	7,700
Roof replacement and wall cladding project to New Services Hangar, Heathrow, to improve building efficiency and insulation	2,800
New high efficiency lighting system for New Services Hangar, Heathrow	270
Support for improvements to the 105 bus route (frequency doubled and rerouted past Compass Centre)	83
Support for improvements to bus services serving the Cargo Centre and southside Heathrow (H26, 140, 555, 556, 557)	300
Support (with HAL) for free travel zone for everyone between Heathrow Central Terminal Area and northside	25
Contribution to the costs of two feasibility studies into western rail links from Terminal 5, Heathrow	20
Soil and ground water investigation at Heathrow maintenance base as part of the British Airways Maintenance Base Initiative (BAMBI) to return leased areas back to HAL	107
Waste disposal costs at Heathrow and Gatwick	1,775
Improvements to drainage system at Heathrow maintenance base	158
Money generated by recycling programme	98
Value of travel assistance to individuals through British Airways Assisting Conservation	450
Administration charge for British Airways Assisting Conservation	20
Donations to environmental organisations	50
Contribution to UNEP publication on children's conference	5

Table 8.10   Cont'd

Project/Initiative	£000s
Recycling revenues donated to staff charities	25
8 additional GPUs have been replaced in 1996–97 at Heathrow and Gatwick to meet the higher noise performance standard. All new models meet Stage 1 of the European Directive for Engine Emissions. Purchases for other locations will also meet these standards	200

Source: BA (1997 : 12).
Reproduced courtesy of British Airways.

Figure 8.2   The structure of environmental management at British Airways (redrawn from BA 1997)

Environmental Assessment and tourist transport

An understanding of the past and present effect of tourist transport systems on the environment is critical to the long-term management of environmental resources, but there is also a need to consider the likely effect of future transport development projects. It is within this context that the significance of research methodologies such as Environmental Assessment (EA) can be examined to show how future tourist transport infrastructure projects may be evaluated. Within the existing literature on the environmental impact of tourism and transport (see Farrington and Ord 1988), a number of research methodologies exist, which are documented by Williams (1987) in terms of their analytical function and the techniques they employ.

There are three levels at which EA of tourism and transport projects can be undertaken; 'identification', 'prediction' and 'evaluation'. Williams (1987) summarises five main methodologies used to assess the impact of tourism on the environment, in which transport is a significant component. These range from 'ad hoc' teams of specialists describing impacts within their professional field of study, through the 'map overlay approach' frequently employed in land use planning, to 'check-lists' of different impacts associated with physical development related to tourism, 'networks' to assess the secondary and tertiary effects associated with action relating to tourism projects and lastly, more sophisticated matrices of impacts within the confines of EA (see Wathern 1990 for a more detailed discussion). Although EA was not specifically designed with tourist transport projects in mind, it is a useful methodological tool to examine the direct and indirect effects of a project on the existing and future tourism environment within an integrated research framework. (See Department of the Environment 1989 for a guide to the scope and complex range of issues which EA in the UK must address as a legal requirement.)

A recent study by Perl et al (1997) moves the EA research frontier forward in the methodology it devised for pricing aircraft emissions at Lyon-Satolas airport. Without reiterating the technical aspects of the study, Perl (1997) highlights the three principal environmental impacts associated with airport operation: air, noise and water pollution. As Perl et al rightly argues:

> One important variation concerns the degree to which these impacts mobilise public participation in, or demands for influence over, airport planning and development. . . . an impact like aircraft noise, which is spatially concentrated in certain areas, has motivated much greater public protest than the air or water pollution impacts from airports, which diffuse more broadly and mix with pollutants from other sources. (Perl et al 1997 : 89)

By linking EA techniques with economic cost evaluations, Perl et al (1997) estimated the cost of air pollution for 1987, 1990, 1994 and 2015. The methodology involved pricing the pollution cost from the landing–takeoff cycle, which includes taxiing, idling, queuing, takeoff and climbout. For 1994, Perl et al (1997) estimated the cost of air pollution at US$3.6–6.6 million. This was projected to rise to US$9.5–17.4 million in 2015 (assuming that aircraft engineering technology did not improve). Such research can be extended to other airports and can certainly assist in scenario planning for possible environmental costs of pollution. It certainly has the potential to make EA a more systematic rather than descriptive method in dealing with tourist transport impacts. To illustrate how an EA has considered a tourist transport system and some of its potential shortcomings (see Ross 1987 for a discussion of how to evaluate EAs), the case of the Channel Tunnel is now examined.

CASE STUDY   The environmental impact of a new tourist transport infrastructure project – the Channel Tunnel

The Channel Tunnel is currently the largest tourist transport infrastructure project in Europe, which cost in excess of £8 billion at 1990 prices. According to SERPLAN (1989), the Channel Tunnel had the potential to generate an additional 450,000

Plate 8.1 Competition on short sea crossing over the English Channel led to the introduction of hovercraft in the 1970s. The time savings achieved by this form of technology have been superseded by those offered by the Channel Tunnel. The hovercraft are themselves being replaced by catamarans (see http://www.hoverspeed.co.uk) (Source: S.J. Page)

tourists for the South East of England, using various modes of transport to travel through the tunnel to mainland Europe. It is evident that the tunnel project has created a new tourist gateway between the United Kingdom and mainland Europe, thereby facilitating more choice in the available modes of cross-Channel travel for tourists. The extent to which the opening of the Channel Tunnel has directly and indirectly affected physical and man-made tourism environments in the 1990s (Plates 8.1 and 8.2) is largely overlooked in recent research on the Channel Tunnel (Page 1994b; Gibb 1996; Vickerman 1995).

The environmental dimension of the project has not received a great deal of attention, with the exception of the controversy surrounding the routing of the Channel Tunnel high-speed rail link through Kent (see Goodenough and Page 1994 for more detail). In the UK, the environmental lobby is particularly concerned with the physical impact, although the majority of such studies have focused on a specific impact rather than the tunnel's effect in different tourism environments in the 1990s.

Environmental Assessment and the Channel Tunnel
In the UK the EC Directive (85/337/EEC) on EA coincided with the government's 'Invitation to Promoters for the Development, Financing, Construction and Operation of a Channel Fixed Link between France and the United Kingdom'. Wathern (1990) examines the recalcitrance of certain member states, particularly the UK, towards adopting the directive since the government's commitment to push the Channel Tunnel

Plate 8.2   The construction of the Channel Tunnel terminals in the UK and France produced major impacts on the environment, as shown by this photograph of the construction site at Coquelles in 1992 (reproduced courtesy of the Channel Tunnel Group)

project to fruition as quickly as possible in 1985. By using the parliamentary device of the hybrid bill, no public planning inquiry was needed, which avoided any obligatory participation in an EA to comply with the impending EC legislation, since 'the Directive does not apply to projects ... which are authorised by a private or Hybrid Bill' (Department of the Environment 1989 : 23), although 'the promoter of such a Bill should provide an environmental statement which can be considered by the select or standing committees ... on the Bill' (Department of the Environment 1989 : 23). The latter situation applied to the four shortlisted promoters of the Fixed Link project (Channel Tunnel Group, Channel Expressway, Eurobridge and Euroroute) which were required to comply with the EC Directive in 1985. This meant that a detailed EA rather than a simple environmental statement was required from each promoter, minimising the cost to the government by obliging the private sector to fund a detailed environmental analysis. As a result, the EAs of the Fixed Link project are a landmark in the UK since they were the first to comply with the EC Directive 85/337.

The four shortlisted promoters' EA reports submitted in 1985 were subsequently appraised by Land Use Consultants (1986) who reviewed both the content, coverage, accuracy and presentation of the reports in relation to their ability to meet the requirements of the draft EC Directive 85/337. Evaluating EAs is a complex process in view of the problems of understanding and forecasting the secondary effects and consequential development such projects may generate. For example, the Channel Tunnel Group's EA (Channel Tunnel Group 1985) underestimated the potential impact of a new mode of tourist transport on tourism, arguing that the tunnel would not in

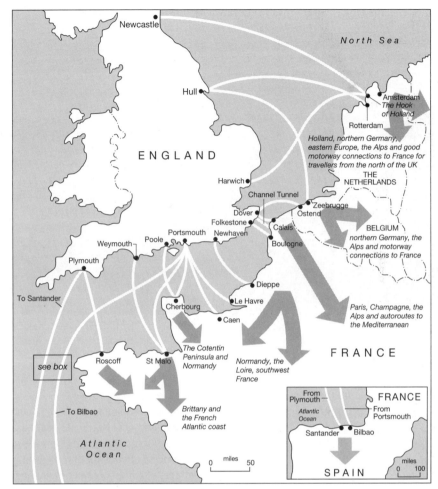

Figure 8.3   The pattern of Channel crossings, 1998

itself directly stimulate a growth in the demand for cross-Channel travel. Planners failed to recognise how the cross-Channel ferry industry would respond to the competition through mergers, acquisitions and the development of new routes (see Figure 8.3). The promoters also overlooked the new tourism markets which will be more accessible to the UK and Europe as a result of the tunnel and improvements to the high-speed European rail network and road (Page and Sinclair 1992b; Page 1993d). Thus, while the Channel Tunnel Group's EA dealt with the physical impacts involved with construction of the tunnel, it failed to make a detailed assessment of the consequences of a growth in visitor arrivals induced by the Fixed Link. Criticisms of the Channel Tunnel Group's EA have also pointed to the voluminous and unintelligible nature of the study (Lee and Wood 1988). Although the EA failed to consider the potential environmental impacts associated with a sustained growth in visitor arrivals and departures once the tunnel opened, it is possible to identify a number of ways in which tourist use of the tunnel would affect the environment.

Plate 8.3 The construction of the Channel Tunnel terminals in the UK and France produced major impacts on the environment at Coquelles. This shows the construction of the loading ramps for vehicles (reproduced courtesy of the Channel Tunnel Group)

The environmental impact on existing tourism resources
The construction of the tunnel has aroused the concerns of amenity bodies such as the Council for the Protection of Rural England and the Nature Conservancy Council. Ardill (1987) examines the conservation lobby's concern for the impact on the landscape, and the land damaged or lost in the process of constructing the tunnel, the terminals and associated infrastructure. The effects of consequential development on the South East resulting from selective land releases for tourism or motorway service area provision at interchanges are also considered, as are concerns that the visual amenity of the Kent landscape would be affected by the tunnel and that tourism resources might be affected by consequential development. The Channel Tunnel Joint Consultative Committee (1986) argues that only a limited environmental impact would result from consequential development associated with the tunnel, based on the fact that since 1970 there has been a threefold growth in the demand for cross-Channel travel but this has not generated any consequential development or led to any dramatic change in Kent's economic geography.

Tourist use of the tunnel and the potential effects on the environment
According to Land Use Consultants (1986 : 49) the tunnel could affect tourism in terms of employment and induced development, but tourism and recreation would need to be controlled and managed to reduce their impact on the environment. Their general assessment is important in terms of the scale of the potential impact of tourism in relation to how many tourists will travel through the tunnel once it is

open. According to SETEC (1989), the traffic forecasting consultants to Eurotunnel, they expected up to 15.8 million road-borne passengers to travel through the tunnel via the shuttle service (see Vickerman 1995 for detail on actual arrivals). This would virtually double the capacity for Channel crossings, and the potential environmental impacts largely depend on the extent to which demand grows to fill this increased capacity and the degree of market capture by Eurotunnel of existing ferry and air traffic. This raises the question of how tourist use of the tunnel can be accommodated within the existing environment and the cumulative effects of additional tourists travelling through Kent.

Direct environmental costs associated with tunnel-related tourist transport infrastructure
The actual impact of major road and rail infrastructure to serve tunnel traffic, was assessed within Belgium as part of its EA for the proposed high-speed TGV link from Lille to Brussels. This examines some of the real environmental costs of both road and rail travel by tourists and non-tourists. The purpose of the EA by the European Centre of Regional Development of the Walloon Region (CEDRE 1990) was to assess the 'micro-ecological effects' – the environment in which the infrastructure is to be established. The 'micro-ecological effects' are of interest in this context since they considered the:

- abiotic impact (i.e. the effect on geology, hydrology, noise and vibration)
- the biological impact (i.e. the effect on flora, fauna and the interactions between the two)
- the human impact (i.e. the effect on agriculture, residential areas, traffic, transport: the human elements in the landscape).

Their assessment of pollution reveals the potential impact tourist travel may have on the environment through which they travel – in this case a dedicated transport corridor (Table 8.11). Tourist travel by rail and road produces pollutants, though on balance, rail is considered to be the most 'environmentally friendly option', as it is more energy-efficient and less intrusive, though it still generates a degree of noise pollution.

The environmental costs and benefits of tourist use of the international passenger terminals (IPTs)
Visitors using the Channel Tunnel who travel by rail or road generate an environmental impact on their destinations. In the case of London, this impact is largely related to the international passenger terminal (IPT) at Waterloo (Page and Sinclair 1992a). In Kent, many of the direct environmental impacts occur at the tunnel terminal and at the planned Ashford IPT, which is the focus for rail and car-borne travellers from the South East wishing to board the Eurostar rail services. A further environmental impact is that of increased tourist traffic in Ashford town centre. Consultants for BR identified a problem of potential congestion in and around Ashford town centre related to traffic generated by the IPT, which is likely to intensify in view of the predicted 45 per cent growth (1990–2006) in rail travel from Ashford to London and European destinations. Congestion and the provision of car parking may concentrate the environmental problems, especially pollution, in a small area of the town, and

Table 8.11 A comparison of pollution from the TGV and car-borne traffic using the
motorway in Belgium[1]

Pollutant	Mass of pollutant (g) per kilometre travelled	
	TGV	Car
Sulphur dioxide[2]	0.124	0.090
Nitrous oxide[3]	0.071	1.460
Aerosol	0.044	0.049
Carbon monoxide[4]	0.005	1.109
Hydrocarbon[5]	0.002	0.179
Carbon dioxid[6]	228.907	135.000
Safety:		
Number of persons	0.8 (train)	20.0 (roads)
killed per billion	0 (French TGV)	6.7 (motorway)
km travelled		

[1] The accuracy of the pollution measurements listed in the table will depend on the meteorological conditions and prevailing winds as to whether pollutants concentrate at particular locations or disperse over a wider area.
[2] Sulphur dioxide may impair health and it contributes to acid rain as sulphuric acid.
[3] Nitrous oxide is a major component in photochemical smog and nitric acid contributes to acid rain. The major emission source is motor vehicles and power stations.
[4] Carbon monoxide directly causes health-related problems and it can induce complications among people suffering from cardiac-related diseases. Concentrations of carbon monoxide in confined areas are harmful as they can reduce the oxygen-carrying capacity of the blood.
[5] Partially burnt hydrocarbons are carcinogenic (cancer-forming) agents and they also contribute to photochemical smog.
[6] Carbon dioxide contributes to the 'greenhouse effect'.
Source: Cited in Page (1992c) and based on CEDRE (1990).

forecast demand for 5,000 car parking spaces in the town illustrates the scale of the IPT development. However, since such forecasts have subsequently been found to be too optimistic, such an expansion is not as problematic as was once thought.

The indirect environmental impact of increased numbers of car-borne tourists on tourist destinations
An associated problem relates to additional car-borne visitors who may stop off in Ashford en route to the tunnel for leisure shopping and accommodation, adding to the potential congestion. Although Ashford is unlikely to suffer what Romeril (1989) calls 'saturation tourism' (i.e. seasonally induced peaks in flows of tourists), it is evident that planned visitor management strategies will not be able to overcome all of the problems, particularly once the tunnel is open. The tourism carrying capacity in Kentish towns like Canterbury has reached saturation point in the peak season and key attractions like Canterbury Cathedral (see Page 1995b) now employ visitor management tools, such as charging and monitoring visitors who use the cathedral. The tunnel has certainly generated the potential for more tourist visits to the county of Kent, but

assessing where, when, and the duration of these visits among domestic and overseas tourists using the Fixed Link remains a difficult process. Whatever locations the potential tourists visit, the existing environmental pressures posed by tourist travel to certain destinations in South East England are unlikely to be reduced without some attempts to spread the seasonal distribution of visitors.

The potential problems resulting from the development of a new tourist transport infrastructure project and the implications for destination areas highlight the significance of focusing on the necessity for tourist transport to be developed and managed in a sustainable framework.

Towards sustainable tourist transport systems

As mentioned earlier, 'sustainability' is a new-found term within tourism and transport literature: for services to be attractive to consumers they must now be 'sustainable' or 'green', though much of the rhetoric associated with sustainability has not led to radical changes in the operation and management of tourist transport systems – merely some readjustment to accommodate green issues in most cases. As transport is fundamental to tourist travel, some researchers argue that it is not possible to make tourism sustainable without a fundamental revision of the concept of tourism, holidaymaking and the role of travel in modern society. Therefore, without a re-evaluation of pleasure travel, measures designed to introduce sustainability into the tourist transport environment debate are unlikely to address the root cause of the problem: the demand for tourism. However, since this is unlikely to be influenced in the short term, the immediate issue is to address the environmental impact of existing tourist travel.

The motivation to achieve sustainable tourist travel has resulted from the actions of pressure groups (e.g. Greenpeace, Friends of the Earth and Transport 2000 in the UK), and their views have permeated national governments as such groups have harnessed grassroots pressure from consumers to develop a greener economy and improve the quality of the environment. But it is at government level that commitment needs to be made to formulate, implement and resource policies to facilitate sustainable transport options. Little attention is given to the issue of tourist transport systems as it is often subsumed in the general theme of transport, which has a bias towards domestic concerns and the effect on economic development. The UK Tourism Society's response to the initial findings of the Government Task Force (the final report is Department of Employment/English Tourist Board 1991) suggests that:

> no analysis of the relationship between tourism and the environment can ignore transportation. Tourism is inconceivable without it. Throughout Europe some 40% of leisure time away from home is spent travelling, and the vast majority of this is by car . . . Approaching 30 per cent of the UK's energy requirements go on transportation . . . [and] . . . the impact of traffic congestion, noise and air pollution . . . [will] . . . diminish the quality of the experience for visitors. (Tourism Society 1990)

How can the sustainability concept be incorporated into the tourist transport system? According to Barbier (1988 : 19, cited in Newson 1992), sustainability needs to be viewed as a process in terms of how different systems interact as:

> the wider objective of sustainable economic development is to find the optimal level of interaction among three systems – the biological and resource system, the economic system and the social system, through a dynamic and adaptive process of trade offs.

This means that economic activity, such as tourism, must try to achieve a balance with the natural environment so that the environment can support the activity without generating unacceptable impacts which affect the future resource base. To achieve this objective, the concept of sustainability needs to be built into the operation of tourist transport systems, and the following action is needed in terms of policy making and management:

- policy formulation
- policy implementation
- facilitating good practice in tourist transport
- the evaluation of sustainable transport practices.

A systems approach is useful in this context as it helps one to understand how the decision-making process associated with the regulation, organisation and management of tourist transport systems affects different elements within the system. In terms of the sustainability concept, actions in one part of the system (e.g. policy formulation) will have repercussions for other parts of the system.

Policy formulation for sustainable tourist transport

Banister and Button (1992 : 2) recognise that the 'whole question of sustainable development . . . is – and likely to remain – a central concern of policy-makers and transport is but one element of this'. The rapid growth in long-distance passenger transport and its dominance by aviation at the international scale, together with the rapid expansion in car ownership within countries poses many problems for policy makers attempting to pursue sustainable transport options. Moreover, the underlying demand for travel seems set to continue to expand, as forecasts for the year 2000 suggest (Edwards 1992). The social and psychological demand for travel and holidays remains a potent force in developed countries. One result of the sustainability debate for policy makers is that the environmental impact of transport is not just a local issue: it is also a global problem, as the case study of BA indicated. This is confirmed by Banister and Button (1992 : 5) who argue that 'transport is an important contributor [to the sustainable development debate] at three levels (local, transboundary and global)'. Policy formulation therefore needs to be undertaken in a context where national governments develop transport policies and coordinate their responses at a transnational and global level through agencies such as the United Nations.

However, political commitment to formulating sustainable transport policies at national level may not be compatible with other political priorities. For example, many governments have facilitated the development of tourist transport infrastructure to foster regional tourism development (e.g. Ireland – see

Page 1993b) and to encourage outbound travel (e.g. Japan). In fact, Wahab and Pigram (1997 : 285) argue that 'a growing trend in policy making in many countries is to leave tourism to private enterprise, and current economic conventional thinking supports the role of market mechanisms'. Sustainable transport policies may require a re-evaluation of these national transport policy objectives in relation to tourism, transport and the cost to the environment. In the context of the UK (see Banister 1992 for a discussion of national transport policies in the UK), D. Hall (1993) argues that sustainable transport is neglected in policy making since the Government's White Paper *This Common Inheritance* (Department of the Environment 1991) and accompanying policies have paid little attention to transport and the environment.

One recent development which is worthy of discussion in this context is Agenda 21. Agenda 21 'is a comprehensive programme of action adopted by 182 governments at the 1992 United Nations Conference on Environment and Development (UNCED), known as the Earth Summit. It provides a blueprint for securing the sustainable future of the planet' (Wahab and Pigram 1997 : 284). While it is the first document to gain widespread international commitment towards conserving the world's resources, it 'did not mention travel and tourism except in a few sections' (Wahab and Pigram 1997 : 284). However, the World Travel and Tourism Council et al (1997) report *Agenda 21 for the Travel and Tourism Industry: Towards Environmentally Sustainable Development* does examine transport as one of the 10 areas of priority action for companies involved in tourism. The report identifies transport as a central feature of Agenda 21 in terms of controlling or reducing harmful emissions into the atmosphere as well as other adverse environmental impacts from tourism-related transport. In fact, the World Travel and Tourism Council et al (1997 : 63) argues that 'Transport is the lifeblood of the travel and tourism industry and failure to take action and improve performance in this area could result in harsh penalties for travel and tourism companies and increased costs for travellers.' To avoid such penalties, the report advocates that companies should:

- use well-maintained and modern transport technology, which may reduce emissions especially in the airline sector but also in other land and sea-based transport sectors
- assist less developed countries to acquire technology and skills to reduce environmental emissions from tourist transport
- develop and manage car-share, cycle or walk-to-work schemes for employees and provide incentives for successful implementation
- provide information for tourists to encourage the use of public transport, cycle ways and footpaths
- work with government to implement measures to reduce congestion in air transport and in urban tourism environments
- work with governments to achieve a greater integration in planning transport modes which not only reduce reliance on the private car, but reduce energy consumption in linking tourists to onward destinations
- use demand management tools to assist in reducing the need for polluting modes of transport in preference for more environmentally friendly modes of transport
Source: Modified and developed from World Travel and Tourism Council (1997 : 63–4).

Table 8.12   Possible components of a strategic, integrated approach to transport planning and investment at various organisational levels

National/European
- A new integrated approach to transport planning in which the case for rail investment is evaluated on a basis comparable to that for roads, taking into account the full environmental costs and benefits.
- Within the framework of this integrated approach, the preparation of a long-term expansion plan for rail, with the aim of at least doubling passenger kilometres carried (restoring the situation to that which applied pre-Beeching), increasing substantially rail's share of freight transport, both internationally and domestically, and in general exploiting fully the potential benefits of the Channel Tunnel link to Europe.

The region/county
- Through land use planning at the regional and/or county level, the maintenance of an appropriate balance of homes and employment to secure local job opportunities and reduce average commuting distance.
- The location, as far as possible, of new settlements and other major developments along railway corridors to secure the best possible public transport access to other major centres.
- The development at regional level of long-term investment strategies for integrated public transport, using resources which would otherwise be spent on new roads.

New business developments
- The development of new policy guidelines for the location of businesses and public offices (together with suitable parking standards), with the aim of minimising total commuting mileages – such policies, which could perhaps be developed along the lines of the Dutch 'right business in the right place' strategy might then be set out in a future planning policy guidance note.
- The introduction of mileage reduction plans, to be developed and implemented by businesses and public organisations within government-issued guidelines and supported through financial incentives.

Shopping
- Continued strong support for city and town centre shopping (with new development permitted only where it reinforces existing provision) and for the maintenance of local shopping facilities.
- Major new retail developments to be subjected to a full shopping impact analysis, which would include an assessment of the traffic likely to be generated, both in terms of the absolute numbers of vehicles and total vehicle mileage, and an assessment of the public transport provision.

Accessibility to schools
- Requirements on local planning and education authorities to place maximum emphasis on safe routes to school by foot, by bicycle and by public transport, and to ensure that new primary schools in urban areas are generally no more than a five-minute walk away from the children's homes.
- A requirement upon highway and education authorities, together with the schools concerned, to carry out comprehensive reviews of the adequacy of public transport and school buses servicing secondary schools, with a view, where necessary, to reorienting spending programmes to upgrade these services.
- The publication of an advice and good practice guide for local authorities on the development of safe routes to school, and the allocation of some initial central funding for such schemes.

Table 8.12   Cont'd

Access to local facilities

- Commitments through local planning policies to secure a full range of local facilities which are readily accessible by foot or bicycle, both in new residential areas and, when the opportunity arises, in older ones.
- The planning of future residential areas of any significant size around high-quality public transport routes, with homes generally no more than a five-minute walk away from a bus or light rail stop.

Integrated transport strategies at the urban level

- Requirement upon those responsible for land use and transport planning within our major cities to prepare integrated transport strategies, the aims of which would include the development of high-quality public transport services, significant reductions in overall vehicle emissions and, generally, a vastly improved environment for those who live and work there.

Source: D. Hall (1993 : 12).

While such suggestions are helpful, in the self-regulation era companies need to be given incentives as such measures require more commitment than conducting an Environmental Audit. It may be that individual countries need to formulate an environmentally based tourist transport strategy with which companies can comply. In fact, D. Hall (1993) suggests that a general environmental transport strategy needs to be formulated for the UK (see Table 8.12). Although tourist transport is subsumed within the wider category of transport systems in Table 8.12, Hall does suggest that coordinated action is needed in relation to:

- regulatory mechanisms (e.g. by setting a ceiling for emissions)
- financial mechanisms (e.g. incentives to favour energy-efficient modes of travel)
- the introduction of technological advances in transport to encourage the use of more fuel-efficient engines
- the development of an integrated and coordinated planning response to transport where land use and transport planning should minimise the distance to travel for economic and leisure activities (e.g. work and shopping).

How does this affect international tourist travel? It would appear that the likely outcome of D. Hall's (1993) strategy would be the promotion of environmentally friendly modes of travel for tourists. Yet the real issue of existing tourists' travel habits is absent from the policy objectives as it is often perceived as an international problem rather than one nation's sole responsibility.

Implementation of sustainable tourist transport policies

A range of government transport policy responses to sustainability issues are discussed in Banister and Button (1992) and one recurrent theme is the need to adopt economic policies to price transport activities so that they reflect the environmental cost. There is growing evidence that countries such as the UK are now looking at controversial measures such as road pricing, which has been developed in Singapore. One approach widely used in developed countries is the differential pricing of petrol through the level of taxation it attracts, to reduce the use of leaded petrol and to increase the consumption of unleaded petrol. A more radical solution advocated by the EU is the introduction of a carbon tax on energy production so that more environmentally sound energy sources are developed to reduce pollution. Yet this has been fiercely resisted by governments such as the UK because they feel it would add additional costs to the price of energy and thereby increase the costs of production. The basis of their argument in the area of tourist transport is it could make UK transport operators uncompetitive on a global basis. In this respect, concerted international government action is needed to reduce levels of pollution from transport, with certain countries taking a lead while others are forced to follow suit through international pressure. For example, in the UK the deregulation of bus services in metropolitan areas initially led to new operators using aged vehicles which contributed higher levels of pollution compared to the former metropolitan Passenger Transport Executives (PTEs) where grants were provided to update fleets, thereby resulting in the use of more energy-efficient vehicles (Knowles and Hall 1992). Yet as Button and Rothengatter (1992) acknowledge, the global nature of transport's impact on the environment is likely to intensify. The implementation of sustainable transport policies needs to be accompanied by changes in the lifestyles of tourists so that they recognise the environmental degradation which their process of travel induces.

Good practice in sustainable tourist transport

The real debate over achieving sustainable tourist transport options is usually focused on the outcome: can such options really be put into practice or do they remain a stated policy objective of environmental planning which is little more than a paper exercise? There are various examples of good practice cited in the tourism literature where transport is a core component of tourism planning, so that conservation and interpretation of the environment raises tourists' awareness of natural habitats and the need for a delicate balance to be achieved between tourist use and preservation. The Tarka Project in Devon is one example where a tourism strategy has achieved these objectives (see Department of the Environment/English Tourist Board 1991 and Charlton 1998 for more details). However, the reliance on public and private sector transport operators to implement sustainable tourism is questioned by Wood and House (1991, 1992). Although Wood and House (1991) acknowledge that transport operators need to pursue good environmental practices, they also advocate that the onus should be placed on the tourist. Their central argument is that tourists should 'environmentally audit themselves' before and during their holiday and this principle could also be applied to aspects of business travel. The environmental audit is based on a number of simple questions:

- Why go on holiday? – consider your motivations and whether you really need to travel.
- Choose the right type of holiday to meet your needs.
- Consider travelling out of season to less well-known destinations.

- Choose the right travel method and tour operator after asking what the company is doing to minimise environmental impacts.
- Consider the form of transport you will use to get to the point of departure.
- Does the tour operator contract transport companies with new energy-efficient vehicles and aircraft or are they old, noisy and less efficient?
- Is public transport, cycling (Scottish Tourist Board 1991) or walking a feasible option when you are at the destination as opposed to hiring a car?

Wood and House's (1992) *The Good Tourist in France* illustrates how tourists can make their trip sensitive to the environment, especially in their use of transport. Wood and House (1992) provide information on 'how to get there' but more importantly they undertake detailed research on each region of France so that tourist travel in the destination area can be based on sustainable options (i.e. forms of transport which do not have major environmental impacts). They outline details of operators and locations where you can hire or purchase travel services based on:

- rail travel
- bus/coach travel
- car travel
- boating
- cycling
- walking
- riding

as well as contact addresses of local groups which encourage and support sustainable development.

Research on fragile tourism environments such as the Arctic has advocated the need for visitor codes of practice (Mason 1994). Marsh and Staple (1995) reinforce such arguments. They argue that cruise ship passengers to such environments need to be educated about the impacts they can cause.

Probably the most sustainable form of transport which tourists can engage in is cycling. Cycling is comparatively neglected in the tourism literature, being discussed in generic transport studies research and in leisure contexts. For this reason, the following case study examines cycling as a sustainable means of tourist transport with a relatively low impact on the environment.

CASE STUDY   Cycling as an environmentally friendly form of tourist transport: the case of the UK

According to Lumsdon (1996a), the market for recreational and tourist cycling can range from day trips to part-time casual usage through to long-distance touring holidays. Tourist use is most likely to involve the occasional casual usage by tourists visiting a destination who may hire a cycle for a day (see Page 1998 for a discussion of such usage on the Norfolk Broads in the UK) or the more determined tourist which undertakes long-distance cycling holidays. Lumsdon (1996b : 5) defines cycle tourism as cycling which is 'part of or the primary activity of, a holiday trip . . . it falls within a categorisation of activity holidays'. Beyond the seminal study which incorporates cycling (Tolley (1990), there are only a limited number of studies on cycle tourism (Beoiley 1995; Schieven 1998; Sustrans 1997; Ritchie 1997). In a review of leisure cycling, Lumsdon (1997b) observes that the Department of Transport (1996) statistics suggest that up to 40 per cent of cycle journeys are for leisure purposes and if other personal trips are included, up to half of all trips are for leisure. Yet as Lumsdon shows, the prevailing literature on cycle transport pays little attention to the leisure dimension (in which tourist use is subsumed). Even the UK's *National Cycling Strategy* (Department of Transport 1996) highlights the significance for non-leisure use, though, as Lumsdon (1997b : 115) shows, leisure is discussed:

> Leisure cycling has great potential for growth, it can be a stimulus to tourism, it is a high-quality way to enjoy the countryside and a good way to introduce people to cycling for their everyday transport needs. To encourage leisure cycling there need to be small scale improvements, especially near where people live, followed by better signposting, marketing and information. Flagship leisure routes, using quiet roads or disused railway paths, can increase the profile and boost leisure cycling in town and countryside. (Department of Transport 1996 : 13 cited in Lumsdon 1997b : 115)

But who are the typical cycle tourists and what motivates them to use this form of transport? The Scottish Tourist Board's (1991) innovative study *Tourism Potential of Cycling and Cycle Routes in Scotland* indicated that cycling had grown in popularity as a recreational activity in the 1970s and 1980s, with the membership of the Cyclists' Touring Club standing at 40,000 in the UK, having grown 10 per cent in the previous decade. The more recent study by the Countryside Commission (1995) *The Market for Recreational Cycling in the Countryside* identified some of the main motivations for cycling, including

- keeping fit
- fun
- fresh air
- access to the countryside

Lumsdon (1996b) simplifies the market segments involved in cycle tourism to include:

- *holiday cyclists,* comprising couples, families or friends who seek a holiday where they can enjoy opportunities to cycle but not necessarily every day. They seek traffic-free routes and are free – independent travellers not seeking a package holiday. While they are likely to take their own bikes on holiday, a proportion will hire bikes and are likely to cycle 15–25 miles each day, a feature examined in New Zealand by Ritchie (1997).
- *short-break cyclists,* who seek to escape and select packages which will provide local knowledge (with or without cycle hire) and comfortable accommodation. They are likely to travel in groups and will cycle 15–25 miles a day.
- *day excursionists* are casual cyclists who undertake leisurely circular rides of 10–15 miles and are not prepared to travel long distances to visit attractions or facilities. They prefer to seek quiet country lanes which are signposted. They tend to comprise 25–30 per cent of the market for cycling and are increasingly using their own bikes rather than hiring them.

However, Lumsdon (1996b) also provides a more detailed analysis of the market for cycling as Table 8.13 shows. Lumsdon (1997b) cites the continued rise of adult cycle sales in Europe as evidence for the growth of interest in cycling for recreational purposes. Lumsdon (1997b) indicates that in Austria, Denmark, Germany, the

Table 8.13 Segmentation of the cycle market

Type	Profile/Nature	Use of infrastructure	Trend	Spend in local economy	Potential for growth
Day excursion					
1 Half-day and day Casual home-based tourer	Occasional rider from home base. Single and couple, age 24–45. Also families. Increasingly using cars to transport bikes. Cycling approx. 10–20 miles. Socioeconomic spread.	Using back lanes or recognised cycle trails.	Sustained increase	Estimated little expenditure	High
2 Half-day and day Casual mountain biker	Occasional rider from home base. Age 24–45. Higher proportion of males and fewer families. Cycling approx. 10–20 miles. Increasingly using cars to transport bikes. Socioeconomic spread.	Seeking off-road routes of easy to moderate terrain. Potential to saturate popular routes in National Parks, etc.	Sustained increase	Estimated little expenditure	Moderate/High
3 Half-day and day Cycle hire	Infrequent rider – more likely not to have bicycle or use when on holiday. Wider age profile of 18–55. Families strong market. Cycling 10–20 miles. Socioeconomic spread.	Seeking publicised off-road and quiet country routes or historic town trails, (such as Oxford, York).	Strong upward trend in late 1980s. Static at present with growth of cycling ownership	Spend in local facilities more likely	Low in most localities, high in key tourist cones, high potential in historic towns if traffic calming introduced.
Holiday market					
4 'Do it yourself' cycle tourer	Organises day rides or cycling tours from an independent base, Keener cyclists, young people, hostellers increasingly using car to transport bikes. More likely to be professional/ managerial. Use of guidebooks.	Mainly country lanes	Slow growth	Higher spend in local facilities than day market	Moderate

Table 8.13 Cont'd

Type	Profile/Nature	Use of infrastructure	Trend	Spend in local economy	Potential for growth
5 'Do it yourself' mountain biker	As in (4) but seeking more strenuous routes. Fewer families and slightly younger age profile. Use of leaflets and guidebooks. More likely to be professional and managerial.	Heavier impact on off-road routes in sensitive areas.	Moderate and sustained growth	Not quite so high as tourers given nature of activity means less time at attractions, tea rooms, etc.	Moderate/High
6 Organised independent self-guided, cycling holidays/tourers, mountain bikes	Participants book an organised holiday, (routes, accommodation, etc.) but travel as couple or group of friends. They are more likely to be professional and managerial.	Companies offer towns, country lanes or mountain bike options. Impact minimal at present.	Moderate growth	High spend in local economy	Moderate
7 Organised group cycling	As above, but participants make up a group for a guided tour.	As above	Static	High spend in local economy	Low
8 Group holidays	As above but booking made for group as part of multi-activity or cycling holiday. Incorporates day hire of cycle fleets by school and youth clubs.	Usually minimal as leaders choose specific routes e.g. through YHA.	Static	High spend in local economy	Low
9 Club riders	Keen riders; knowledgeable, self-arranged, long-distance day rides and holidays.	Mainly touring, minimal impact.	Static	High spend on holidays	Low
10 Sports competitors	Mainstream cycling as a sporting activity.	Heavy impact, e.g. Kellogg's Tour of Britain, Milk Race.	Static	Limited potential for spend by spectators, media, back-up teams	Spectator sport
11 Events riders	Cycling for charity mainly.	As above	Increasing	Greater potential for spend	Moderate

Note: Estimates in the table are based on evaluation of cycle hire holiday company brochures, qualitative comment by companies.

Netherlands and Switzerland, tourism and recreational networks are now developing which also enhance the image of cycling. The Scottish Tourist Board (1991) outlines some of the constraints on and needs of cycling tourists in Scotland (Table 8.14). Table 8.14 highlights a range of needs and constraints, but probably the most important issue is that of appropriate infrastructure and opportunities for cycle tourism, an issue recently reviewed in New Zealand by Ritchie (1997). Although the Royal Commission on Environmental Pollution (HMSO 1994) identified the impact of other forms of tourist transport on the environment and the role of cycling as a mode of personal transport, it was recognised that it has a limited environmental impact.

The UK's national cycle network

The Royal Commission on Environmental Pollution (HMSO 1994) recommended that cycle trips should be quadrupled to 10 per cent of all journeys in the UK by 2005. Wardman et al (1997) review some of the measures needed to achieve the target of 10 per cent by 2005, using behavioural model-based research. This research has important implications for infrastructure provision. One of the important findings of the Royal Commission (HMSO 1994) was that local authorities in the UK should have a central role in meeting the 2005 targets and in infrastructure provision. In a planning context, this was to be achieved through the existing planning mechanism – the local authority's annual Transport Policies and Programme (TPP) submissions. While the purpose of this was to improve the level of cycle use, it has implications for tourism, which can utilise any infrastructure put in place for residents and leisure users in local areas. It may also assist in reducing fatalities among cyclists (McClintock and Cleary 1996). A number of UK local authorities have appointed cycling officers, who have developed strategy documents for local use, but one of the principal catalysts for facilitating the development of a national cycle network in the UK is Sustrans.

Sustrans is a national sustainable transport and construction company operating as a charity 'which designs and builds routes for people' (Lumsdon 1996b : 10). One of its early aims was to develop a 2,000-mile national cycle network to link all the main urban centres in the UK, using a combination of traffic-calmed roads, cycle paths and disused railway lines and river/canal paths. This aim was realised in 1996 by a grant of £42.5 million from the Millenium Commission to create a 6,500-mile route on the basis of Sustrans' original vision, which would become the UK's National Cycle Network (Figure 8.4). Initial estimates seem to indicate that the network has the potential to generate 100 million trips per annum, 45 million of which will be cycle-based, of which 40 per cent will be leisure-based (18–20 million journeys a year). Sustrans (1995) argues that the network has the potential to generate £150 million in tourist receipts annually and to create 3,700 jobs. This has to be viewed within the context of cycle tourism, since Beoiley (1995) estimated that it generates £535 million a year from leisure day trips, domestic holidays and overseas trips. The C2C route illustrates the generative effect which new cycle routes can have on tourism. It is a 170-mile coast-to-coast route in Northern England which Sustrans (1995) estimates attracted 15,000 mainly cycle tourists in an economically marginally area (West Cumbria and the North Pennines). Sustrans (1995) also has a pan-European perspective on cycle tourism, with its European Cycle Route Network (Figure 8.5). Some of the principal routes are:

Table 8.14   Characteristics and needs of different types of recreational and tourism cycling

Category of cycling activity	Characteristics of users	Main constraints	Main needs	Growth potential
Day touring	Home (or holiday based) excursions for whole or part of day. Trips of 20 miles upwards. Mainly experienced users.	Few constraints although safety reaching minor road network may be a problem. Design of roads a problem in some areas. Rail travel can be restricted.	Safe town/country links, alternatives to busy main trunk roads. Improved access to rail network. Off-road cycleways.	Medium/High
Cycle hire	Casual cycling usually holiday-based for whole or part of day. Experienced and inexperienced cyclists.	Lack of cycle hire centres in some areas. Problems of catering for diverse cycle types and sizes. Only a short season.	Off-road cycle routes in popular areas. Improved publicity and marketing. Need for information on where to cycle.	High
Cycle touring	Extended day touring requiring overnight accommodation. Mainly experienced cyclists with good knowledge.	Difficulties of transporting cycles by rail. Need for alternative routes in town centres/on trunk roads. Accommodation sometimes a problem and conflict with cars in the summer. Cycle repair shops infrequent.	Good rural road network. Varied accommodation from campsites to hostels. Some off-road routes. Improved tourist information.	Medium/High
Organised cycle touring	Extended day touring requiring overnight accommodation. Less experienced cyclists and overseas visitors.	Difficulties of transporting cycles by rail.	Need for back-up services. Quiet rural road network.	High
Mountain bikes	Major growth; car-based and hire-based activity.	Availability of off-highway facilities. Cost of bikes and hire. Lack of certainty about where cyclists can and cannot cycle. Conflict with other users.	Extensive network of off-road routes, e.g. forestry tracks. Improved information on rights of access. Signed trails. Cycle hire.	High

Source: Scottish Tourist Board (1991 : 5).

Figure 8.4   The national cycle network in the UK (redrawn from Sustrans 1997)

- the 5,000 km Atlantis route (Isle of Skye in Scotland to Cadiz in Spain)
- the 470 km Noordzee route (Den Helder in the Netherlands to Boulogne-sur-Mer in France)

Some commentators might view Sustrans' (1996) work as making a valid contribution to local Agenda 21 initiatives, with its close working relationship with UK local authorities. It is also argued that since almost 75 per cent of leisure trips on the National Cycle Network are expected to be new or switched from other modes of transport (Sustrans 1995), it can make an important contribution to sustainable tourism and community-based strategies for environmental management. In fact,

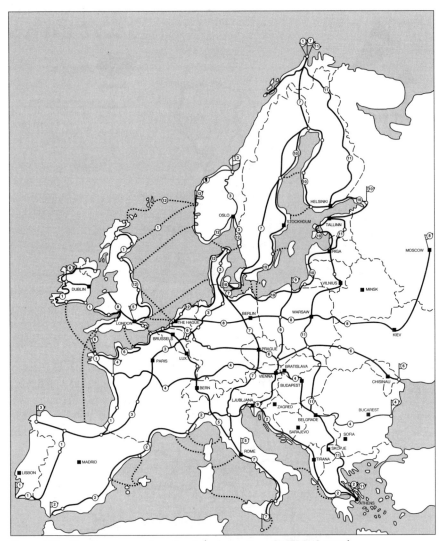

Figure 8.5 The European cycle network (redrawn from SUSTRANS 1997)

the launch of the Kingfisher Trail in Northern Ireland in 1998, as part of the National Cycle Network, is evidence of how the network can also contribute to rural tourism initiatives. The Kingfisher Trail evolved from a desire to harness the popularity of the Shannon–Erne Waterway, using a network of quiet country roads. As a result the trail is marketed as an activity-based rural tourism corridor with cycling as the vital transport link. It is similar in many ways to the widely cited Tarka Trail in Devon (Charlton 1998), which is a 180-mile walking route with a 30-mile cycling route incorporated within it. The scale of cycling on the Tarka Trail was estimated to be 75,000 cyclists in a four-month period, based on evidence from cycling businesses. Therefore, how sustainable is cycle tourism?

According to Lumsden (1996b : 10–12), there are three ways in which the National Cycle Network may contribute to sustainable tourism:

- by encouraging tourists to switch from cars to cycles at their destination, although it needs a cycle-friendly culture to implement such changes in tourist attitudes. Lumsdon (1996b) argues that this could reduce recreational car journeys at the destination by 20–30 per cent.
- by reducing car-based day excursions, particularly at honeypot attractions or sites near to resorts and urban areas. Lumsdon (1996b) views the National Cycle Network as offering tourists 'escape routes' as evidence from the UK's Forest of Dean and Wye Valley implies (Lumsdon and Speakman 1995).
- a growth in cycle-based holidays in both the short break and longer duration category by UK residents and overseas visitors.

Lumsdon (1996b) also provides a detailed study of:

- the market for cycling opportunities
- the supply of cycling opportunities

which is an excellent analysis of the marketing issues that need to be addressed to assist in promoting cycle tourism. However, Lumsdon (1997b : 126) views the development of a cycle culture as vital to encourage the growth of recreational and tourist cycling. Cycle tourism, as the examples of the Tarka Project and Kingfisher Project suggest, is able to make a valid contribution to sustainable tourism development, encouraging less environmentally damaging forms of activity to be developed. Cycle tourism is certainly beginning to assume a much higher profile in the UK, and if leisure use encourages people to become more avid cyclists and to reduce car usage, it will certainly make a valid contribution to Agenda 21 objectives in transport and tourism at a variety of spatial scales.

But how can one evaluate the extent to which sustainable tourism and tourist travel are realistic propositions in the next millenium?

The evaluation of sustainable principles for tourist transport

During the 1980s the concept of mass tourism came under greater scrutiny as a range of influential books questioned whether the economic benefits of tourism were adequately compensating for the increasing environmental impact. As Wahab and Pigram (1997 : 287) argue, 'mass tourism, with the detriments it may inflict on the environment, has been severely criticised as a major environmental predator. It is therefore necessary that tourism adopts a different perspective that should be compatible, for all practical purposes, with the environment and the community in which it is active.' This close scrutiny of mass tourism was followed by the development of 'sustainable', 'responsible', 'green' or 'soft' tourism, and a growing recognition that tourism cannot easily be managed where the carrying capacity (see Pigram 1992 for a discussion of this term) of the environment is greatly exceeded. Marketing strategies with 'sustainable' in their title have emerged as a response to this interest in the environment, but all too often they have failed to grasp the carrying capacity and absolute numbers of tourists which different locations can support. Consequently, tourist transport has contributed to growing pressure on tourism environments by the provision of services to locations that have outgrown their carrying capacity. Therefore, it is not surprising to find criticisms of the sustainable tourism movement, which has been manipulated by certain commercial interests as a new trend they can use to sell tourism and transport services to the more discerning and environmentally aware tourists. Yet as Wahab and Pigram argue,

> Tourism sustainability is a byproduct of a multitude of factors that contribute to the successful present integration and future continuity of tourism at the macro and micro level in the destination. As all socioeconomic, cultural, political and environmental factors are subject to change in time and space, sustainability is therefore a relative term and not an absolute fact. (Wahab and Pigram 1997 : 289)

Despite sustainability being a relative term, there is growing evidence that tour operators and tourist transport providers have seen the positive benefits of appearing to offer sustainable products.

Wheeler (1992a) argues that it is difficult to visualise sustainable tourism as a realistic solution as the world is now experiencing 'megamass tourism', which is viewed as the next stage on from mass tourism. Although sustainable tourism (Wheeler 1992b) is emphasising small-scale individual tourist activities at specific locations and the substitution of the term 'traveller' for 'tourist', a rather elitist movement has developed, supported by a small number of more 'progressive tourists'. Herein lies a major contradiction in the sustainable debate: the insatiable demand for tourist travel is incompatible with the rather up-market, small-scale and expensive form of tourism which only a limited number of tourists are likely to be able to afford. As a concept, sustainability is still in its early stages of development and is unlikely to lead to major changes in the tourist transport system, being more appropriate as a marketing tool for 'new tourism' (Poon 1989). In all probability, sustainable tourist travel cannot be achieved until the concept has been researched further and the fundamental problem of megamass tourism is addressed.

Although the design of resort areas and man-made tourist attractions able to meet the demands of megamass tourism may be able to deal with a high throughput of tourists in a restricted geographical area (e.g. Disneyland – Plate 8.4) there will be a growing demand for tourist transport to reach these artificial and synthetic tourism environments. This may have the temporary effect of reducing pressure on other more fragile tourism environments while the needs of tourists are met by these staged tourist attractions. But the real prospect of environmental damage will emerge if mass tourism trends are based on the search for a more authentic experience (MacCannel 1976). The fundamental problem of environmental impacts is likely to remain and future technological advances may offer a lifeline for the environment if staged tourism can be developed further for the mass tourist, using new ideas such as 'virtual reality' to meet the tourists' need for entertainment, excitement and pleasure.

Plate 8.4 Mass tourism can be managed in man-made environments such as Disneyland, Los Angeles, using visitor management techniques to enable large numbers of people to queue (Source: S.J. Page)

As tourist transport operations are usually characterised by private sector ventures, voluntary agreements have typically been the basis for environmental management policies. Government organisations can assist in this process by ensuring that legislation is in place to encourage a reduction in environmental pollution from transport. Government commitment to sustainable transport policies in developed countries seems to have foundered as the decision to deregulate transport is unlikely to see such options implemented, given the reliance on profitability in tourist transport provision rather than environmental issues. Pigram (1992) argues that in the process of policy formulation and implementation of sustainable tourism options, it is important to recognise the role of major decision makers such as transport operators in influencing the long-term success of such schemes. Even so, Wahab and Pigram (1997) identify a very worrying trend which may militate against developing sustainable tourist transport systems because

> a growing trend in policy making in many countries is to leave tourism to private enterprise, and current economic conventional thinking supports the role of market mechanisms, the pendulum shows signs of change in theory and in application . . . countries such as Canada and the United States have reduced or abolished the role of the public sector in national tourism administration in favour of private enterprise. State tourism bodies remain as promotion agencies at the macro level. (Wahab and Pigram 1997 : 289)

In other words, those public sector bodies able to understand and make the links between transport, tourism and sustainability at a policy-making level are no longer able to fulfil that role. This is certainly the case in New Zealand in relation to sustainable tourism (Page and Thorn 1997).

Ultimately, the tourists' desire for international and domestic travel may need to be the focus of long-term educational strategies to identify some of the problems travel, tourism and transport pose for the environment, and codes of practice may assist in that respect. One radical solution may be to increase the cost of travel and introduce government regulations to restrict the demand in order to reduce the impact on the environment. Yet there are many political and ethical objections to such an approach since it is reminiscent of the situation in some Eastern European bloc countries before the collapse of communist rule and it would be tantamount to an infringement of individual freedom in democratic societies. Increasing the cost and restricting the opportunity to travel has other social implications because it may run contrary to the objectives of social tourism which aims to make travel and holidays accessible to all social groups. A partnership approach between responsible transport and tour operators and governments committed to making tourists more aware of their own actions may be one way forward. Instead of tourist travel being regulated, tourists should be encouraged to exercise greater restraint in their demand for travel, though it is evident that there is no short-term solution to preventing the environmental impact of tourist transport systems.

Summary

The human effects and environmental consequences of tourist transport have led to a greater awareness of how tourist transport systems interact with the human and physical environment. The concept of 'sustainable tourism has burdened itself with incompatible conflicting objectives – small scale sensitivity and limited numbers to be achieved in tandem with economic viability and significant income and employment impacts' (Wheeler 1991 : 95). In other words, sustainable tourism's implicit assumption that smaller-scale tourist activities will result from such developments could pose threats to the economic threshold at which tourist travel services are provided. If sustainable tourism were viewed as the only legitimate form of tourism, it would have unrealistic social impacts by limiting travel to a privileged minority. This

> appeases the guilt of the thinking tourist while simultaneously providing the holiday experience they or we want. The industry is happy because the more discerning (and expensive) range of market can be catered for by legitimately opening up new areas to tourism. (Wheeler 1991 : 96)

This highlights the rather superficial nature of the sustainability concept (Wheeler 1994), which does not really offer any long-term solutions to the tourist and the transport provider because it fails to address the global impact of tourism, which is too large a problem for governments and transport operators to address in isolation. Even if tourist transport providers and tour operators withdrew from carrying tourists to sensitive environments, the competitive nature of tourist transport provision in market economies would mean that another rival operator, with less interest in environmental issues, might enter the market. Environmental Auditing and Environmental Impact Assessment are moving the transport business towards considering the consequences of transporting tourists to different environments but the reliance on private sector cooperation in minimising their impact may only result in action:

> where the benefits [of environmental auditing] are largely enjoyed by third parties or the general public . . . [but] . . . if consumers' current search for quality embraces an increasing environmental awareness, the tourism industry would face demand-led pressure to adopt environmental auditing more widely. (Goodall 1992 : 73)

Even so, the growing interest in alternative modes of transport, such as cycling and government commitment to fund infrastructure such as the National Cycle Network, are encouraging signs for more sustainable forms of tourism transport for the next millenium. Change does not take place rapidly, as a major culture shift is needed in tourists' and transport providers' attitudes, so that more environmentally sensitive and sustainable tourist activities are promoted.

Questions

1. How would you classify the environmental impacts of tourist transport?
2. What is sustainability? How has it been used by tourist transport operators to market their services?
3. Is cycling the only sustainable form of tourist transport with a minimal impact on the environment?
4. What is an Environmental Audit? How can it help a tourist transport operator to assess impacts on the environment?

Further reading

The literature on sustainability and tourism has expanded rapidly in the 1990s. An interesting array of articles can be found in the *Journal of Sustainable Tourism*. The following books are also a good introduction to the topic:

Hall, C.M. and Lew, A. (eds) (1998) *The Geography of Sustainable Tourism*, London: Addison Wesley Longman.
Middleton, V. (1998) *Sustainable Tourism: A Marketing Perspective*, Oxford: Butterworth-Heinemann.

On the environmental practices of tourist transport operators, see:

BA (1997) *Annual Environmental Report*, Harmondsworth: British Airways.

Internet sites

The range of sites with material of interest is expanding rapidly, but the following offer a range of commercial, charitable and critical perspectives on sustainability and the role of transport.

On the development of sustainable transport and cycling see:

http://www.sustrans.org.uk

A critical analysis of tourism can be found at the site developed by Tourism Concern, a charity set up to promote awareness of the impact of tourism on people and their environments:

http://www.gn.apc.org/tourismconcern

Chapter 9

Prospects and challenges for tourist transport systems: towards quality service provision

Introduction

This book aims to raise awareness of the relationship between tourism and transport by developing the concept of a tourist transport system as a means of analysing the processes shaping the provision and consumption of transport services by tourists. Throughout the book, transport is emphasised as a dynamic and active element in the tourist's experience of travelling because it is a vital part of the process of tourism. Some of the first-generation tourism textbooks (e.g. Mathieson and Wall 1982) regarded tourist transport as an essential part of tourism but not worthy of study in its own right. In fact, a number of subsequent texts (e.g. Cooper et al 1993, 1998) continue to view transport as a passive element in the tourist experience (Ryan 1996) and it remains a descriptive feature of most texts.

In Chapters 1 and 2, the scope of multidisciplinary research on tourism and transport is reviewed in terms of the concepts and methods of each discipline (economics, geography, marketing and management) used to analyse tourist transport. However, the different philosophical backgrounds of researchers from these disciplines mean that their approach to tourist transport is not easy to synthesise into a holistic framework. Moreover, the tendency for researchers to retain their disciplinary training – whether in economics, geography, marketing or management – has simply contributed to the growing body of knowledge on transport and tourism. For our understanding of tourist transport systems and the tourist's experience of travel to grow, a greater degree of coherence and a theoretical basis needs to be developed. This means that research will need to be interdisciplinary in nature. Interdisciplinary research requires people from different disciplines to collaborate and focus on a specific research problem, where different questions are asked about the topic without each researcher losing sight of the problem under consideration. This may help to integrate the contributions which different disciplines can make to the analysis of tourist transport systems in order to achieve a more holistic understanding of the operation, management and use of transport services by tourists.

Although there is not space within this introductory book to undertake a comprehensive review of transport and tourism, it has sought to focus on how the consumer, provider and other agencies (e.g. national governments) interact in different transport systems. The concept of a tourist transport system was developed as a framework in which to understand the interrelationships between different elements in such systems. Using a systems approach to the analysis of tourist transport also highlighted the importance of *inputs* to the system (e.g. the demand and supply) as well as *controlling influences* (e.g. government policy) and *outputs* (the tourist travel experience) and the effect on the environment. The book has also sought to identify a number of processes which characterise the tourist transport system. For example, deregulation and privatisation is a process now affecting tourist transport systems in North America, Western Europe and Australasia (Button and Gillingwater 1991) as well as communist states such as China (Taplin 1993). Within the existing literature, the discussion of tourist transport systems has remained fragmented and dependent upon generalised and empirical studies or extremely specialised studies of both tourism and transport. The interface between tourism and transport has not been integrated into a holistic framework. Whilst tourism is now regarded as a complex phenomenon by educators and researchers, its frequent association with transport has meant that social science researchers have failed to integrate these issues in a framework where the complementarity between tourism and transport could be explored further. The tendency within tourism research to focus on typologies of tourism and tourists has led to a critical separation of the tourist from the mode of transport they use. This has the effect of contributing to the separation of tourism and transport research, with tourist motivation to travel viewed in isolation from the process of travelling. The result is that tourist travel is divided into two discrete elements (transport and the tourist) rather than being conceptualised as a continuous process using a systems approach. But what are the processes shaping tourist transport in the new millennium?

Tourist transport provision in the late 1990s and the new millenium

One of the overriding themes affecting the tourist transport system is globalisation, especially in those sectors which deal with the management and logistics of international travel (Lovelock and Yip 1996). Globalisation inevitably produces 'winners and losers' in the pursuit of business and four distinct processes are associated with it. These are:

- *Deregulation*, where the entry barriers to many sectors of the tourist transport business have been removed and large oligopolies are challenged by new entrants (Pearce 1995b). As the example of the US domestic airline industry illustrates, in newly deregulated industries, competition increased at a rapid pace. However, there is debate within the American domestic airline industry as to whether consumers have been the main beneficiaries, with lower prices. Goetz and Sutton (1997) explain that the benefits of deregulation have accrued to those passengers travelling on trunk routes, while business travellers and passengers travelling to/from more peripheral locations have experienced higher fares.

- *Technological change*, which has revolutionised the organisation, management and day-to-day running of tourist transport businesses with the introduction of information technology (IT). IT has also helped reduce some of the costs of business operations. The introduction of CRSs and GDSs have certainly assisted with the globalisation of the supply of tourist transport services. The introduction of the Internet has also had a major impact on the supply of transport services (Macdonald-Wallace 1997). In fact many of the world's airlines now have Internet sites (see Appendix 3) and as Whitaker and Levere (1997) show, some are being used for bookings, but 'the scope and standard of airline-related material on the Internet varies dramatically'. In fact the evolution of Internet sites and their use in marketing have now moved beyond a tool simply to advertise and sell tourist transport services. This traditional use, based on sales and marketing, is reflected in the UK express coach network site – http://www.nationalexpress.co.uk. However, there is evidence that some companies (i.e. Red Funnel Ferries in Southampton) are developing a more holistic approach to transport and tourism and using the Internet to address the impact of competitive forces such as rival carriers. The company's Internet site – http://www.redfunnel.co.uk contains the traditional sales and marketing function. But it also moves into a tactical marketing role where bookings can be actioned and place-marketing is undertaken in relation to the main destination they serve – the Isle of White. The website provides ideas for themed itineraries and the main attractions to visit which complement the tourism marketing activities of the public sector (e.g. the Southern England Tourist Board). This is certainly leading the way in providing a seamless tourism experience facilitated by technology and the activities of the transport operator. Some airlines also offer sophisticated systems allowing passengers to plan, book and pay for their flights; others can master little more than sketchy corporate information (Whitaker and Levere 1997 : 27).
- *Regional change*: the highest costs for air travel remain in Europe and North America whereas in other trading blocs such as ASEAN, lower costs exist. For tourist transport providers in the global economy, it can mean airlines are competing on a different cost basis as Hanlon (1996) observes in terms of regional wage rates and remuneration of airline employees.
- *Hypercompetition*: Within the global marketplace, tourist transport providers are facing pressures continually to improve products and to remain competitive. In some cases, organisations are constantly struggling to remain in business as experience in the international airline industry suggests. As the privatisation characteristic of the 1990s and deregulation (see Meersam and van de Voorde 1996) seem set to continue, established industry leaders and oligopolies find their position challenged or destroyed by fierce competition. According to D'Aveni (1998), this hypercompetition is typified by:

  - rapid product innovation
  - aggressive competition
  - shorter product life cycles
  - businesses experimenting with meeting customers' needs
  - the rising importance of alliances
  - the destruction of norms and rules of national oligopolies.

  D'Aveni (1998) identifies four processes which are fuelling hypercompetition:

  - customers requiring better quality at lower prices. One of the innovations airlines have pursued to develop improved quality at lower prices is in-flight catering (Jones 1995)

- rapid technological change, especially the use of IT
- the rise of aggressive large companies willing to enter markets for a number of years with a loss-leader product in the hope of destroying the competition and capturing the market in the long term
- government policies towards barriers to competition are being progressively removed. This is evident in the tourist transport sector throughout the world, albeit to differing degrees depending on the political persuasion and commitment to deregulation.

At first sight, D'Aveni's (1998) processes are not particularly different from those listed under globalisation (i.e. deregulation, technological change, consumer preferences and regional change). But the fundamental difference lies in the business strategy of hypercompetitors. As D'Aveni (1998) argues, hypercompetitors tend to destroy the existing competencies of businesses. Those affected by such change are often trapped by an inability to think laterally and to adopt new competencies. Even when new competencies are introduced, businesses often have difficulty in diffusing them throughout their organisation. Some belatedly look towards the concept of 'change management' but this can sometimes be too little action too late. Often firms are so severely affected by hypercompetitors and their action, that their responses are bound by age-old reactions based on previous rules of competition. However, the hypercompetitor can only remain in a competitive position while it retains the advantage.

According to D'Aveni (1998), hypercompetitors enter the market by disrupting the competition in some of the following ways:

- by redefining the product market, thereby redefining the meaning of the quality while offering it at a lower price. This is the strategy adopted by EasyJet in the UK which entered the market with low-cost air travel from Luton Airport to challenge the market leaders (e.g. BA, British Midland and KLM UK).
- by modifying the industry's purpose and focus by bundling and splitting industries. BA's response to EasyJet was to reduce fares in the short term, but then it provided a splitting action by establishing a similar low-cost operation based at London Stansted, with lower landing fees. This avoids eroding profit margins and using high-cost airline capacity from Heathrow and Gatwick. In other words, BA can operate a loss-leader small business to compete head on with EasyJet on equal terms. A similar response occurred in New Zealand in the mid-1990s when Air New Zealand established a low-cost airline (Freedom Air) to compete with the Hamilton-based airline Kiwi Air.
- by disrupting the supply chain by redefining the knowledge and know-how needed to deliver the product to the customer.
- by harnessing the global resources from alliances (see Dresner and Windle 1996) to compete with the non-aligned businesses. This is particularly acute in the airline industry although to date the term 'hypercompetitor' has not been used to describe the business strategy of key players.

The process of globalisation and hypercompetition are powerful forces affecting the tourist transport sector and a number of themes emerge which are worthy of further discussion:

- the role of the consumer
- the growing significance of service quality
- the introduction of Total Quality Management Systems.

The tourist as a consumer

Much of the rhetoric and hype associated with the rapid expansion of popular business books and the elevation of individuals to 'guru' status in the 1980s and 1990s is characterised by one consistent theme: that businesses need to understand the customer and to get near to them as 'end-users'. The tourist transport system is no exception to this as Chapter 1 has shown.

Swarbrooke (1997 : 67) reiterates the importance of consumer behaviour research in tourism, since from a tourist transport perspective it allows businesses to plan infrastructure developments, identify product opportunities, set price levels for products and identify market segments and the best marketing medium to promote the product. Consumer behaviour research also allows businesses to modify their product and its delivery to align it more closely with consumer expectations. For the tourist transport business, understanding how tourists make their purchasing decisions and the factors affecting their choice of product is critical. In particular, the travellers' predisposition towards certain forms of transport will obviously affect their overall satisfaction with the product. For the tourism sector in general, Swarbrooke (1997) identifies a number of weaknesses in consumer behaviour research in the UK which are particularly relevant to the transport sector (although the exception may be the major airlines who commission in-house research that remains confidential and commercially sensitive). The main weaknesses are:

- an absence of reliable and up-to-date data, a feature emphasised in Chapter 4
- a lack of longitudinal studies to trace the evolution of consumer behaviour in tourism through time
- the methodologies and techniques used to collect data on consumer behaviour in tourism remain relatively crude and unsophisticated
- the most robust data collated by private sector companies remains inaccessible to researchers
- methods of segmenting the market remain outdated due to a reliance on the life-cycle concepts and age, despite major societal and value changes which have questioned their validity in the late 1990s
- cross-cultural differences in tourism markets and a predisposition towards using specific tourist transport modes remain poorly understood. The research identified by Lumsdon (1997), in part, addresses some of these issues in relation to cycling
- there are few media available to disseminate results to the practitioner audience.

As a result, consumer behaviour is one area which tourist transport operators will need to focus on if they seek to understand what motivates tourists to travel, and to select specific modes of transport.

Tourist transport systems are likely to be affected by various opportunities and constraints on tourist travel in the late 1990s and beyond. For example, congestion of airspace in developed countries such as North America and Western Europe (French 1994, 1997b) will remain a persistent problem for policy makers and transport planners in the late 1990s and new millennium. At the same time the demand for long-haul travel in developed countries is set to expand in the late 1990s, which may pose opportunities for transport providers and tour operators if constraints cannot be overcome. Environmental issues will also feature more prominently in tourist transport systems as a new generation of travellers, having become familiar with green issues in the 1980s, emerge as consumers of tourist transport services. Understanding the relative importance of these factors in shaping the tourist's desire to travel on different modes of transport will be a major challenge for service providers, as the sustainability debate (see Weiler 1993) focuses on more environmentally sensitive and novel modes of transport.

Increasingly, the patronage of tourist transport services is going to depend upon the ability of providers to differentiate their services on the basis of image, market positioning and reputation for service quality. The 1990s are emerging as the decade of the consumer in relation to tourist travel, with providers responding to legitimate requests for higher standards of comfort, reliability and courtesy as part of the travel experience. The new millennium is also set to see a continuity and intensity of these processes of change, while the discussion of globalisation and hypercompetition indicates that the pressure on transport providers will intensify. Passengers are now recognised as customers and their rights and needs are beginning to gain a higher profile in the provision, quality and management of tourist transport services.

Service quality issues in tourist transport

In Chapter 2, the concept of service was introduced in the context of marketing. While that discussion provided a broad overview of the importance of service issues in tourist transport, it is evident from the processes affecting the tourist as a consumer, that service quality is assuming a greater role in their purchasing decisions and travel behaviour. Irons (1994) argues that services are relationships and that whether that relationship is a transient one or a longer-term proposition, it needs to be conducted in a professional and consistent manner. As Irons (1994 : 13) shows,

> Such a relationship will be based on a series of contacts or interactions. It is from these interactions with the organisation that consumers form their perceptions . . . to assess value, decide to buy, repeat purchase or recommend to others.

Such interactions are also repeated within the organisation and Irons (1994) expresses this process as a triangle (Figure 9.1). Irons explains the triangle in the following way:

- An organisation need to associate its internal culture with the one it portrays externally and this underpins the relationships evident in Figure 9.1.
- Within the organisation, power needs to be devolved so that the relationships can be developed and the appropriate skill and know-how is provided at the point where customer satisfaction is met.
- The organisational values and culture need to be clearly understood by all employees so that they affect their actions and activities in relation to customers.
- Managers need to lead the process, empowering people at the various levels in the organisation to achieve customer-related targets. In other words, managers need not only to exercise a degree of control in the management function, but also to lead the organisation in this era of the consumer.

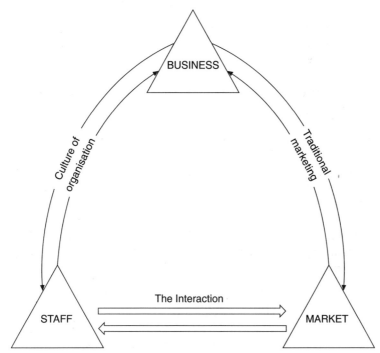

Figure 9.1   Iron's service quality triangle (redrawn from Irons 1994)

- A customer focus is critical rather than a focus first on the product and then its purchasers.

To create a services culture in an organisation, Irons (1997) identified the following key points:

- Service businesses need to identify what the priorities are for the customer. Irons (1997) cites the example of Southwest Airlines in the USA which saw a set of priorities – reliability, low fares, personal treatment – and set about 'rigorously building the airline around meeting these needs and cutting out those things the customer did not want' (Irons 1997 : 8).
- Organisations need to develop a clear vision of 'what they stand for and where they aspire to go . . . This vision should be for the customer, for the staff and for the owners' (Irons 1997 : 9).
- Organisations and employees need to communicate so that they understand what is to be achieved, why, how and the role of employees in the corporate vision.
- The organisation needs to learn from its experience through problem-solving and how this can benefit its vision.
- The service culture needs to be led from the top in the organisation rather than through passive forms of managerialism.
- It is at the point of interaction between the market and the consumer that value can be created.
- Service delivery is an integral part of the process for service organisations and it should drive the business.

While the principles outlined by Irons (1994, 1997) may be useful in outlining how businesses may create a service culture, at a practical level the service requirements of the tourist transport sector need to be examined in more detail. This is because in certain sectors of the tourist transport business, service qualities offer particular challenges to operators because of the nature of the service interaction. It should also be emphasised that in some cases, tourists' expectations are rising beyond the reach of mass transport providers and their ability to meet these needs.

Within the literature on tourism and transport there are comparatively few systematic reviews of service quality. While studies reviewed in Chapter 3 on rail travel highlighted the experience of InterCity prior to privatisation, few other reviews exist. Those studies which have been undertaken have largely focused on the airline sector (e.g. Ostrowski et al 1993, 1994; Van Borrendam 1989). Probably the most influential publication to date is that by Witt and Mühlemann (1995) which not only reviews the previous research in the area, but also identifies the idiosyncrasies and conditions which influence service quality in airlines.

Service quality: conceptual issues

Ostrowski et al (1993) argued that service quality issues were comparatively poorly developed in the airline industry, based on a survey of 6,000 travellers using two US airlines. They concluded that there was considerable scope for improvement. Hamill (1993) places this in the context of changes in the aviation sector in the 1990s, with deregulation and privatisation forcing companies to become more customer-oriented. This process is continuing in the late 1990s in Europe with liberalisation (Graham 1998), since Lufthansa prepared itself for privatisation in 1997. By January 1998, at least four other European flag carriers had prepared for privatisation (Air France, TAP Air Portugal, Alitalia and LOT of Poland). Hamill (1993) also points to the strategies pursued by some airlines, where the use of computer reservation systems (CRSs) was seen as improving service. Yet this development has no direct impact on the actual service encounter and such a perception is further evidence of the need for a focus on service quality issues.

Witt and Mühlemann (1995) explore the problem of establishing a working definition of service quality. Gronroos (1984) introduced the idea of a *technical quality* dimension (the customer interaction with the service organisation) and a *functional quality* dimension (the process through which the technical quality is delivered). As a result, the consumer's perception of service is a result of the service dimensions combining technical and functional aspects. In contrast, Gummesson (1993) argued that four qualities affected customer perceived satisfaction. These were:

- design quality
- delivery quality
- relational quality
- production quality.

A further model of quality was developed by Zeithmal et al (1990) which was based on gap analysis and focused on four dimensions:

- customers not knowing what to expect
- inappropriate service quality standards
- a service performance gap
- company promises not matched by delivery

To evaluate quality, Zeithmal et al (1990) used ten dimensions which were reduced to five elements:

- tangibles
- reliability
- responsiveness
- assurance
- empathy

which are combined in the SERQUAL model used to evaluate customers' perceptions of quality. While SERQUAL and measures of perceived quality were certainly dominant elements in the research agenda in the late 1980s and early 1990s, Witt and Mühlemann (1995 : 34) argue that 'the successful organisation will be one which establishes a total quality culture' based on total quality management (TQM). But what is TQM, where does it originate from and how will it affect tourist transport providers?

Total Quality Management

It is widely acknowledge that the 1980s saw many service providers in North America respond to a perceived 'quality' crisis posed by products and services offered by rivals in the Pacific Rim (Deming 1982). Many service providers responded with corporate strategies focused on quality issues as a method of retaining market share. Yet if the late 1980s were characterised by a business environment committed to quality, the 1990s were dominated by total quality management (TQM) as a more sophisticated form of recognising customers' needs as an integral part of an organisation's goals. TQM developed as a corporate business management philosophy and it even has an academic journal – *TQM* – devoted to research in this area. Why should this be of interest to the tourist transport system in the 1990s? The growing concern for consumers, quality and total supply management in the tourist transport system is part of the move towards TQM among service providers. Furthermore, TQM is likely to assume a greater role in academic and commercial research on tourist transport in the 1990s.

TQM is an all-embracing approach which enables an organisation to develop a more holistic view of consumers, quality issues and service provision as an ongoing process. Yet one of the principles of TQM – the concern for quality – is explicitly dealt with in detail in this book. One difficulty is in establishing a universal definition of quality which could be applied to tourist transport systems. Dotchin and Oakland (1992) provide an excellent review

of this issue, citing the work by Townsend and Gebhart (1986) which distinguishes between the subjective evaluation of quality by the customer (quality
of perception) and the provider's more objective assessment (quality of fact).
Clearly the meaning of quality will vary according to the context and the perception of who is establishing what can be deemed as quality, as the discussion
of conceptual issues of quality showed. While the journal *TQM* contains
many interesting discussions of this issue, operationalising TQM in a tourist
transport context requires organisations to work towards specific goals focused
on an agreed concept of quality. Corporate commitment is required so that
TQM permeates all areas of the company's business. TQM also provides an
organisation with the opportunity to monitor and implement internal procedures and to control suppliers using established quality standards such as
BS 5750, as discussed in Chapter 6 in relation to the use of logistics solutions,
and supply chain management, as discussed in Chapter 5.

One of the real challenges for TQM in tourist transport systems is to
establish what the customer considers as excellence in service provision and
the design of service delivery systems to deal with individual tourists' requests,
requirements and needs. Many corporations involved in tourist transport provision are trying to make individual tourists feel more valued as customers
but, until delivery systems are able to deal fully with this issue, operators will
be unable to claim success in TQM. It is at the strategic policy and planning
stage that organisations may need to agree on how to improve continuously
and strive for quality in service provision so that the tourist's travel experience
is enhanced. One challenge is to ensure that the process of travel is not perceived as such a mundane and stressful experience for some tourists.

Implementing a TQM strategy is no easy task for organisations where it
may involve a change in corporate culture. Nevertheless, a number of critical
factors characterise success in TQM in service provision. As Table 9.1 shows,
senior management set on developing a policy for TQM will need to follow
certain principles and management strategies. Many of the principles discussed
in Table 9.1 expand the ideas developed by Irons (1994, 1997) on developing
a service culture, while TQM is a more systematic attempt to ensure quality is
dealt with in a consistent manner. Witt (1995) cites Oakland's (1989) route
to implementation as a series of steps which are outlined in Figure 9.2. Oakland
(1989) explains that the CEO of any organisation must begin by *understanding*
the concepts of TQM and the route to implementation. This then needs to be
followed by *commitment and policy* to set out what the organisation hopes to
achieve from its quality strategy. Following this, it may be necessary to alter
the *organisational structure* to fit with the new ethos. In terms of *measurement*,
the inputs (raw materials), output (product), performance of employees and any
costs of failure need to be quantified. Even though it is often hard to measure
intangible elements in a service, the SERVQUAL survey tool might be used.
The process of *planning* is the next step, to assess the nature of the service
process, who it serves, when and where. This is a good point to use the results
of the SERVQUAL survey to plan changes to the delivery of the service. The
next stage is called *systems*, where a quality manual is produced to explain how

Table 9.1   Implementing a Total Quality Management programme

Senior management in an organisation seeking to implement a TQM programme should consider:

- an organisation needs long-term commitment to constant improvement
- culture of 'right first time' is required
- employees need to be trained to understand customer–supplier relationships
- purchasing practices need to consider more than just the price – they must also consider the total cost
- improvements in delivery systems need to be managed
- the introduction of methods of supervision and training needs to be explained to avoid fear and intransigence
- breaking down interdepartmental barriers by managing the service process to improve communications and teamwork
- eliminating
  - goals without methods
  - standards based only on numbers
  - fiction, get facts by using the correct tools (e.g. by using appropriate research techniques)
- developing an ongoing human resource management strategy to develop experts and 'gurus'
- developing a systematic approach to managing the implementation of TQM.

The implementation of a TQM programme can be shaped using these principles to achieve: *outcomes* which involve:

- the identification of customer–supplier relationships
- managing processes
- cultural changes
- commitment

which may need to be accompanied by management necessities including:

- systems based on international standards
- teams to monitor and improve quality throughout the systems
- tools to analyse and predict what type of corrective action is needed to improve quality.

Source: Modified from Dotchin and Oakland (1992 : 142).

the company undertakes its quality policies, with the management systems in place. This is also an opportunity to specify the nature of the product being delivered and how it is produced. The term *capability* refers to the next stage where the organisation can assess whether it has the ability to meet each customer's set of requirements or if modifications are needed. This is followed by a *control* function to ensure the service is delivered in a consistent manner, within acceptable tolerance levels, on each occasion. Since service delivery often involves more than one person, the role of *teamwork* needs to be considered. This may also involve the use of quality circles in the organisation, where employees work in teams to solve problems and promote a commitment to quality. To ensure a continuous improvement in quality, *training* is essential. At the top of the steps is TQM *implementation*. Porter and Parker (1992) note that management behaviour and their willingness to carry through such programmes is often the key to the successful implementation of TQM.

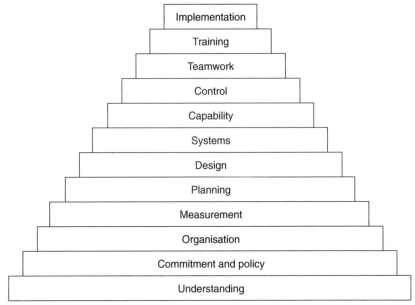

Figure 9.2   Implementation of a TQM programme for a tourist transport operator

However, interest in TQM is no substitute for the organisational and logistical skills involved in coordinating and managing tourist transport systems. Conveying large numbers of people over short and long distances for pleasure and business is a complex process requiring a great deal of planning and organisation on a day-to-day basis as well as in the longer term. Adding a concern for quality provision in this process makes the delivery of services a more complex undertaking and it is not surprising that service interruptions occur due to the sheer volume and scale of people handled in tourist transport systems. But when things do go wrong, companies and their front-line staff must be empowered to deal with incidents, or systems must be in place to deal with crises when they occur. Whether a tourist transport provider prefers a gradual improvement approach to quality or a TQM approach, it is worth considering some of the impediments to quality improvements in relation to the airline sector.

Quality issues in the airline industry

Witt and Mühlemann (1995) identify two persistent problems to meeting travellers' requirements:

- How may a quality service be defined, what factors influence the customer's experience and how may these factors be identified?
- How may performance or delivery of the product be measured or monitored, given the intangible nature of services? (Witt and Mühlemann 1995 : 35)

Witt and Mühlemann (1995 : 35) identify the following five factors which may pose particular problems

- the mixed nature of airline markets, where leisure and business travellers may be mixed on any flight, each with different requirements
- lack of direct control over factors contributing to the traveller's experience, including:

  - ticket purchases from travel agents, which can involve mistakes in ticketing
  - experiences at the airport (e.g. air traffic control problems and weather conditions)
  - the impact of airline alliances, where partner airlines may not have harmonised standards of service to ensure a consistent quality throughout the journey regardless of the carrier.

- congestion and slot availability, where large carriers dominate the main slots at a time when air travel in Europe and North America is becoming more congested
- restrictions versus deregulation, where spatial inequalities occur in service provision depending upon the traveller's location in the system and choice of route, as Goetz and Sutton (1997) observe in the USA
- differentiation in the product, where the airlines seek to segment the market and attract more travellers through the use of marketing tools such as frequent flyer programmes (Mason and Barker 1996; Beaver 1996).

Some airlines, such as SAS, have implemented quality management systems moving near to TQM and BA is a further example of an airline which has attracted a great deal of attention in the research literature for its focus on quality (see Hamill 1993 for example). KLM is also implementing a full TQM scheme (Van Borrendam 1989). However, as Witt and Mühlemann (1995 : 39) suggest, 'Singapore Airlines is probably the best known for customer focus', as confirmed in Chapter 5. Aside from quality issues, there are a range of other themes likely to affect tourist transport in the next millennium.

Government policy, planning and investment in infrastructure assume a significant role in facilitating the efficient movement of people for the purpose of tourism. In this context, the London Tourist Board's (1990) *At the Crossroads: The Future of London's Transport* reaffirms the essential relationship between transport and tourism dealt with in Chapter 1. The London Tourist Board study is unique in this respect since it recognised that:

- an efficient transport network is necessary for tourists to gain access to a destination such as London; tourism would not exist without a transport network as it is part of the tourism infrastructure
- an integrated transport network with convenient transfers between different modes of transport is essential, with reasonably priced travel options
- within the destination, tourists need a choice of transport to transfer between the port of arrival and their final destination
- investment in public transport provides social, economic and environmental benefits for both residents and tourists alike. Investment in transport infrastructure is a long term proposition and is unlikely to yield tangible benefits in market-led economies in relation to tourism. Yet without it, tourism would not be able to develop.

As the London Tourist Board Study notes, the development and long-term prosperity of tourism depends on transport both to make destinations accessible and to facilitate tourist travel within the destination area. Efficiency, safety and ease of travel and convenient interchanges are likely to be viewed as important performance indicators by users of tourist transport systems. These principles apply to the wider context of tourist travel and making the travel experience more rewarding is one major challenge for all parties involved in providing tourist transport systems.

Within the airline sector, concerns with dropping yields and moves to secure the loyalty of economy travellers continue to face many companies. In terms of consumer behaviour, economy class travellers tend to seek the cheapest fare, which means a greater emphasis on securing the loyalty of commercial passengers, regardless of whether they travel economy or business class. Even some of the British railway companies, such as Midland Mainline, which operates London to the East Midlands/South Yorkshire services, have recognised this. Their introduction of a premium business service follows the same principle for securing the commercial traveller. Many of the world's airlines have turned to the creative flair of advertising agencies to appeal to their prestige markets such as the business traveller. For example, in August 1997 Air New Zealand launched a new advertising campaign designed by Saatchi and Saatchi. At the same time, a database marketing company worked with Air New Zealand to identify its Koru Club and Air Points members. A 'teaser' was then set to select a few members and encourage them to watch the advertisement when it was first screened, motivated by a prize opportunity. The chosen few then received a follow-up mailing some weeks after the advertisement extolling the virtues of flying Air New Zealand. At the same time Air New Zealand customer support staff also received a newsletter to explain the focus of the campaign on business travellers, highlighting what they needed to do to show that Air New Zealand is unique.

What such campaigns show is that the marketing activities are selective and based on the concept of hand-picking. While the airline continued to use powerful national icons, such as the Koru image which is displayed on the tail of every aircraft, it is apparent that the targeting of high-yield travellers is now becoming the key focus for airlines. Such activities are likely to continue in the transport sector, as businesses seek to improve yields.

In a similar vein, airlines are also securing software which will improve yields. For example, Air New Zealand already uses Sabre Technology solutions in yield management and flight planning systems. In December 1997, Air New Zealand was in the process of signing a contract with Sabre to build a new Internet site and to introduce new software to replace its flight operation and crew management systems. This illustrates that IT is enabling tourist transport operators to remain competitive and hopefully to reduce operational costs.

Transport operators are also turning to new solutions to reduce other components of their operational costs. For example, in May 1997 Cathay Pacific launched new Airbus A340 services on its Auckland–Hong Kong route – Airbus Industries claims that the fuel cost per seat of the A340 is 40 per cent of that for a Boeing 747. This was part of a $US9 billion fleet replacement programme for Cathay Pacific. Fleet replacement costs represent a perennial

Table 9.2   The potential for cost reductions among airlines

Cost items	Cost drivers		
	Route network	Fleet composition	Company policies
Aircraft crew costs	XXX	XXX	XXX
Engineering overheads	X	XXX	
Direct engineering costs	X	XXX	X
Marketing	XXX		X
Aircraft standing	XXX	X	
Station and ground services	X		X
Passenger services	X		X
General and administrative costs	X		X
Fuel		X	
Airport and en route costs	X		
Direct passenger service			X

Notes
XXX Significant cost reduction potential
X   Some cost implications
Source: Modified from Seristö and Vepsäläinen (1997 : 21).

problem for many airlines. The capital cost of fleet replacement means more innovative solutions need to be sought such as lease-buy schemes, manufacturer funding and straight lease schemes. This frees airlines from major sunk capital costs over and above those needed to service debt repayments on leases. Even so, cash-rich airlines such as Singapore Airlines continue to purchase aircraft and options on future aircraft rather than seeking leasing options (see Chapter 5).

A recent study by Seristö and Vesäläinen (1997) offers a number of insights into the actual cost and revenue factors associated with airline operations. This is important in an age of cost-competitiveness, especially when airlines have been trimming staffing levels (Alamdari and Morrell 1997) and salaries in the 1990s to remain afloat. Yet as Seristö and Vepsäläinen (1997 : 11) argue, 'for many a carrier ever more critical measures will be needed to achieve sustainable profitability', which is also relevant to the wider tourist transport sector. In the analysis of cost drivers in 42 of the world's airlines, a number of variables were examined:

- the fleet composition of airlines
- the flying personnel used, particularly the number of flight crew per aircraft
- the route network
- cost drivers, operating expenses and profitability in terms of:
  - the composition of traffic
  - route structure
  - salaries/remuneration levels.

Using quantitative research methods (e.g. factor analysis), the variables were analysed and a model was built (Figure 9.3). This model highlights how various factors and variables were interrelated and as a result, it identifies the cost items and the factors where cost reductions were possible. Such analyses highlight that transport operators will need to focus on systematic appraisals of costs in a climate of increasing customer expectations, competitiveness amongst providers and a declining yield per passenger through time. One strategy which airlines have followed is the pursuit of cost savings by divesting themselves of non-core activities such as in-flight catering operations. In June 1997, for example, Air New Zealand sold its catering business and planned to involve IBM in running its computer centre in a contracting-out of specialist non-core activities. These changes were identified in the company's 'Project Save' in the 1997/98 financial year, which is expected to save up to NZ$100 million in operating costs. Such savings are also expected to liberate capital to be reinvested in core business activities (Hanning 1997).

Interest in environmental factors such as sustainability in tourist transport seems set to continue as a powerful theme embracing tourism well into the next millennium. A growing interest in the use of public transport infrastructure to support tourist travel (Charlton 1998) is evident, with initiatives such as the Devon and Cornwall Rail Partnership, which has attracted leisure travel to offset losses in the non-leisure local rail market. Such rail tourism projects certainly have the potential to offer an alternative to the ingrained role of the car in recreational and tourist travel (Page 1998). Public transport certainly has a valid role to play in achieving sustainable tourism objectives in local areas. Such initiatives not only make a contribution to the reduction of congestion and environmental pollution in areas of natural beauty, but also offer access opportunities for the disabled, cyclists and casual travellers in place of the car. Even in urban areas, the development of public transport systems may offer the tourist more opportunities to enjoy the urban environment without the stress of parking and driving a car in congested cities. Brooks (1995) documented the reintroduction of historic Victorian trams in Christchurch, New Zealand, where the five vintage trams cover a 2.5 km inner city track. By 1997 it was obvious that they were not profitable. While the trams undoubtedly offer an attraction for the tourist, like those used in Blackpool and Fylde in Lancashire, UK, it is evident that transport systems may sometimes need to be subsidised to generate tourist business for other sectors of the urban economy (as is the case with the tourist tram service in Melbourne) (Page 1993a, 1995b). Yet this seems to run somewhat contrary to the political policy-making environment of the late 1990s where transport users, particularly tourists, need to pay the economic cost of transport.

Within an international context, there is also evidence to suggest that with the globalisation of the airline industry and other transport sectors, there is a growing need for an agency to ensure fair competition. According to Downes and Tunney (1997), European competition law for the air transport industry may 'become the foundation stone for global competition rules for aviation' (Downes and Tunney 1997 : 76). The World Trade Organisation is seen

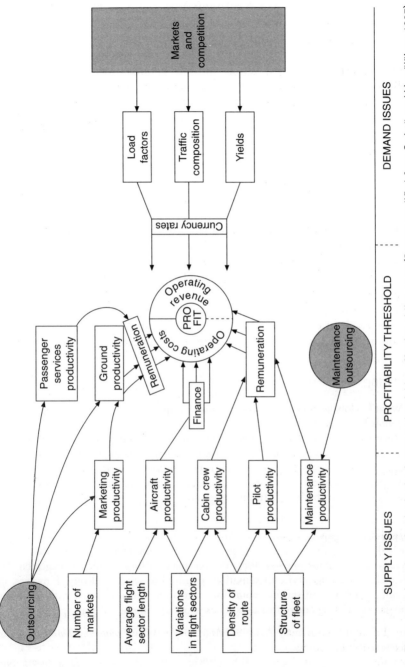

Figure 9.3 Interdependencies in the factors and variables affecting airline costs (Source: modified from Seristö, and Vepsäläinen 1997)

as the most likely body to ensure competition rules are upheld. At the same time tourist transport providers are facing an operating environment where increased health and safety regulations (see Caves 1996) and airport risk controls as well as measures such as the EU's (1995) *Protection of Tourists* (European Parliament 1995) place a greater onus on the operator and packager to provide accurate information to travellers. These types of measure are likely to encourage transport operators to consider the tourist experience within the context of ever-increasing demands for quality improvements.

One worrying trend in the airline sector is the rise in 'on-board incidents' where violent passengers disrupt the flight. Skapinker (1998b) reports that while BA only encounters 10 to 15 such incidents a year, they can endanger the lives of hundreds of people. It is likely that this is only the tip of the iceberg and it is certainly a concern for both airline employees and passengers alike, as the author found when observing such an incident. One of the contributory factors is alcohol consumption and BA now empower the staff to prevent drunk passengers from boarding and to stop serving passengers who appear drunk on board. Airport delays and inadequate information may lead to increased alcohol consumption and at least one Asian airline has even issued cabin staff with restraints to prevent disorderly conduct. Although such events usually occur on long-haul flights, airlines are imposing heavy fines on passengers, especially where pilots divert to eject passengers. It is just one additional issue with which airlines now have to deal. Managing such situations in an appropriate manner can also result in commitment and customer loyalty if the emphasis is on the enhancement of a quality experience for all passengers.

Questions

1. What is globalisation and how are transport operators responding to operating in a global tourism economy?
2. What types of problem can affect service quality in tourist transport systems?
3. What is hypercompetition? How are transport operators responding to such issues?
4. Are travellers developing unrealistic expectations of tourist transport providers? What can transport operators do to avoid a continued gap between expectation and delivery in the tourist travel experience?

Further reading

There is an enormous business literature on quality issues, quality management and the challenges facing service organisations involved in service delivery. However, the following offer a good starting point for a literature which is laden with acronyms, hype, jargon and repetitive messages focused on a simple set of principles.

Berry, L.L. and Parasuraman, A. (1991) *Marketing Services: Competing Through Quality*, New York: Free Press.
Fitzsimmons, J.A. and Sullivan, R.S. (1982) *Service Operations Management*, New York: McGraw-Hill.

Gnoth, J. (1995) 'Quality of service and tourist satisfaction', in S. Witt and L. Moutinho (eds), *Tourism Marketing and Management Handbook: Student Edition*, Hemel Hempstead; Prentice Hall, 243–54.
This is a good analysis of service encounters in tourism.
Ilinitch, A., Lewin, A. and D'Aveni, R. (eds) (1998) *Managing in Times of Disorder: Hypercompetition Organisational Responses*, London: Sage.
Townsend, P.L. and Gebhart, J.E. (1986) *Commit to Quality*, New York: John Wiley.
Zeithmal, A. and Berry, L.L. (1985) 'A conceptual model of service quality and its implications for future research', *Journal of Marketing*, Fall: 41–50.

Internet sites

See Appendices 1 and 2. A good example of innovation in the development of an integrated tourism and transport internet site is Red Funnel Ferries:

http://www.redfunnel.co.uk

Appendix 1

Guidelines for the carriage of elderly and disabled passengers by air

1. States should ensure that airports and airline operators maintain in the airport premises and in aircraft, when necessary, adequate medical service and equipment to assist the sick and disabled.
2. Provision must be made to provide infants' seats in aircraft for the carriage of infants.
3. Special provision must be made to store powered wheelchair batteries of passengers in aircraft, in accordance with acceptable policies of security and safety. Carriers should therefore have in their aircraft a kit for the safe transport of wet-cell batteries that are used to power wheelchairs.
4. Special attention should be given to the retrieval of mishandled baggage of elderly and disabled passengers.
5. Baggage and equipment of elderly and disabled passengers must be loaded on the aircraft in such a manner as to retrieve them expeditiously upon unloading.
6. At an appropriate time, airline operators must render advice and all possible assistance to blind passengers with the veterinary requirements of their guide-dogs.
7. At an appropriate time and if necessary, airline operators should assist elderly and disabled passengers to obtain necessary official clearance for their journeys (e.g. by assisting with the obtaining of visas).
8. Airport and airline staff should be specially trained to handle emergency situations involving elderly and disabled passengers such as the evacuation of passengers.
9. Elderly and disabled passengers must be treated equally with other passengers at all times and should not be required to sign any documentation such as waivers of liability.
10. Exit row seating policies for blind or otherwise disabled passengers must be consistent among airlines of contracting states and should not derogate the right of such passengers to request seating in aircraft on the same lines as other passengers, except in instances where it is considered justifiable that the airline take other appropriate measures.
11. At least 3% of the seats in an aircraft (not less than two seats) should be primarily intended for the use of disabled travellers. These seats should be designed for physically disabled persons so that sufficient space is available to facilitate the act of sitting down, standing up, and being generally comfortable throughout the journey. These seats should have movable armrests next to the aisle and the optional center armrests must also be movable. They should also be equipped with signaling devices that can be used by the passenger to summon the cabin crew.
12. Seats dedicated to disabled travellers should have a wider seat pitch than others in the aircraft.

13. When a person identifies the nature of his or her disability, an air carrier should, before assigning a seat to that person, inform the passenger of those seats in the aircraft that are most accessible to him or her. As far as possible, the carrier should bear in mind and apprise the passenger of the advantage to him or her of being seated near a toilet during the journey.

14. Elderly and disabled passengers must be provided air transport upon reasonable request.

15. National facilitation committees of states must ensure that airports and airlines act jointly to ensure the provision and operation of effective loading equipment for elderly and disabled passengers and their equipment.

16. An air carrier should accept as priority baggage, without charge and in addition to the free baggage allowance permitted to a passenger: an electric wheelchair, a scooter or manually operated rigid-frame wheelchair; a manually operated folding wheelchair; a walker, cane, crutches or braces; any prosthesis or medical device; and any other device that assists a passenger to communicate better.

17. Where a person in a wheelchair that is not independently mobile awaits boarding, the carrier shall inquire periodically about the needs of that person while he or she is awaiting the flight.

18. An air carrier should, whenever possible, indicate in the record of a person's reservation any services that the air carrier will provide to the person and provide the person with a written confirmation of the services that will be provided.

19. The carrier will also transmit to the appropriate personnel handling the person requiring the services previously mentioned a list of the services that the carrier has undertaken to provide at the time of reservation.

20. Airline reservations offices should also maintain a form containing special handling requests of elderly and disabled passengers, and this should be handed over to the cabin crew with other special instructions.

21. Information containing the number of elderly and disabled passengers on board, together with their details as referred to previously, should be given to the aircraft commander prior to the flight.

22. Airline reservations officers must display material containing details of hotels at destination points that provide special services/rates for the disabled.

23. Air carriers should inform elderly and disabled passengers within reasonable time whether their reservations and, if relevant, those of their escorts are confirmed. This would enable these passengers to seek reservations elsewhere if the carrier cannot offer accommodation in its flights.

24. In the case of aircraft chartered, sold or leased for the carriage of a group of disabled persons, the number of escorts must at least be equal to the following:

    (a) one escort for each nonambulatory disabled passenger. The airline may, however, for reasons of safety or weight control, require two escorts per passenger to permit the rapid evacuation of a disabled person;

    (b) one escort for each ambulatory disabled person who is able to move about unaided but with difficulty;

    (c) one escort for up to 12 ambulatory passengers in other cases, with the exception of blind persons;

    (d) one escort for every two blind persons. (A blind person may travel alone if accompanied by a guide [seeing-eye] dog.)

25. Any person accompanying a disabled person must be over 18 years of age (except in the case of blind passengers).

26. When a particular flight is dedicated to carrying disabled passengers and tickets have been sold accordingly for that flight, no person other than a disabled passenger or an escort may be allowed on board.

27. Airline manuals must clearly specify the conditions and requirements necessary for disabled passengers to obtain a confirmed reservation on a flight.

28. If medical clearance is necessary for a passenger to travel by air, the carrier must ensure that the document that evidences such clearance is attached to the passenger's airline ticket. The carrier must also ensure in such instances that a document indicating the state of the passenger's health is attached to the ticket.

29. On request, air carriers should arrange for special transportation of a passenger, such as an ambulance at the passenger's expense.

30. Information desks at airline offices, airline check-in counters, and customs and immigration inspection counters at airport terminals must be adjustable to a height of 1.2 m. Wherever possible, a wheelchair person's bypass must be used at these counters.

31. Toilets of aircraft and airport terminal buildings must be equipped with door expansion and grab bars. In larger aircraft, at least one toilet should be easily accessible by wheelchair or transport chair.

32. Flushing devices, taps for wash basins, door locks, and signaling devices in the toilets both in the terminal buildings and in the aircraft must be easily accessible and easy to handle. These devices should be marked in an easily identifiable manner.

33. The lighting in the toilets must be antidazzling so the symbols therein are clearly visible to a vision-impaired person.

34. Flooring material in aircraft must facilitate good foothold and the floor should contrast in colour with the rest of the decor.

35. All aircraft windows must be equipped with sunshades or tinted glass.

36. Both general and optional lighting in the aircraft should be antidazzling and the lighting in the seats should provide an illumination intensity acceptable to the vision-impaired at a reading level of one meter above the floor.

37. The furniture in the aircraft should be made of nonallergenic material.

38. Aircraft should have clearly defined smoking and non-smoking areas.

39. Phones at the airport terminal should have a coin slot at a height of 1.2 m. These phones must be equipped with amplifiers for the benefit of the hearing-impaired.

40. Money exchange facilities should also have provision for operation at 1.2 m.

41. Airports should also install in their terminal buildings induction loops connected to the public address system for the use of hearing aids.

42. Airport operators should ensure the availability of car rentals with special facilities for the disabled in the airport premises. They should also ensure that buses operating from and to their airport are equipped to carry disabled passengers.

43. Medical services such as ambulance services should be available at the airport.

44. Air carriers and their staff must be specially aware of medical insurance availability and inform elderly and disabled passengers of these details at the time a reservation for travel is made. Carriers must also be equipped with the facilities needed to obtain such insurance if the passenger so requests.

45. Ideally, states should maintain special transport services free of charge for elderly and disabled passengers to and from the airport.

46. States should encourage their carriers to permit access to disabled passengers claiming to be self-reliant, unaccompanied by an attendant, with the only exception being in the consideration of safety (already suggested in assembly working paper).

47. A passenger wheelchair should be the last item to be stored in the aircraft and the first to be removed.

48. Escorts and disabled passengers must be seated next to each other.

49. Disabled passengers travelling with guide-dogs should be assigned seats that allow as much space as possible for the dog.

50. There should be wider aisles in aircraft, at least near the bulkhead seats, so that wheelchairs can move more freely to these seats.

51. Airports should have a clearly marked accessible entrance for disabled passengers with a doorway controlled electronically and activated by either pressure pads or magic eyes. There should be no stairs between this entrance and the check-in counter.

52. Airports should have wheelchairs or other motorized equipment at the arrival and departure areas of the terminal for the use of walk-impaired passengers. These wheelchairs should have the following features: high backrests; vertically adjustable footrests, detachable armrests; self-propelled wheels; and at least two different seat widths.

53. Accessible toilets should be available after the check-in process so that wheelchair passengers can prepare for their journey.

54. Facilities for parents with infants, such as diaper changing and nursing areas, should be available between the check-in area and the sterile lounge of airport terminal buildings.

55. Elderly and disabled passengers who may need assistance on arrival to get to the baggage claim area or to customs control should be handed over to ground staff after they leave the aircraft.

56. Immigration and customs forms should be easy to read, simple, and contain large-type instructions or notices so that travellers with limited vision can easily decipher their contents.

57. Signs indicating baggage claim areas must be clearly displayed and preferably be accompanied by verbal announcements.

58. There should be special staff deployed by the airport authorities to assist pregnant women, parents with infants, and elderly and other disabled persons in the baggage claim areas.

59. Exits from baggage claim areas should be made possible through gates that are large enough for a wheelchair or trolley to pass through. The passage from the baggage claim area to the exit area should be flat and clearly marked with directions (i.e. indicators to taxi stands, buses, car parks). Alternatively, there should be suitable elevators and ramps for the use of disabled passengers in these areas.

60. Car parks at airports should be located 200 feet at most from the terminal entrance and have specially assigned places for the disabled. These spaces must be directionally signed.

61. The path leading to the area assigned to the disabled in the car park should be protected from the weather.

62. Parking meters and tickets at the areas designated for disabled persons at airport car parks must be accessible to the driver.

63. There should be stairs and ramps at all changes of level at airports. All ramps must be wide enough to accommodate a wheelchair or trolley.

64. There must be level vehicle loading and unloading areas close to the entrances of airport terminal buildings. These areas must be protected from the weather.

65. All public areas in the airport terminal building must be accessible by level or ramped route.

66. There must be public elevators at all levels of the terminal building including the garage.
67. There must be a ramp or loading bridges to the aircraft.
68. Rest rooms at airport terminal buildings must be accessible to the elderly and disabled. These rest rooms must have at least 5 feet by 5 feet turning space, with the doors having a clearing space of at least 3 feet. Doors should be equipped with lever handles (not knobs). Paging phones should be wall mounted.
69. Where airports have more than one building with facilities, there should be appropriate connecting transport that could also be used by elderly and disabled passengers.
70. Vending machines at airports should have their controls identified with raised lettering and the controls should be within 4 feet of the floor.
71. Where walking distances between the check-in counters and the departure gates are too long for elderly and other medically disadvantaged passengers, sufficiently wide conveyor belts should be installed at airport terminal buildings.
72. Airports should be equipped with vehicles with the capability to transport handicapped persons with or without a wheelchair within the airport.
73. Airports should have vehicles with a ramp or lifting device to transport a wheelchair on the apron.
74. Airports should have ambulift vehicles, with a lifting device capable of taking a wheelchair up to the aircraft door, that transfer disabled passengers from the departure lounges to the aircraft.
75. Since security screening of disabled passengers is more sensitive and complicated than the process employed on other passengers, airports should be equipped with special rooms for the carrying out of security checks on these passengers so the general flow of the screening process is not hampered, and so the screening is carried out without causing the passenger indignity or embarrassment. Officials must be specially trained by the appropriate authorities for carrying out these tests.
76. Airports should make as much use as possible of the pictogram signs that have been recommended for use, particularly to assist illiterate passengers.

Source: Abeyratne (1995 : 56–59)
© *Journal of Travel Research*

World Wide Web addresses for world airports and associated organisations

Amsterdam/Schiphol	www.schiphol.nl
Atlanta	www.gatech.edu/3020/travelink/airport.homepage.html
Baltimore–Washington	www.haltwashintairport.com
Boston	www.massport.com
Calgary	www.calgary.sh.ca
Chicago/O'Hare	www.cl.chl.ll.us/worksmart/aviation/charhome.html
Dallas–Fort Worth	www.dfwairport.com
Denver	infodenver.denver.co.us/-aviation/dlaintro.html
Detroit	www.waynecounty.com/-world/wayne/alt.shtml
Frankfurt	www.frankfurt-airport.de
Hong Kong	www.hkairport.com
Houston	www.clearlake.lbm.com/visitor/airports.html
Las Vegas	www.mccarran.com
London	www.baa.co.uk
Malaysia Airports	www.jaring.my/airport
Memphis	www.macaa.com
Miami	www.Miami-Airport.com
Nashville	www.nahinti.com
Osaka/Kansai	www.klls.or.jp/kixinfo/kix.html
Paris	www.pairs.org/accuell/airport
Portland	www.portofportiandor.com
Reno	www.renoairport.com
St Louis	www.st-louis.mo.us
San Francisco	www.Airportsintl.com
Seoul	www.airport.or.kr
Singapore	www.Changi.Airport.com.sg
Tampa	www.tampaairport.com
Tokyo/Narita	www.narita-airport.or.jp/airport
Toronto	www.lbpia.toronto.on.ca/menu.htm
Washington	www.metwashairports.com

Also consult the Air Transport Action Group's Worldwide Web site for a state-by-state series of home pages for US airports on:

http://www.ebb.con/ebb/airports/airports.htm

or via the general home page:

http://www.atag.org

There are also a range of associations and government bodies with World Wide Web pages that may be of interest, including:

American Association of Airport Executives	www.airportnet.org
Airports Council International	www.airports.org
Air Transport Association of America	www.air-transport.org
US Bureau of Transport Statistics	www.bts.gov
US Department of Transport	www.dot.gov
UK Department of Transport	www.coi.gov.UK/coi/depts/GDT/GDT.html
European Regional Airlines Association	www.eraa.org
US Federal Aviation Administration (FAA)	www.faa.gov/faahome.htm
International Air Transport Association (IATA)	www.iata.org
US National Air Traffic Control Association	www.natca.org
US Regional Airlines Association (RAA)	www.raa.org/index.htm

Appendix 3

World Wide Web addresses for a selection of world airlines

European airlines

Adria Airways	www.kabl.sl/sl21/aa
Aer Lingus	www.aerlingus.ie
Aero Lloyd	www.aerolloyd.de
Aeroflot RIA	www.aeroflot.org
Air Europa	www.irinfo.es/aviacion/aireuropa.html
Air France	www.airfrance.fr
Air UK	www.airuk.co.uk
Alitalia	www.zenonet.com/itllatour/alitalia.html
AOM French Airlines	www.aom.ch
Austrian Airlines	www.aua.co.at/aua
Aviaco	www.irinfo.es/aviacion/aviaco.html
Braathens Safe	www.nettvlk.no/terminalen/braathens
Britannia Airways	www.flightconnection.co.uk/BRITANNI/BRITANNI.TYM
British Airways	www.british-airways.com
British Midland	www.Iflybritishmidland.com
Cargolux	www.traxon.com/clx.htm
Condor	www.condor.de
EasyJet Airlines	205.164.229.4/easyjet
Eurowings	www.eurowings.de
Finnair	www.finnair.fl
Iberia	www.irinfo.es/aviacion/iberia/iberia.html
Icelandair	www.centrum.is/icelandair
Jersey European Airways	www.jea.co.uk
KLM Royal Dutch Airlines	www.klm.nl
Lauda Air	www.laudaair.com
LOT Polish Airlines	www.lot.com
LTR Airlines	www.irinfo.es/aviacion/lte.html
LTU International Airways	www.ltu.com
Lufthansa	www.lufthansa.com
Luxair	www.sita.int/luxair/index.html

Maiev Hungarian Airlines	www.osgweb.com/airlines/malev/Index.html
Malmo Aviation	www.maviation.se
Monarch Airlines	www.cuug.ab.ca:8001/-busew/monarch.html
Oasis	www.irinfo.es/aviacion/oasis.html
Olympic Airways	agn.hol.gr/info/olympic1.htm
Portugalla Airlines	emporium.turnpike.net/~portugalla/index.htm
Sabena	www.sabena.com
SAS	www.sas.se
Spanair	www.irinfo.es/aviacion/spanair.html
Swissair	www.swissair.com
TAP Air Portugal	www.Tap-Air PORTUGAL.pt
TAT European Airlines	www.british-airways.com/banetac/mworld/allies/tat.htm
THY Turkish Airlines	www.turkishairlines.com
Transwede Airways	www.transwede.se
Virgin Atlantic Airways	www.fly.virgin.com/atlantic
Vivo Air	www.irinfo.es/aviacion/viva.html

North American airlines

Air Canada	www.aircanada.ca
Alaska Airlines	www.alaskaair.com/hom.html
Aloha Airways	www.alohaair.com/aloha.air
America West Airlines	www.americawest.com
American Airlines	www.americanair.com
American Transair	www.halycyon.com/integra/ata.html
Canada 3000 Airlines	www.2000.com/c3000.html
Canada Airlines International	www.CdnAir.CA
Carnival Air Lines	www.carnivalair.com
Comair	www.fly-comair.com
Continental Airlines	www.flycontinental.com:80/index.html
Delta Air Lines	www.delta-air.com/index.html
Northwest Airlines	www.nwa.com
Reno Air	www.renoair.com
Sky West Airlines	www.skywest-air.com/index.html
Southwest Airlines	www.Iflyswa.com
Sun County Airlines	www.suncounty.com
Trans World Airlines (TWA)	www2.twa.com/TWA/Airlines/home/home.htm
United Airlines	www.ual.com
USAir	www.usair.com
ValuJet	www.valujet.com
Western Pacific Airlines	www.ipo-network.com/members/ipos/Western_Airlines.html
World Airways	www.teaminteract.com/worldair/world.html

Latin American and Caribbean airlines

Aerolineas Argentinas	www.pinos.com/Aero/aero.html
Aeromexico	www.wotw.com/aeropmexico
ALM Antillean Airlines	www.empg.com/alm/index.html
Avianca	www.avianca.com
BWIA International Airways	rum.bajan.com/caribbean/bwia/bwiainfo.htm
Cayman Airways	www.caymans.com/-caymans/CaymanAirways.html
Cubana	www.cubaweb.cu/turismo/cubana/indice.html
Ladeco Chilean Airlines	www.ladeco.com
LanChile	www.lanchile.com
Mexicana Airlines	www.mexicana.com/index.html
Taesa	www.wotw.com.wow.mexico/city/taesa.html
TAM	www.tamairlines.com.br
Transbrasil	www.transbrasil.com.br
Varig	www.varig.com.br
Vasp	www.vasp.com.br

African and Middle Eastern airlines

Air Afrique	www.sinergia.lt/air/afrique.htm
Air Mauritius	www.cyber.be/air-mauritius
Air Namibia	www.ipi.co.uk/docs/air_namibia/airnam.htm
El Al	www.elal.co.ll
Emirates	www.ekgroup.com
Gulf Air	www.gulfair.net
Kuwait Airways	www.travelfirst.com/sub/kuwaitair.html
Royal Air Maroc	www.xbrcom.qc.ca/RAM/RAM_home.html
Royal Jordanian	www.rja.com.jo
Saudi Arabian Airlines	www.saudiarabian-airlines.com
South African Airways	www.saa.co.za/saa/yesnet2.htm

Asia–Pacific airlines

Air China	www.airchina.com
Air New Zealand	www.airnz.com
Air Pacific	www.singapore.com/pata/airpac.htm
All Nippon Airways	www.ana.co.jp/index-e.html
Ansett Australia	www.ansett.com.au
Asiana Airlines	www.kumho.co.kr/asiana
Cathay Pacific Airways	www.cathaypacific-air.com
China Airlines	metrotel.co.uk/travlog/ca.html
China Eastern Airlines	206.170.104.72
Garuda Indonesia	www.gob-usa.com/gob-usa/f3.html
Japan Air System	www.jas.co.jp
Japan Airlines	www.jal.co.jp

Korean Air	www.mabuhay.com/Korean_Air
Malaysia Airlines (MAS).	www.MalaysiaAir.com
Miat Mongolia Airlines	www.arpnet.it/-mongolia/viaggi/aereo.htm
Philippine Airlines	www.sequel.net/PAL
Qantas, Australia	www.qantas.com.au
Singapore Airlines	www.singaporeair.com
Thai International Airways	metrotel.co.uk/travelog/thaiair.html

Bibliography

Abeyratne, R. (1993) 'Air transport tax and its consequences on tourism', *Annals of Tourism Research*, 20 (3): 450–60.

Abeyratne, R. (1995) 'Proposals and guidelines for the carriage of elderly and disabled persons by air', *Journal of Travel Research*, 33 (3): 52–59.

Abeyratne, R. (1998) 'The regulatory management in air transport', *Journal of Air Transport Management*, 4 (1): 25–37.

Adams, J. (1981) *Transport Planning: Vision and Practice*, London: Routledge and Kegan Paul.

Adamson, M., Jones, W. and Platt, R. (1991) 'Competition issues in privatisation: lessons for the railways', in D. Banister, K.J. Button (eds) *Transport in a Free Market Economy*, Basingstoke, Macmillan: 49–78.

Air Transport Action Group (ATAG) (1993) *Surface Transportation and Airports*, Geneva: ATAG.

Air Transport Action Group (ATAG) (1994) *North American Traffic Forecasts 1984–2010*, Geneva: ATAG. (http://www.atag.org/NATF/Index.htm)

Air Transport Action Group (ATAG) (1997a) *The Economic Benefits of Air Transport*, Geneva: ATAG. (http://www.atag.org/ECO/Index.htm)

Air Transport Action Group (ATAG) (1997b) *Asia Pacific Air Traffic Growth and Constraints*, Geneva: ATAG. (http://www.atag.org/ASIA/Index.htm)

Alamdari, F. and Morrell, P. (1997) 'Airline labour cost reduction: post-liberalisation experience in the USA and Europe', *Journal of Air Transport Management*, 3 (2): 53–66.

Alamdari, F. and Morrell, P. (1997) 'Airline alliances: a catalyst for regulatory change in key markets', *Journal of Air Transport Management*, 3 (1): 1–2.

Allen, D. and Williams, G. (1985) 'The development of management information to meet the needs of a new management structure', in K.J. Button and D. Pitfield (eds) *International Railway Economics: Studies in Management and Efficiency*, Aldershot, Gower: 85–100.

Anon (1992) 'Japan', *International Tourism Reports*, 4: 5–35.

Anon (1997) *The Europa World Yearbook 1997*, Woking: Gresham Press.

Anon (1998) 'Scenic wonder pays price of popularity', *New Zealand Herald*, 26 February, A24.

Annals of Tourism Research (1991) 'Special issue: Tourism social science', *Annals of Tourism Research*, 18, 1.

Archdale, G. (1991) 'Computer Reservation Systems – the international scene', *INSIGHTS*, November: D15–20.

Archer, B.H. (1987) 'Demand forecasting and estimation', in J.R.B. Ritchie and C.R. Goeldner (eds) *Travel, Tourism and Hospitality Research*, New York: Wiley, 77–85.

Archer, B.H. (1989) 'Tourism and small island economies', in C.P. Cooper (ed) *Progress in Tourism, Recreation and Hospitality Management*, Volume 1, London: Belhaven, 125–35.

Archer, B. (1995) 'The impact of international tourism on the economy of Bermuda', *Journal of Travel Research*, 34 (2): 27–30.

Ardill, J. (1987) 'The environmental impact', in B. Jones (ed) *The Tunnel: The Channel Tunnel and Beyond*, Chichester: Ellis Harwood, 177–212.

Ashford, H. Stanton, H. and Moore, C. (1991) *Airport Operations*, London: Pitman.

Ashford, H. and Moore, C. (1992) *Airport Finance*, New York: Van Nostrand Reinhold.

Ballantyne, T. (1996) 'Airlines in China', *Travel and Tourism Analyst*, 2: 4–20.

Bania, N., Bauer, P. and Zlatoper, T. (1998) 'US air passenger service: a taxonomy of route networks, hub locations and competition', *Transportation Research E*, 34 (1): 53–74.

Banister, D. (1992) 'Policy responses in the UK', in D. Banister and K.J. Button (eds) *Transport, the Environment and Sustainable Development*, London: E & FN Spon, 53–78.

Banister, D. (1994) *Transport Planning in the UK, USA and Europe*, London: E & FN Spon.

Banister, D. (ed) (1998) *Transport Policy and the Environment*, London: E & FN Spon.

Banister, D. and Button, K.J. (eds) (1991) *Transport in a Free Market Economy*, Basingstoke: Macmillan.

Banister, D. and Button, K.J. (eds) (1992) *Transport, the Environment and Sustainable Development*, London: E and FN Spon.

Banister, D. and Hall, P. (eds) (1981) *Transport and Public Policy Planning*, London: Mansell.

Barbier, E.B. (1988) *New Approaches in Environmental and Resource Economics: Towards an Economics of Sustainable Development*, London: International Institute for Environment and Development.

Barke, M. (1986) *Transport and Trade*, Edinburgh: Oliver and Boyd.

Barlay, S. (1995) *Cleared for Take-off: Behind the Scenes of Air Travel*, London: Kyle Cathie Limited.

Barnes, C. (1989) *Successful Marketing for the Transport Operator: A Practical Guide*, London: Kogan Page.

BaRon, R. (1997) 'Global tourism trends to 1996', *Tourism Economics*, 3 (3): 289–300.

Baum, T. (ed) (1993) *Human Resource Issues in International Tourism*, Oxford: Butterworth-Heinemann.

Beaver, A. (1996) 'Frequent flyer programmes: the beginning of the end?', *Tourism Economics*, 21: 43–60.

Belobaba, P. (1987) *Air Travel Demand and Airline Seat Inventory Management*, Cambridge, MA.: MIT Flight Transportation Laboratory Report R87-7.

Belobaba, P. and Wilson, J. (1997) 'Impacts of yield management in competitive airline markets', *Journal of Air Transport Management*, 3, 1: 3–10.

Beesley, M.E. (1989) 'Transport research and economics', *Journal of Transport Economics and Policy*, 23: 17–28.

Bell, G., Blackledge, D. and Bowen, P. (1983) *The Economics of Transport and Planning*, London: Heinemann.

Beioley, S. (1995) 'On yer bike – cycling and tourism', *INSIGHTS*, A17–A31.

Beioley, S. (1997) 'Four weddings, a funeral and the visiting friends and relatives market', *INSIGHTS*, B1–B6.

Bennett, M. (1997) 'Strategic alliances in the world airline industry', *Progress in Tourism and Hospitality Research*, 3: 213–23.

Berry, L.L. and Parasuraman, A. (1991) *Marketing Services: Competing Through Quality*, New York: Free Press.

Bertolini, L. and Spit, T. (1998) *Cities on Rails: The Redevelopment of Station Areas*, London: E & FN Spon.

Bertrand, I. (1997) 'Airlines in the Caribbean', *Travel and Tourism Analyst*, 6: 4–24.

Bird, P. and Rutherford, B.A. (1989) *Understanding Company Accounts*, London: Pitman.

Bitner, M.J. (1992) 'Servicescapes: the impact of physical surroundings on customers and employees', *Journal of Marketing*, 56 (2): 57–71.

Bitner, M.J., Booms, B.H. and Tetreanit, M.S. (1990) 'The service encounter: diagnosing favourable and unfavourable incidents', *Journal of Marketing*, 1: 71–84.

Boeing Commercial Airplane Group (1994) *1994 Current Market Outlook*, Seattle: Boeing Commercial Airplane Group.

Boeing Commercial Airplane Group (1996) *1996 Current Market Outlook*, Seattle: Boeing Commercial Airplane Group.

Bóte Gomez, V. and Sinclair, M.T. (1991) 'Integration in the tourism industry: a case study approach', in M.T. Sinclair and M.J. Stabler (eds) *The Tourism Industry: An International Analysis*, Oxford: CAB International, 67–90.

Bowen, J. (1997) 'The Asia Pacific airline industry: prospects for multilateral liberalisation', in C. Findlay, L. Chia and K. Singh (eds) *Asia Pacific Air Transport: Challenges and Policy Reforms*, Singapore: Institute of South East Asian Studies, 123–53.

Braithwaite, G., Caves, R. and Faulkner, J. (1998) 'Australian aviation safety – observations from the lucky country', *Journal of Air Transport Management*, 4 (1): 55–62.

Briggs, S. (1997) 'Marketing on the Internet – tangled webs or powerful network?', *INSIGHTS*, A27–A34.

British Airports Authority (1997) *Investing for Growth: BAA Annual Report 1996/97*, London: British Airports Authority.

British Airways (1992) *Annual Environmental Report*, Hounslow: British Airways.

British Airways (1997) *Annual Environmental Report*, Hounslow: British Airways. (http://www.british-airways.com)

British Railways Board (1992) *The British Rail Passenger's Charter*, London: British Railways Board.

British Railways Board (1996) *National Conditions of Carriage*, London: British Railways Board.

British Tourist Authority (1991) *1993 Cross-Channel Marketing Strategy*, London: British Tourist Authority.

British Travel Association and the University of Keele (1967 and 1969) *Pilot National Recreation Survey*, Keele: British Travel Association and the University of Keele.

Britton, S. (1982) 'The political economy of tourism in the Third World', *Annals of Tourism Research*, 9 (3): 331–58.

Broads Authority (1997) *Broads Plan*, Norwich: Broads Authority.

Brooks, K. (1995) 'Old trams creak back to streets', *New Zealand Herald*, 16 January: 9.

Buckley, P.J. (1987) 'Tourism – an economic transaction analysis', *Tourism Management*, 8 (3): 190–4.

Buckley, P.J. and Casson, M. (1987) *The Economic Theory of the Multinational Enterprise: Selected Papers*, London: Macmillan.

Bull, A. (1991) *The Economics of Travel and Tourism*, London: Pitman.

Bull, A. (1996) 'The economics of cruising: an application to the short ocean cruise market', *Journal of Tourism Studies*, 7 (2): 28–35.

Burkart, A. and Medlik, S. (1974) *Tourism, Past, Present and Future*, Oxford: Heinemann.

Burkart, A. and Medlik, S. (eds) (1975) *The Management of Tourism*, Oxford: Heinemann.

Burns, P. (1996) 'Japan's Ten Million programme: the paradox of statistical success', *Progress in Tourism and Hospitality Research*, 2: 181–92.

Burt, T. (1998) 'Europeans cruise to luxury contracts', *Financial Times*, 5 February, 4.

Burton, T. (1966) 'A day in the country: a survey of leisure activity at Box Hill in Surrey', *Journal of the Royal Institute of Chartered Surveyors*, 98: 378–80.

Burton, J. and Hanlon, P. (1994) 'Airline alliances: cooperating to compete?', *Journal of Air Transport Management*, 1 (4): 209–28.

Bus and Coach Council (1991) *Buses and Coaches: the Way Forward*, London: Bus and Coach Council.

Button, K.J. (1982) *Transport Economics*, London: Heinemann.

Button, K.J. (1990) 'The Channel Tunnel – economic implications for the South East', *Geographical Journal*, 156, 2: 87–99.

Button, K.J. (1996) 'Aviation dergulation in the European Union: do actors learn in the regulation game?', *Contemporary Economic Policy*, 14: 70–80.

Button, K.J. (1997) 'Developments in the European Union: lessons for the Pacific Asia region', in C. Findlay, L. Chia and K. Singh (eds) *Asia Pacific Air Transport: Challenges and Reforms*, Singapore: Institute of South East Asian Studies, 170–80.

Button, K.J. and Gillingwater, K. (eds) (1983) *Future Transport Policy*, London: Routledge.

Button, K.J. and Gillingwater, K. (eds) (1991) *Airline Deregulation: International Experiences*, London: Fulton.

Button, K.J. and Rothengatter (1992) 'Global environmental degradation: the role of transport', in D. Banister and K.J. Button (eds) *Transport, the Environment and Sustainable Development*, London: E and FN Spon, 19–52.

Bywater, M. (1990) 'Japanese investment in South Pacific tourism', *Travel and Tourism Analyst*, 3: 51–64.

Cannon, T. (1989) *Basic Marketing Principles and Practice*, London: Holt, Rinehart and Winston, 3rd edition.

Carpenter, T. (1994) *The Environmental Impact of Railways*, Chichester: Wiley.

Carter, D. (1996) 'Broader horizons beckon', *Financial Times*, 18 November, (World Airports Survey), II.

Caves, R. (1994) 'A search for more apron capacity', *Journal of Air Transport Management*, 1 (2): 109–20.

Caves, R. (1996) 'Control of risk near airports', *Built Environment*, 22 (3): 223–33.

CEDRE (1990) *Transports à grande vitesse: développement régional et ménagement du territoire, rapport de synthèse*, Strasburg: Centre Européen du Développement Régional.

Channel Tunnel Group (1985) *The Channel Tunnel Project: Environmental Effects in the UK*, London: Channel Tunnel Group.

Channel Tunnel Joint Consultative Committee (1986) *Kent Impact Study*, London: Department of Transport.

Charlton, C. (1998) 'Public transport and sustainability: the case of the Devon and Cornwall Rail Partnership', in C.M. Hall and A. Lew (eds) *The Geography of Sustainable Tourism*, London: Addison Wesley Longman.

Charlton, C., Gibb, R. and Shaw, J. (1997) 'Regulation and continuing monopoly on Britain's railways', *Journal of Transport Geography*, 5 (2): 147–53.

Charlton, C. and Gibb, R. (1998) 'Transport deregulation and privatisation', *Journal of Transport Geography*, 6 (2): 85.

Chartered Institute of Public Finance and Accountancy (CIPFA) (1996) *The UK Airports Industry: Airport Statistics 1994/95*, London: CIPFA.

Chew, S. (1997) 'SIA eyes more ventures overseas to raise bottom line', *Straits Times* (Singapore) 17 February.

Chou, Y.H. (1993) 'Airline deregulation and nodal accessibility', *Journal of Transport Geography*, 1 (1): 36–46.

Chubb, M. (1989) 'Tourism patterns and determinants in the Great Lakes region: Population, resources and perceptions', *GeoJournal*, 19 (3): 291–6.

Christopher, M. (1994) *Logistics and Supply Chain Management*, New York: Irwin.

Civil Aviation Authority (CAA) (1997) *Reports and Accounts 1997*, London: Civil Aviation Authority.

Clift, S. and Page, S.J. (eds) (1996) *Health and the International Tourist*, London: Routledge.

Cline, R., Ruhl, T., Gosling, G. and Gillen, D. (1998) 'Air transportation demand forecasts in emerging market economies: A case study of the Kyrgyz Republic in the former Soviet Union', *Journal of Air Transport Management*, 4 (1): 11–23.

Coase, R. (1937) 'The nature of the firm', *Economica*, 4: 386–405.

Cockerell, N. (1994) 'Europe to the Asia Pacific region', *Travel and Tourism Analyst*, 1: 65–82.

Cockerell, N. (1997) 'Growth of river cruising', *Travel Industry Monitor*, October, 7–8.

Cohen, E. (1972) 'Towards a sociology of international tourism', *Social Research*, 39: 164–82.

Cole, S. (1994) 'Rail privatisation and its impact on the local authority's transport role', *INSIGHTS*, A73–A77.

Collier, A. (1989) *Principles of Tourism*, Auckland, Longman Paul.

Collier, A. (1994) *Principles of Tourism*, Auckland, Longman Paul, 3rd edition.

Conlin, M. (1995) 'Rejuvenation planning for island tourism: The Bermuda example', in M. Conlin and T. Baum (eds) *Island Tourism*, Chichester: Wiley, 181–202.

Cooper, C., Fletcher, J., Gilbert, D. and Wanhill, S. (1993) *Tourism, Principles and Practice*, London: Pitman.

Cooper, C., Fletcher, J., Gilbert, D. and Wanhill, S. (1998) *Tourism, Principles and Practice*, London: Addison Wesley Longman, 2nd edition.

Cooper, J. (ed) (1988) *Logistics and Distribution Planning: Strategies for Management*, London: Kogan Page.

Coppock, J.T. (ed) (1977) *Second Homes: Curse or Blessing*, Oxford: Pergamon.

Coppock, J.T. and Duffield, B. (1975) *Recreation in the Countryside*, Macmillan: London.

Cossar, J. (1993) 'Travellers' health – a review', *Travel Medicine International*, 17: 2–6.

Cossar, J., Reid, D., Fallon, R., Bell, E., Riding, M., Follett, E., Dow, B., Mitchell, S. and Grist, N. (1990) 'A cumulative review of studies of travellers, their experience of illness and the implications of these findings', *Journal of Infection*, 21: 27–42.

Countryside Commission (1992) *Trends in Transport and the Countryside*, Cheltenham: Countryside Commission.

Countryside Commission (1995) *The Market for Recreational Cycling in the Countryside*, Cheltenham: Countryside Commission.

Cowell, D.W. (1986) *The Marketing of Services*, London: Heinemann.

Craven, J. (1990) *Introduction to Economics*, Oxford: Blackwell, 2nd edition.

D'Aveni, R. (1998) 'Hypercompetition closes in', *Financial Times*, 4 February (Global Business Series: Part 2, 12–13).

Dann, G. (1994) 'Travel by train: keeping nostalgia on track', in A. Seaton (ed) *Tourism: State of the Art*, Chichester: Wiley.

David, F.R. (1989) 'How companies define their mission statements'; *Long Range Planning*, 22 (1): 90–7.

De Kadt, E. (ed) (1979) *Tourism: Passport to Development?*, New York: Oxford University Press.

Deming, W.E. (1982) *Quality Productivity and Competitive Position*, Cambridge: Massachusetts Institute of Technology.

Dempsey, P., Goetz, A. and Szyliowicz, J. (1997) *Denver International Airport: Lessons Learned*, New York: McGraw-Hill.

de Neufville, R. (1991) 'Strategic planning for airport capacity: an appreciation for Australia's process for Sydney', *Australian Planner*, 29: 174–80.

de Neufville, R. (1994) 'The baggage system at Denver: prospects and lessons', *Journal of Air Transport Management*, 1 (4): 229–36.

Department of Employment/English Tourist Board (1991) *Tourism and the Environment: Maintaining the Balance*, London: HMSO.

Department of the Environment (1989) *Environmental Assessment: A Guide to the Procedures*, London: HMSO.

Department of the Environment (1991) *This Common Inheritance*, London: HMSO.

Department of the Environment, Transport and the Regions (1997) *Transport Statistics Great Britain 1997*, London: HMSO.

Department of Transport (1987), *Tourism, Leisure and Roads*, London: HMSO.

Department of Transport (1992a) *New Opportunities for the Railways*, Cmmd. 2012, London: HMSO.

Department of Transport (1992b) *The Franchising of Passenger Rail Services: A Consultative Document*, London: Department of Transport.

Department of Transport (1993) *Railways Bill 1993*, Bill 117, London: HMSO.

Department of Transport (1994) *Britain's Railways: A New Era*, London: Department of Transport.

Department of Transport (1996) *Transport Statistics Report: Cycling in Great Britain*, London: HMSO.

Dickinson, R. (1993) 'Cruise industry outlook in the Caribbean', in D.J. Gayle and J.N. Goodrich (eds) *Tourism Marketing and Management in the Caribbean*, London: Routledge, 113–21.

Dickinson, R. and Vladmir, A. (1997) *Selling the Sea: An inside look at the Cruise Industry*, Chichester: Wiley.

Doganis, R. (1991) *Flying Off Course: The Economics of International Airlines*, London: Routledge.

Doganis, R. (1992) *The Airport Business*, London: Routledge.

Doganis, R. and Graham, A. (1987) *Airport Management. The Role of Performance Indicators*, London: Polytechnic of Central London.

Dotchin, J.A. and Oakland, J.S. (1992) 'Theories and concepts in Total Quality Management', *Total Quality Management*, 3 (2): 133–45.

Douglas, N. and Douglas, N. (1996) 'P&O's Pacific', *The Journal of Tourism Studies*, 7 (2): 2–14.

Downes, J. and Tunney, J. (1997) 'The legal framework for airline competition', *Travel and Tourism Analyst*, 3: 76–92.

Dresner, M. and Windle, R. (1995) 'Are US air carriers to be feared? Implications of hubbing to North Atlantic competition', *Transport Policy*, 2 (3): 195–202.

Dresner, M. and Windle, R. (1996) 'Alliances and code sharing in the international airline industry', *Built Environment*, 22 (3): 201–11.

Dunning, J. (1977) 'Trade, location of economic activity and the MNE: a search for an eclectic approach', in B. Ohlin, P. Hesselborn and P. Wijkman (eds) *The International Allocation of Economic Activity*, London: Macmillan.

Dwyer, L. and Forsyth, P. (1996) 'Economic impacts of cruise tourism in Australia', *Journal of Tourism Studies*, 7 (2): 36–45.

Eadington, W.R. and Redman, M. (1991) 'Economics and tourism', *Annals of Tourism Research*, 18 (1): 41–56.

Eaton, A. (1996) *Globalization and Human Resource Management in the Airline Industry*, Aldershot: Avebury Aviation.

Eaton, B. and Holding, D. (1996) 'The evaluation of public transport alternatives to the car in British national parks', *Journal of Transport Geography*, 4 (1): 55–65.

Edwards, A. (1991) *European Long Haul Travel Market: Forecasts to 2000*, London: Economist Intelligence Unit.

Edwards, A. (1992) *Long Term Tourism Forecasts to 2005*, London: Economist Intelligence Unit.

English Historic Towns Forum (1997) *Code of Practice for Coach Tourism*, Bristol: English Historic Towns Forum.

English Tourist Board/Employment Department (1991) *Tourism and Environment: Maintaining the Balance*, London: English Tourist Board.

European Conference of Ministers of Transport (1992) *Privatisation of Railways*, Brussels: European Conference of Ministers of Transport.

European Parliament (1995) *Protection of Tourists*, Luxembourg: European Parliament.

Eurostat (1987) *Transport, Communications, Tourism*, Luxembourg: Office for Official Publication of the European Communities.

Farrington, J.H. (1985) 'Transport geography and policy – deregulation and privatisation', *Transactions of the Institute of British Geographers*, 10 (1): 109–19.

Farrington, J.H. (1995) 'Transport, environment and energy', in B.S. Hoyle and R.D. Knowles (eds) *Modern Transport Geography*, London: Belhaven, 51–66.

Farrington, J. and Ord, D. (1998) 'Bure Valley Railway: an EIA', *Project Appraisal*, 3 (4): 210–18.

Faulks, R. (1990) *Principles of Transport*, Maidenhead: McGraw-Hill, 4th edition.

Feiler, G. and Goodovitch, T. (1994) 'Decline and growth, privatisation and protectionism in the Middle-East airline industry', *Journal of Transport Geography*, 2 (1): 55–64.

Findlay, C. and Forsyth, P. (1984) 'Competitiveness in internationally traded services: the case of air transport', *Working Paper No. 10*, ASEAN-Australian Joint Research Project, Kuala Lumpur and Canberra.

Findlay, C., Chia, L. and Singh, K. (eds) (1997) *Asia Pacific Air Transport*, Singapore: Institute of South East Asian Studies.

Fitzsimmons, J.A. and Sullivan, R.S. (1982) *Service Operations Management*, New York: McGraw-Hill.

Flipo, J. (1988) 'On the intangibility of services', *Service Industries Journal*, 8 (3): 286–98.

Forer, P.C. and Pearce, D.G. (1984) 'Spatial patterns of package tourism in New Zealand', *New Zealand Geographer*, 40 (1): 34–42.

Forsyth, P. (1997a) 'Privatisation in Asia Pacific aviation', in C. Findlay, L. Chia and K. Singh (eds) *Asia Pacific Air Transport: Challenges and Policy Reforms*, Singapore: Institute of South East Asian Studies, 48–64.

Forsyth, P. (1997b) 'Price regulation of airports: principles with Australian applications', *Transportation Research E*, 33 (4): 297–309.

Forsyth, P. and King, J. (1996) 'Competition, cooperation and financial perform-ance in South Pacific aviation', in G. Hufbaner and C. Findlay (eds) *Flying High: Liberalising Aviation in the Asia Pacific*, Washington and Canberra: Institute for International Economics and Australia–Japan Research Centre, 99–176.

Forsyth, T. (1997) 'Environmental responsibility and business regulation: the case of sustainable tourism', *Geographical Journal*, 163 (3): 270–80.

Foster, D. (1985) *Travel and Tourism Management*, London: Macmillan.

Frapin-Beange, A., Bennett, M. and Wood, R. (1994) 'Some current issues in airline catering', *Tourism Management*, 15 (4): 295–305.

Frechtling, D. (1996) *Practical Tourism Forecasting*, Oxford: Butterworth-Heinemann.

French, T. (1994) 'European airport capacities and airport congestion', *Travel and Tourism Analyst*, 5: 4–19.

French, T. (1995) *Regional Airlines in Europe, Strategies for Survival*, London: Economist Intelligence Unit.

French, T. (1996a) 'World airport development plans and constraints', *Travel and Tourism Analyst*, 1: 4–16.

French, T. (1996b) 'No frills airlines in Europe', *Travel and Tourism Analyst*, 3: 4–19.

French, T. (1997a) *Airports in Europe: Meeting the Market Challenge*, London: Travel and Tourism Intelligence.

French, T. (1997b) 'Global trends in airline alliances', *Travel and Tourism Analyst*, 4: 81–101.

Gallacher, J. and Odell, M. (1994) 'Airline alliances, tagging along', *Airline Business*, July.

Gant, R. and Smith, J. (1992) 'Tourism and national development planning in Tunisia', *Tourism Management*, 13 (3): 331–6.

Gayle, D.J. and Goodrich, J.N. (eds) (1993) *Tourism Marketing and Management in the Caribbean*, London: Routledge.

Getz, D. and Page, S.J. (eds) (1997) *The Business of Rural Tourism: International Perspectives*, London: International Thomson Publishing.

Gibb, R. (ed) (1996) *The Channel Tunnel*, Chichester: Wiley.

Gibb, R. and Charlton, C. (1992) 'International surface passenger transport: prospects and potential', in B.S. Hoyle and R.D. Knowles (eds) *Modern Transport Geography*, London: Belhaven, 215–32.

Gibb, R., Lowndes, T. and Charlton, C. (1996) 'The privatisation of British Rail', *Applied Geography*, 16 (1): 35–51.

Gibb, R., Shaw, J. and Charlton, C. (1998) 'Competition, regulation and the privat-isation of British Rail', *Environment and Planning C: Government and Policy*.

Gilbert, D.C. (1989) 'Tourism marketing – its emergence and establishment', in C.P. Cooper (ed) *Progress in Tourism, Recreation and Hospitality Management*, Volume 1, London: Belhaven, 77–90.

Gilbert, D.C. (1990) 'Conceptual issues in the meaning of tourism', in C.P. Cooper (ed) *Progress in Tourism, Recreation and Hospitality Management*, Volume 2, London: Belhaven, 4–27.

Gilbert, D.C. and Joshi, I. (1992) 'Quality management and the tourism and hospit-ality industry', in C.P. Cooper and A. Lockwood (eds) *Progress in Tourism, Recreation and Hospitality Management*, Volume 4, London: Belhaven, 149–68.

Gillen, D. and Waters, W. (1997) 'Introduction: airport performance measurement and airport pricing', *Transportation Research E*, 33 (4): 245–47.

Glaister, S. (1981) *Fundamentals of Transport Economics*, Oxford: Blackwell.

Glaister, S. and Mulley, C.M. (1983) *Public Control of the Bus Industry*, Aldershot: Gower.

Glyptis, S. (1981) 'People at play in the countryside', *Geography*, 66 (4): 277–85.

Glyptis, S. (1991) *Countryside Recreation*, Harlow: Longman.

Gnoth, J. (1995) 'Quality of service and tourist satisfaction', in S. Witt and L. Moutinho (eds) *Tourism Marketing and Management Handbook: Student Edition*, Hemel Hempstead: Prentice Hall, 243–54.

Go, F. (1993) 'International airline trends', in F. Go and D. Frechtling (eds) *World Travel and Tourism Review: Indicators, Trends and Issues*, Volume 3, Oxford: CAB International, 178–83.

Go, F. and Murakami, M. (1990) 'Transnational corporations capture Japanese travel market', *Tourism Management*, 11 (4): 348–58.

Goetz, A. and Sutton, C. (1997) 'The geography of deregulation in the US airline industry', *Annals of the Association of American Geographers*, 87 (2): 238–63.

Goetz, A. and Szyliowicz, J. (1997) 'Revisiting transportation planning and decision making theory: the case of Denver International Airport', *Transportation Research*, A31 (4): 263–80.

Golich, V.L. (1988) 'Airline deregulation: economic boom or safety bust', *Transportation Quarterly*, 42: 159–79.

Goodall, B. (1991) 'Understanding holiday choice', in C.P. Cooper (ed) *Progress in Tourism, Recreation and Hospitality Management*, Volume 3, London: Belhaven, 58–79.

Goodall, B. (1992) 'Environmental Auditing for tourism', in C.P. Cooper and A. Lockwood (eds) *Progress in Tourism, Recreation and Hospitality Management*, Volume 4, London: Belhaven, 60–74.

Goodall, B. and Stabler, M. (1997) 'Environmental awareness, action and performance in the Guernsey hospitality sector', *Tourism Management*, 18 (1): 19–34.

Goodenough, R. and Page, S.J. (1993) 'Tourism training and education in the 1990s', *Journal of Geography in Higher Education*, 17 (1): 57–75.

Goodenough, R. and Page, S.J. (1994) 'Evaluating the environmental impact of the Channel Tunnel rail-link', *Applied Geography*, 14 (1): 26–50.

Gormsen, E. (1995) 'International tourism in China: its organisation and socio-economic impact', in A. Lew and L. Yu (eds) *Tourism in China: Geographical, Political and Economic Perspectives*, Colorado: Westview Press, 63–88.

Government of Ireland (1990) *Operational Programme on Peripherality: Roads and Other Transport Infrastructure 1989–1993*, Dublin: Stationery Office.

Graham, A. (1992) 'Airports in the United States', in R. Doganis, *The Airport Business*, London: Routledge, 188–206.

Graham, B. (1995) *Geography and Air Transport*, Chichester: Wiley.

Graham, B. (1997a) 'Air transport liberalisation in the European Union: an assessment', *Regional Studies*, 31: 807–12.

Graham, B. (1997b) 'Regional airline services in the liberalised European Union Single Aviation Market', *Journal of Air Transport Management*.

Graham, B. (1998) 'Liberalisation, regional economic development and the geography of demand for air transport in the European Union', *Journal of Transport Geography*, 6 (2): 87–104.

Grant, M., Human, B. and Le Pelley, B. (1997) 'More than getting from A to B: transport strategies and tourism', *INSIGHTS*, A43–A48.

Gray, H. (1982) 'The contribution of economics to tourism', *Annals of Tourism Research*, 9: 105–25.

Great Eastern Railways Ltd. (1997) *Passenger's Charter*, London: Great Eastern Railways Ltd.

Green, C. (1994) 'The future for InterCity: 1994 and beyond', in M. Vincent (ed) *The InterCity Story*, Somerset: Oxford Publishing Company, 142–52.

Griffith, L. (1989) 'Airways sanctions against South Africa', *Area*, 21 (3): 249–59.

Gronroos, C. (1980) *An Applied Service Marketing Theory*, Helsinki: Working Paper No. 57, Swedish School of Economics and Business Administration.

Gronroos, C. (1984) 'A service quality model and its marketing implications', *European Management Journal*, 11 (3): 332–41.

Gummesson, E. (1993) *Quality Management in Service Organisations*, Stockholm: International Service Quality Organisation.

Gunn, C.A. (1988) *Tourism Planning*, New York: Taylor and Francis, 2nd edition.

Hall, C.M. (1991) *Introduction to Tourism in Australia: Impacts, Planning and Development*, Melbourne: Longman Cheshire.

Hall, C.M. (1997) *Tourism in the Pacific Rim*, Melbourne: Addison Wesley Longman, 2nd edition.

Hall, C.M. and Jenkins, J. (1995) *Tourism and Public Policy*, London: Routledge.

Hall, C.M. and Lew, A. (eds) (1998) *The Geography of Sustainable Tourism*, London: Addison Wesley Longman.

Hall, C.M. and Page, S.J. (eds) (1996) *Tourism in the Pacific: Issues and Cases*, London: International Thomson Business Press.

Hall, C.M. and Page, S.J. (1999) *The Geography of Tourism and Recreation: Environment, Place and Space*, London: Routledge.

Hall, D. (1993) 'Getting around – transport and sustainability', *Town and Country Planning*, 62 (1/2): 8–12.

Hall, D.R. (1993a) 'Transport implications of tourism development', in D.R. Hall (ed) *Transport and Economic Development in the new Central and Eastern Europe*, London: Belhaven, 206–25.

Hall, D.R. (1993b) 'Key themes and agendas', in D.R. Hall (ed) *Transport and Economic Development in the new Central and Eastern Europe*, London: Belhaven, 226–36.

Hall, D.R. (1997) personal communication.

Halsall, D. (1982) *Transport for Recreation*, Lancaster: Institute of British Geographers Transport Study Group.

Halsall, D. (1992) 'Transport for tourism and recreation', in B.S. Hoyle and R.D. Knowles (eds) *Modern Transport Geography*, London: Belhaven, 155–77.

Hamill, J. (1993) 'Competitive strategies in the world airline industry', *European Management Journal*, 11 (3): 332–41.

Hamilton, J. (1988) 'Trends in tourism demand patterns in New Zealand', *International Journal of Hospitality Management*, 7 (4): 299–320.

Handy, C. (1989) *The Age of Unreason*, London: Business Books Ltd.

Hanghani, A. and Chen, M. (1998) 'Optimising gate assignments at airport terminals', *Transportation Research A*, 32 (6): 437–54.

Hanlon, P. (1996) *Global Airlines: Competition in a Transnational Industry*, Oxford: Butterworth-Heinemann.

Hannegan, T. and Mulvey, F. (1995) 'International airline alliances: an analysis of code-sharing's impact on airlines and consumers', *Journal of Air Transport Management*, 2 (2): 131–37.

Hanning, P. (1997) 'Caterers cut as airline goes leaner', *New Zealand Herald*, 14 June, A15.

Harding, R. (1994) 'Aeromedical aspects of commercial air travel', *Journal of Travel Medicine*, 1 (4): 211–15.

Harper, K. (1998a) 'Channel rail link collapses', *The Guardian*, 29 January, 1.

Harper, K. (1998b) 'BA to suffer £200 m hit from strong pound', *The Guardian*, 10 February, 16.

Harris, R. and Masberg, B. (1997) 'Factors critical to the success of implementing vintage trolley operations', *Journal of Travel Research*, 35 (3): 41–5.

Harrison, C. (1991) *The Countryside in a Changing Society*, London: TML Partnership.

Hay, A. (1973) *Transport for the Space Economy: A Geographical Study*, London: Macmillan.

Hayashi, Y., Yang, Z. and Osman, O. (1998) 'The effects of economic restructuring on China'a system for financing transport infrastructure', *Transportation Research A*, 32 (3): 183–95.

Henshall, D., Roberts, R. and Leighton, A. (1985) 'Fly-drive tourists: motivation and destination choice factors', *Journal of Travel Research*, 23 (3): 23–7.

Hensher, D., King, J. and Oum, T. (eds) (1996) *Transport Policy*, Proceedings of the 7th World Conference on Transport Research, Oxford: Pergamon.

Hepworth, M. and Ducatel, K. (1992) *Transport in the Information Age: Wheels and Wires*, London: Belhaven.

Heraty, M.J. (1989) 'Tourism transport – implications for developing countries', *Tourism Management*, 10 (4): 288–92.

Hilling, D. (1996) *Transport and Developing Countries*, London: Routledge.

HMSO (1978) *Airports Policy*, London: HMSO.

HMSO (1985) *Airports Policy*, Cmmd. 9542, London: HMSO.

HMSO (1990) *International Passenger Survey*, London: HMSO.

HMSO (1994) *Royal Commission on Environmental Pollution. Eighteenth Report, Transport and the Environment*, Cmmd. 2674, London: HMSO.

HMSO (1997a) *Social Trends 1997*, London: HMSO.

HMSO (1997b) *Royal Commission on the Environment*, London: HMSO.

Hoare, A. (1974) 'International airports as growth poles: a case study of Heathrow Airport', *Transactions of the Institute of British Geographers*, 63 (1): 75–96.

Hobson, J.S.P. and Uysal, M. (1992) 'Infrastructure: the silent crisis facing the future of tourism', *Hospitality Research Journal*, 17 (1): 209–15.

Hodgson, P. (1991) 'Market research in tourism: how important is it?', *Tourism Management*, 11 (4): 274–77.

Hoivik, T. and Heiberg, T. (1980) 'Centre–periphery tourism and self reliance', *International Social Science Journal*, 32 (1): 69–98.

Holder, J. (1988) 'Pattern and impact of tourism on the environment of the Caribbean', *Tourism Management*, 9 (2): 119–27.

Holliday, I., Marcou, G. and Vickerman, R. (1991) *The Channel Tunnel: Public Policy, Regional Development and European Integration*, London: Belhaven.

Hollings, D. (1997) 'Europe's railway in the 21st century', *Travel and Tourism Analyst*, 4: 4–24.

Holloway, J.C. (1989) *The Business of Tourism*, London: Pitman, 3rd edition.

Holloway, J.C. (1998) *The Business of Tourism*, London: Addison Wesley Longman, 5th edition.

Holloway, J.C. and Plant, R. (1988) *Marketing for Tourism*, Pitman: London.

Holloway, J.C. and Robinson, C. (1995) *Marketing for Tourism*, London: Longman, 3rd edition.

Hooper, P. (1998) 'Airline competition and deregulation in developed and developing country contexts – Australia and India', *Journal of Transport Geography*, 6 (2): 105–16.

Hooper, P. and Hensher, D. (1997) 'Measuring total factor productivity of airports – an index number approach', *Transportation Research E*, 33 (4): 249–59.

Horner, A. (1991) 'Geographical aspects of airport and air-route development in Ireland', *Irish Geography*, 24 (1): 35–47.

Horner, S. and Swarbrooke, J. (1996) *Marketing Tourism, Hospitality and Leisure in Europe*, London: International Thomson Business Press.

Horonjeft, R. and McKelvey, F. (1983) *Planning and Design of Airports*, New York: McGraw-Hill.

Hoyle, B.S. and Knowles, R.D. (eds) (1992) *Modern Transport Geography*, London: Belhaven.

Hoyle, B. and Knowles, R. (eds) (1998) *Modern Transport Geography*, Chichester: Wiley, 2nd edition.

Humphries, B. (1992) 'The air transport market', *INSIGHTS*, September: A37–A42.

Humphries, B. (1994) 'The implications of international code-sharing', *Journal of Air Transport Management*, 1 (4): 195–208.

Hunt, J. (1988) 'Airlines in Asia', *Travel and Tourism Analyst*, 5: 5–25.

Inkson, K. and Kolb, D. (1995) *Management: A New Zealand Perspective*, Auckland: Longman Paul.

Irons, K. (1994) *Managing Service Companies: Strategies for Success*, London: Economist Intelligence Unit and Addison Wesley Publishing Company.

Irons, K. (1997) *The World of Superservice: Creating Profit through a Passion for Customer Service*, London: Addison-Wesley.

Jahanshahi, M. (1998) 'The US railroad industry and open access', *Transport Policy*, 5 (2): 73–81.

Javalgi, R.G., Thomas, E.G. and Rao, S.R. (1992) 'US pleasure travellers' perceptions of selected European destinations', *European Journal of Marketing*, 26 (1): 46–64.

Jefferson, A. and Lickorish, L. (1991) *Marketing Tourism: A Practical Guide*, Harlow: Longman.

Jemiolo, J. and Oster, C.V. (1981) 'Regional changes in airline service since deregulation', *Transportation Quarterly*, 41: 569–86.

Johnson, P. (1988) 'The impact of a new entry on UK domestic air transport: a case study of the London–Glasgow route', *Service Industries Journal*, 8 (3): 299–316.

Jones, D. and White, P. (1994) 'Modelling of cross-country rail services', *Journal of Transport Geography*, 2 (2): 111–21.

Jones, P. (1995) 'Developing new products and services in flight catering', *International Journal of Contemporary Hospitality Management*, 7 (2/3): 24–8.

Kihl, M. (1988) 'The impacts of deregulation on passenger transportation in small towns', *Transportation Quarterly*, 42: 27–43.

Killen, J. and Smith, A. (1989) 'Transportation', in R.W.G. Carter and A.J. Parker (eds) *Ireland: A Contemporary Geographical Perspective*, London: Routledge, 271–300.

Knowles, R.D. (ed) (1985) *Implications of the 1985 Transport Bill*, Salford: Transport Geography Study Group, Institute of British Geographers.

Knowles, R.D. (1989) 'Urban public transport in Thatcher's Britain', in R.D. Knowles (ed) *Transport Policy and Urban Development: Methodology and Evaluation*, Salford: Transport Geography Study Group, Institute of British Geographers.

Knowles, R.D. (1993) 'Research agendas in transport geography for the 1990s', *Journal of Transport Geography*, 1 (1): 3–12.

Knowles, R. (1998) 'Passenger rail privatisation in Great Britain and its implications, especially for urban areas', *Journal of Transport Geography*, 6 (2): 117–33.

Knowles, R.D. and Hall, D.R. (1992) 'Transport policy and control', in B. Hoyle and R.D. Knowles (eds) *Modern Transport Geography*, London: Belhaven, 11–32.

Knowles, S.T. and Garland, M. (1994) 'The strategic importance of CRSs in the airline industry', *Travel and Tourism Analyst*, 4: 4–16.

Kosters, M. (1992) 'Tourism by train: its role in alternative tourism', in V. Smith and W. Eadington (eds) *Tourism Alternatives: Potential and Problems in the Development of Tourism*, Philadelphia: University of Pennsylvania Press.

Kotler, P. and Armstrong, G. (1991) *Principles of Marketing*, 5th edition, New Jersey: Prentice Hall.

Lamb, B. and Davidson, S. (1996) 'Tourism and transportation in Ontario, Canada', in L. Harrison and W. Husbands (eds) *Practising Responsible Tourism: International Case Studies in Tourism Planning, Policy and Development*, Chichester: Wiley, 261–76.

Land Use Consultants (1986) *The Channel Fixed Link: Environmental Appraisal of Alternative Proposals: A Report Prepared for the Department of Transport*, London: HMSO.

Langley, R. (1997) *Airports*, London: Keynote Publications.

Latham, J. (1989) 'The statistical measurement of tourism', in C.P. Cooper (ed) *Progress in Tourism, Recreation and Hospitality Management*, Volume 1, London: Belhaven, 55–76.

Latham, J. (1992) 'International tourism statistics', in C.P. Cooper and A. Lockwood (eds) *Progress in Tourism, Recreation and Hospitality Management*, Volume 4, London: Belhaven, 267–73.

Lavery, P. (1989) *Travel and Tourism*, 1st edition, Huntingdon: Elm.

Laws, E. (1991) *Tourism Marketing: Service Quality and Management Perspectives*, Cheltenham: Stanley Thornes.

Laws, E. (1995) *Managing Packaged Tourism*, London: International Thomson Business Press.

Laws, E. and Ryan, C. (1992) 'Service on flights – issues and analysis by the use of diaries', *Journal of Travel and Tourism Marketing*, 1 (3): 61–71.

Lawton, L.J. and Butler, R.W. (1987) 'Cruise ship industry – patterns in the Caribbean 1880–1986', *Tourism Management*, 8 (4): 329–43.

Lee, N. and Wood, C. (1988) 'The European Directive on environmental assessment: implementation at last?', *The Environmentalist*, 9 (3): 177–86.

Leggat, P. (1997) 'Travel health advice provided by in-flight magazines of international airlines in Australia', *Journal of Travel Medicine*, 4 (2): 102–3.

Leggat, P. and Nowak, M. (1997) 'Dietary advice for airline travel', *Journal of Travel Medicine*, 4 (1): 14–16.

Leiper, N. (1990) *Tourism Systems: An Interdisciplinary Perspective*, Palmerston North: Massey University, Department of Management Systems, Occasional Paper 2.

Leiper, N. and Simmons, D. (1993) 'Tourism: a social science perspective', in H. Perkins and G. Cushman (eds) *Leisure, Recreation and Tourism*, Auckland: Longman Paul, 204–20.

Lew, A. (1991) 'Scenic roads and rural development in the US', *Tourism Recreation Research*, 16 (2): 23–30.

Lew, A. and Yu, L. (eds) (1995) *Tourism in China: Geographic, Political and Economic Perspectives*, Boulder, Colorado: Westview Press.

Lickorish, L.J. in association with Jefferson, A., Bodlender, J. and Jenkins, C.L. (1991) *Developing Tourism Destinations: Policies and Perspectives*, Harlow: Longman.

Lipsey, R.G. (1989) *An Introduction to Positive Economics*, London: Weidenfeld and Nicolson.

London Tourist Board (1987) *The Tourism Strategy for London*, London: London Tourist Board.

London Tourist Board (1990) *At the Crossroads: The Future of London's Transport*, London: London Tourist Board.

Lovelock, C. (1992a) 'Seeking synergy in service operations: seven things marketers need to know about service operations', *European Management Journal*, 10 (1): 22–9.

Lovelock, C. (1992b) *Managing Services: Marketing, Operations and Human Resources*, New Jersey: Prentice Hall, 2nd edition.

Lovelock, C. and Yip, G. (1996) 'Global strategies for service businesses', *California Management Review*, 38 (2): 64–86.

Loverseed, H. (1996) 'Car rental in the USA', *Travel and Tourism Analyst*, 4: 4–19.

Lowe, J.C. and Moryadas, S. (1975) *The Geography of Movement*, Boston: Houghton Mifflin.

Lumsdon, L. (1996a) 'Future for cycle tourism in Britain', *INSIGHTS*, A27–A32.

Lumsdon, L. (1996b) *Cycling Opportunities: Making the Most of the National Cycling Network*, Stockport: Simon Holt Marketing Services.

Lumsdon, L. (1997a) *Tourism Marketing*, London: International Thomson Business Press.

Lumsdon, L. (1997b) 'Recreational cycling: is this the way to stimulate interest in everyday urban cycling?', in R. Tolley (ed) *The Greening of Urban Transport Planning for Walking and Cycling in Western Cities*, Chichester: Wiley, 113–27.

Lumsdon, L. and Speakman, C. (1995) *Railways to Cycleways in the Forest of Dean and the Wye Valley*, Ilkley: Transport for Leisure.

Lundberg, D.E. (1980) *The Tourist Business*, New York: Van Nostrand Reinhold.

McClennan, R., Inkson, K., Dakin, S., Dewe, P. and Elkin, G. (1987) *People and Enterprises: Human Behaviour in New Zealand Organisations*, Auckland: Rinehart and Winston.

McClintock, H. and Cleary, J. (1996) 'Cycle fatalities and cyclists' safety. Experience from Greater Nottingham and lessons for future cycling provision', *Transport Policy*, 3 (1/2): 67–77.

McHale, J. (1969) *The Future of the Future*, New York: George Braziller.

McIntosh, I.B. (1989) 'Travel considerations in the elderly', *Travel Medicine International*: 69–72.

McIntosh, I.B. (1990a) 'The stress of modern travel', *Travel Medicine International*, 118–21.

McIntosh, I.B. (1990b) 'Travel sickness', *Travel Medicine International*, 80–3.

McIntosh, I.B. (1995) 'Travel phobias', *Journal of Travel Medicine*, 2 (2): 99–100.

McIntosh, I.B. (1998) 'Health hazards and the elderly traveller', *Journal of Travel Medicine*, 5 (1): 27–29.

McIntosh, R.W. (1973) *Tourism: Principles, Practices and Philosophies*, Columbus Ohio: Grid.

McIntosh, R.W. (1990) *Tourism: Principles, Practices and Philosophies*, New York: Wiley.

McIntosh, R.W. and Goeldner, C.R. (1990) *Tourism: Principles, Practices and Philosophies*, Columbus Ohio: Grid.

MacCannel, D. (1976) *The Tourist: A New Theory of the Leisure Class*, New York: Schocken Books.

Macdonald-Wallace, D. (1997) 'Internet – worldwide road to success', *INSIGHTS*, A1–A4.

Maldonado, J. (1990) 'Strategic planning: an approach to improving airport planning under uncertainty', Master's thesis, Massachusetts Institute of Technology, Cambridge, Massachusetts.

Majumdar, A. (1994) 'Air traffic control problems in Europe: their consequences and proposed solutions', *Journal of Air Transport Management*, 1 (3): 165–78.

Majumdar, A. (1995) 'Commercialising and restructuring air traffic control: a review of the experience and issues involved', *Journal of Air Transport Management*, 2 (2): 111–22.

Mann, J.M. and Mantel, C.F. (1992) 'Travel and health: a global agenda', *Travel Medicine Two*, Proceedings of the Second International Conference on Travel Medicine, Paris, 1–4.

Mansfeld, Y. (1990) 'Spatial patterns of international tourist flows: towards a theoretical approach', *Progress in Human Geography*, 14 (3): 372–90.

Mansfeld, Y. (1992) 'Tourism: towards a behavioural approach', *Progress in Planning*, 38 (1): 1–92.

Marsh, J. and Staple, S. (1995) 'Cruise tourism in the Canadian Arctic and its implications', in C.M. Hall and M. Johnston (eds) *Polar Tourism: Tourism in Arctic and Antarctic Regions*, Chichester: Wiley, 63–72.

Mason, G. and Barker, N. (1996) 'Buy now fly later: an investigation of airline frequent flyer programmes', *Tourism Management*, 17 (3): 219–23.

Mason, P. (1994) 'A visitor code for the Arctic', *Tourism Managememt*, 15 (2): 93–7.

Mathieson, A. and Wall, G. (1982) *Tourism: Economic, Physical and Social Impacts*, Harlow: Longman.

Matthews, L. (1995) 'Forecasting peak passenger flows at airports', *Transportation*, 22 (1): 55–72.

Maunder, D. and Mbara, T. (1995) *The Initial Effects of Introducing Commuter Omnibus Services in Harare*, Zimbabwe: Crowthorne, Transport Research Laborartory Report 123.

May, A. and Roberts, M. (1995) 'The design of integrated transport strategties', *Transport Policy*, 2 (2): 97–106.

Meersam, H. and van de Voorde, E. (1996) 'The privatisation of air transport in Europe: interaction between policy, economic power and market performance', *Built Environment*, 22 (3): 177–91.

Meredith, J. (1994) 'Air traffic management in Europe – can it cope with future growth?', *Journal of Air Transport Management*.

Middleton, V.T.C. (1988) *Marketing in Travel and Tourism*, London: Heinemann.

Middleton, V. (1998) *Sustainable Tourism: A Marketing Perspective*, Oxford: Butterwoth-Heinemann.

Mill, R.C. (1992) *Tourism: The International Business*, New Jersey: Prentice Hall, 2nd edition.

Mill, R.C. and Morrison, A.M. (1985) *The Tourism System: An Introductory Text*, New Jersey: Prentice Hall.

Milman, A. and Pope, D. (1997) 'The US airline industry', *Travel and Tourism Analyst*, 3: 4–21.

Monopolies and Mergers Commission (1989) *Cross-Channel Car Ferries*, Cmmd. 584, London: HMSO.

Moore, A. (1985) 'Japanese tourists', *Annals of Tourism Research*, 12 (4): 619–43.

Morean, B. (1983) 'The language of Japanese tourism', *Annals of Tourism Research*, 10 (2): 93–109.

Morris, S. (1990) *Japanese Outbound Travel Market in the 1990s*, London: Economist Intelligence Unit.

Morris, S. (1994) 'Japan outbound', *Travel and Tourism Analyst*, 1: 40–64.

Morrison, A., Yang, C., O'Leary, J. and Nadkarni, N. (1996) 'Comparative profiles of travellers on cruises and land-based resort vacations', *The Journal of Tourism Studies*, 7 (2): 15–27.

Morrison, S. and Winston, C. (1986) *The Economic Effects of Airline Deregulation*, Washington DC: Brookings Institution.

Moscardo, G., Morrison, A., Cai, L., Nadkarni, N. and O'Leary, J. (1996) 'Tourist perspectives on cruising: multidimensional scaling analyses of cruising and other holiday types', *Journal of Tourism Studies*, 7 (2): 54–64.

Moses, L.N. and Savage, I. (1990) 'Aviation deregulation and safety: theory and evidence', *Journal of Transport Economics and Policy*, 14: 171–88.

Mulligan, C. (1979) 'The Snowdon sherpa: public transport and national park management experiment', in D. Halsall and B. Turton (ed) *Rural Transport Problems in Britain, Papers and Discussion*, Keele: Institute of British Geographers Transport Study Group.

Murphy, P.E. (1985) *Tourism: A Community Approach*, London: Routledge.

National Air Traffic Services Ltd (1997) *Reports and Accounts 1997*, London: Civil Aviation Authority.

National Consumer Council (1991) *Consumer Concerns 1990*, London: National Consumer Council.

National Consumer Council (1992) *British Rail Privatisation*, London: National Consumer Council.

National Tours Association (1992) *Tourism Traveller Index. The Benchmark Study*, USA: National Tours Association.

Newson, M. (1992) 'Environmental economics, resources and commerce', in M. Newson (ed) *Managing the Human Impact on the National Environment: Patterns and Processes*, London: Belhaven, 80–106.

Nozawa, H. (1992) 'A marketing analysis of Japanese outbound travel', *Tourism Management*, 13 (2): 226–34.

Oakland, J. (1989) *Total Quality Management*, Oxford: Heinemann.

O'Connor, K. (1995) 'Airport development in South East Asia', *Journal of Transport Geography*, 3 (4): 269–79.

O'Connor, K. (1996) 'Airport development: a Pacific Asian perspective', *Built Environment*, 22 (3): 212–22.

OECD (1992) *Torurism Policy and International Tourism in OECD Countries*, Pairs: OECD.

O'Hare, G. and Barrett, H. (1997) 'The destination life cycle: international tourism in Peru', *Scottish Geographical Magazine*, 113 (2): 66–73.

O'Kelly, M.E. (1986) 'The location of interacting hub facilities', *Transportation Science*, 20 (2): 92–106.

O'Rourke, V. (1996) 'Queensland rail – a sleeper awakes', *Chartered Institute of Transport* (Queensland section), September.

Ontario Ministry of Culture, Tourism and Recreation (1994) *Ontario's Tourism Industry: Opportunity, Progress, Innovation*, Toronto: Ontario Ministry of Culture, Tourism and Recreation.

Ontario Ministry of Transportation (1993) *Strategic Directions. Draft*, Ontario: Ontario Ministry of Transportation.

Oosterveld, W. (1995) 'Motion sickness', *Journal of Travel Medicine*, 2 (3): 182–87.

Orams, M. (1998) *Marine Tourism*, London: Routledge.

Ortuzar, J. de D. and Willumsen, L.E. (1990) *Modelling Transport*, Chichester: John Wiley.

Ostrowski, P., O'Brien, T. and Gordon, G. (1993) 'Service quality and customer loyalty in the commercial airline industry', *Journal of Travel Research*, 32 (2): 16–24.

Ostrowski, P., O'Brien, T. and Gordon, G. (1994) 'Determinants of service quality in the commercial airline industry: differences between business and leisure travellers', *Journal of Travel and Tourism Marketing*, 3 (1): 19–47.

Oum, T. (1995) 'A comparative study of productivity and cost competitiveness of the world's major airlines', *Discussion Paper No. 363*, Osaka: Institute of Social and Economic Research, Osaka University.

Oum, T. (1997) 'Challenges and opportunities for Asian airlines and governments', in C. Findlay, L. Chia and K. Singh (eds) *Asia Pacific Air Transport: Challenges and Policy Reforms*, Singapore: Institute for South East Asian Studies, 1–22.

Oum, T. and Yu, C. (1995) 'A productivity comparison of the world's major airlines', *Journal of Air Transport Management*, 2 (3/4): 181–95.

Oum, T. and Yu, C. (1998) 'Cost competitiveness of major airlines: An international comparison', *Transportation research A*, 32 (6): 407–22.

Oxley, P. and Richards, M. (1995) 'Disability and transport. A review of the personal costs of disability in relation to transport', *Transport Policy*, 2 (1): 57–66.

Page, C., Wilson, M. and Kolb, D. (1994) *On the Inside Looking In: Management Competencies in New Zealand*, Wellington: Ministry of Commerce.

Page, S.J. (1987) 'London Docklands: redevelopment schemes in the 1980s', *Geography*, 72 (1): 59–63.

Page, S.J. (1989a) 'Changing patterns of international tourism in New Zealand', *Tourism Management*, 10 (4): 337–41.

Page, S.J. (1989b) 'Tourist development in London Docklands in the 1980s and 1990s', *GeoJournal*, 19 (3): 291–5.

Page, S.J. (1989c) 'Tourism planning in London', *Town and Country Planning*, 58 (3): 334–5.

Page, S.J. (1991) 'Tourism in London: the Docklands connection', *Geography Review*, 4 (3): 3–7.

Page, S.J. (1992a) 'Managing tourism in a small historic city', *Town and Country Planning*, 61 (7/8): 208–11.

Page, S.J. (1992c) 'Perspectives on the environmental impact of the Channel Tunnel on tourism', in C.P. Cooper and A. Lockwood (eds) *Progress in Tourism, Recreation and Hospitality Management*, Volume 4, London: Belhaven, 82–102.

Page, S.J. (1993a) 'Regenerating Wellington's waterfront', *Town and Country Planning*, 63 (1/2): 29–31.

Page, S.J. (1993b) 'Tourism and peripherality: a review of tourism in the Republic of Ireland', in C.P. Cooper (ed) *Progress in Tourism, Recreation and Hospitality Management*, Volume 5, London: Belhaven, 26–53.

Page, S.J. (1993c) 'European rail travel', *Travel and Tourism Analyst*, 1: 5–30.

Page, S.J. (1993d) 'Editorial', *Tourism Management*, 14 (6): 419–23.

Page, S.J. (1994a) 'Editorial: the spatial implications of the Channel Tunnel', *Applied Geography*, 14 (1): 3–8.

Page, S.J. (1994b) 'European bus and coach travel', *Travel and Tourism Analyst*, 1: 19–39.

Page, S.J. (1995a) 'Waterfront revitalisation in London: market-led planning and tourism in London Docklands', in S. Craig-Smith and M. Fagence (eds) *Urban Waterfront Development: An International Survey*, New York: Praeger, 53–70.

Page, S.J. (1995b) *Urban Tourism*, London: Routledge.

Page, S.J. (1996) 'Pacific Islands', *International Tourism Reports*, 1: 67–102.

Page, S.J. (1997) *The Costs of Adventure Tourism Accidents for the New Zealand Tourism Industry: Final Report*, Wellington: The Tourism Policy Group, Ministry of Commerce.

Page, S.J. (1998) 'Transport for tourism and recreation', in B. Hoyle and R. Knowles (eds) *Modern Transport Geography*, Chichester: John Wiley, 2nd edition.

Page, S.J. and Fidgeon, P. (1989) 'London Docklands: a tourism perspective', *Geography*, 74 (1): 66–8.

Page, S.J. and Hardyman, R. (1996) 'Place-marketing and town centre management in the UK', *Cities: The International Journal of Urban Policy and Planning*, 13 (3): 153–64.

Page, S.J. and Meyer, D. (1996) 'Tourist accidents: an exploratory analysis', *Annals of Tourism Research*, 23 (3): 666–90.

Page, S.J. and Sinclair, M.T. (1992a) 'The Channel Tunnel: an opportunity for London's tourism industry', *Tourism Recreation Research*, 17 (2): 57–70.

Page, S.J. and Sinclair, M.T. (1992b) 'The Channel Tunnel and tourism markets', *Travel and Tourism Analyst*, 1: 8–32.

Page, S.J. and Thorn, K. (1997) 'Towards sustainable tourism planning in New Zealand: public sector planning responses', *Journal of Sustainable Tourism*, 5 (1): 59–77.

Patmore, J.A. (1983) *Recreation and Resources*, Oxford: Blackwell.

Pearce, D.G. (1979) 'Towards a geography of tourism', *Annals of Tourism Research*, 6 (3): 245–72.

Pearce, D.G. (1985) 'Tourism and environmental research: a review', *International Journal of Environmental Studies*, 25 (4): 247–55.

Pearce, D.G. (1987) *Tourism Today: A Geographical Analysis*, Harlow: Longman.

Pearce, D.G. (1990) *Tourist Development*, Harlow: Longman, 2nd edition.

Pearce, D.G. (1992) *Tourism Organisations*, Harlow: Longman.

Pearce, D.G. (1995a) *Tourism Today: A Geographical Analysis*, London: Longman, 2nd edition.

Pearce, D.G. (1995b) 'CER, Trans-Tasman tourism and a single aviation market', *Tourism Management*, 16 (2): 111–20.

Pearce, D.G. and Butler, R.W. (eds) (1992) *Tourism Research: Critiques and Challenges*, London: Routledge.

Pearce, D.G. and Elliot, J.M. (1983) 'The Trip Index', *Journal of Travel Research*, 22 (1): 6–9.

Pearce, P.L. (1982) *The Social Psychology of Tourist Behaviour*, Oxford: Pergamon.

Pearce, P.L. (1992) 'Fundamentals of tourist motivation', in D.G. Pearce and R. Butler (eds) *Tourism Research: Critiques and Challenges*, London: Routledge, 113–34.

Peisley, T. (1992a) *World Cruise Ship Industry in the 1990s*, London: Economist Intelligence Unit.

Peisley, T. (1992b) 'Ferries, short sea cruises and the Channel Tunnel', *Travel and Tourism Analyst*, 4: 5–26.

Peisley, T. (1995) 'The cruise ship industry to the 21st century', *Travel and Tourism Analyst*, 2: 4–25.

Peisley, T. (1996) *The World Cruise Ship Industry to 2000*, London: Travel and Tourism Intelligence.

Peisley, T. (1997) 'The cross-Channel ferry market', *Travel and Tourism Analyst*, 1: 4–20.

Perkins, H.C. and Cushman, G. (eds) (1993) *Leisure, Recreation and Tourism*, Auckland: Longman Paul.

Perks, A. (1993) 'Tourism and transport issues for the Channel Islands of Guernsey and Alderney', MSc thesis, Department of Management Studies, University of Surrey, Guildford.

Perl, A., Patterson, J. and Perez, M. (1997) 'Pricing aircraft emissions at Lyon-Satolas airport', *Transportation Research D*, 2 (2): 89–105.

Peters, T.J. and Waterman, R.H. (1982) *In Search of Excellence*, London: Harper and Row.

Petrie, J. and Dawson, A. (1994) 'Recent developments in the treatment of jet-lag', *Journal of Travel Medicine*, 1 (2): 19–83.

Pigram, J.J. (1992) 'Planning for tourism in rural areas: bridging the policy implementation gap', in D.G. Pearce and R.W. Butler (eds) *Tourism Research: Critiques and Challenges*, London: Routledge, 156–74.

Polunin, I. (1989) 'Japanese travel boom', *Tourism Management*, 10 (1): 4–8.

Poon, A. (1989) 'Competitive strategies for a new tourism', in C.P. Cooper (ed) *Progress in Tourism, Recreation and Hospitality Management*, Volume 4, London: Belhaven, 91–102.

Poon, A. (1993) *Tourism, Technology and Competitive Strategies*, Wallingford: CAB International.

Porter, L.J. and Parker, A.J. (1992) 'Total Quality Management – the critical success factors', *Total Quality Management*, 4 (1): 13–22.

Potter, S. (1987) *On the Right Lines: The Limits of Technological Innovation*, London: Pinter.

Potter, S. (1997) *Vital Travel Statistics*, London: Landor Publishing Limited.

Prideaux, B. (1997) 'Tracks to tourism: Queensland rail joins the tourism industry', Paper submitted to *Progress in Tourism and Hospitality Research*.

Prideaux, J. (1990) 'InterCity: passenger railway without subsidy', *Royal Society of Arts Journal*, March: 244–54.

Qaiters, C.G. and Bergiel, B.J. (1989) *Consumer Behaviour: A Decision-Making Approach*, Ohio: South Western Publishing.

Quayle, M. (1993) *Logistics: An Integrated Approach*, Kent: Hodder and Stoughton.

Raguraman, K. (1986) 'Capacity and route regulation in international scheduled air transportation: a case study of Singapore', *Singapore Journal of Tropical Geography*, 7 (1): 53–69.

Raguraman, K. (1997) 'Airlines as instruments for nation building and national identity: case study of Malaysia and Singapore', *Journal of Transport Geography*, 5 (4): 239–56.

Reynolds-Feighan, A. and Feighan, K. (1997) 'Airport services and airport charging systems: a critical review of the EU common framework', *Transportation Research E*, 33 (4): 311–20.

Ritchie, B. (1997) 'Cycle tourism in the South Island of New Zealand: infrastructure considerations for the twenty-first century', paper presented at *Trails in the Third Millennium*, Cromwell, New Zealand, 2–5 December, 325–34.

Romeril, M. (1985) 'Tourism and the environment – towards a symbiotic relationship', *Journal of Environmental Studies*, 25: 215–8.

Romeril, M. (1989) 'Tourism and the environment – accord or discord?', *Tourism Management*, 10 (3): 204–8.

Ross, G. (1994) *The Psychology of Tourism*, Melbourne: Hospitality Press.

Ross, W.A. (1987) 'Evaluating environmental impact statements', *Journal of Environmental Management*, 25: 137–47.

Rowe, V. (1994) *International Business Travel: A Changing Profile*, London: Economist Intelligence Unit.

Ryan, C. (1991) *Recreational Tourism: A Social Science Perspective*, London: Routledge.

Ryan, C. (ed) (1996) *The Tourist Experience*, London: Cassell.

Rycroft, R. and Szylowicz, J. (1980) 'The technological dimension of decision-making: the case of the Aswan high dam', *World Politics*, 32: 36–61.

Schieven, A. (1988) *A Study of Cycle Tourists on Prince Edward Island*, unpublished Master's thesis, University of Waterloo, Canada.

Schiffman, L.G. and Kanuk, L.L. (1991) *Consumer Behaviour*, New Jersey: Prentice Hall, 4th edition.

Scottish Tourist Board (1991) *Tourism Potential of Cycling and Cycle Routes in Scotland*, Edinburgh: Scottish Tourist Board.

Sealy, K. (1976) *Airport Strategy and Planning*, Oxford: Oxford University Press.

Sealy, K. (1992) 'International air transport', in B.S. Hoyle and R.D. Knowles (eds) *Modern Transport Geography*, London: Belhaven, 233–56.

Seaton, A. and Bennett, M. (1996) *Marketing Tourism Products: Concepts, Issues, Cases*, London: International Thomson Business Press.

Seaton, A. and Palmer, C. (1997) 'Understanding VFR tourism behaviour: the first five years of the United Kingdom Survey', *Tourism Management*, 18 (6): 345–55.

Seibert, J.C. (1973) *Concepts of Marketing Management*, New York: Harper & Row.

Selman, P. (1992) *Environmental Planning: The Conservation and Development of Biophysical Resources*, London: Paul Chapman.

Seristö, H. and Vepsäläinen, A (1997) 'Airline cost drivers: cost implications of fleet, routes, and personnel policies', *Journal of Air Transport Management*, 3 (1): 11–22.

SERPLAN (1989) *The Channel Tunnel: Impact on the South East*, London: SERPLAN.

SETEC/Wilbur Smith Associates (1989) *Review of Market Trends and Forecasts*, Paris: Eurotunnel.

Sharpley, R. (1993) *Tourism and Leisure in the Countryside*, Huntingdon: Elm.

Shaw, S. (1982) *Airline Marketing and Management*, London: Pitman.

Shaw, S.L. (1993) 'Hub structures of major US passenger airlines', *Journal of Transport Geography*, 1 (1): 47–58.

Sheldon, P. (1997) *Tourism Information Technology*, Wallingford: CAB International.

Shen, Q. (1997) 'Urban transportation in Shanghai, China: Problems and planning implications', *International Journal of Urban and Regional Research*, 21 (4): 589–606.

Shilton, D. (1982) 'Modelling the demand for high speed train services', *Journal of the Operational Research Society*, 33: 713–22.

Sikorski, D. (1990) 'A comparative evaluation of the government's role in national airlines', *Asia Pacific Journal of Management*, 7 (1): 97–120.

Simmons, D. and Leiper, N. (1993) 'Tourism: a social science perspective', in H.C. Perkins and G. Cushman (eds) *Leisure, Recreation and Tourism*, Auckland: Longman Paul, 204–20.

Simon, D. (1996) *Transport and Development in the Third World*, London: Routledge.

Sinclair, M.T. (1991) 'The economics of tourism', in C.P. Cooper (ed) *Progress in Tourism, Recreation and Hospitality Management*, Volume 3, London: Belhaven, 1–27.

Sinclair, M.T. and Page, S.J. (1993) 'The Euroregion: a new framework for regional development', *Regional Studies*, 27 (5): 475–83.

Sinclair, M.T. and Stabler, M. (1991) 'New perspectives on the tourism industry', in M.T. Sinclair and M.J. Stabler (eds) *The Tourism Industry: An International Analysis*, Oxford: CAB International, 1–14.

Sinclair, M.T. and Stabler, M. (1997) *The Economics of Tourism*, London: Routledge.
Singapore Airlines (1991) *Singapore Airlines Annual Report 1991–92*, Singapore: Singapore Airlines.
Singapore Airlines (1997) *Singapore Airlines Annual Report*, 1996–97, Singapore: Singapore Airlines.
Singapore Mass Rapid Transit (SMRT) (1997) *At Your Service: SMRT's Commitment to Passengers*, Singapore: SMRT.
Skapinker, M. (1996) 'Airports are poised and ready for takeoff', *Financial Times*, 18 November (World Airports Survey), I.
Skapinker, M. (1998a) 'Shopping while you wait', *Financial Times*, 5 February (Special Report: The Business of Travel), VI.
Skapinker, M. (1998b) 'When passenger trouble strikes', *Financial Times*, 5 February (Business of Travel Supplement VI).
Smith, C. and Jenner, P. (1997) 'The seniors' travel market', *Travel and Tourism Analyst*, 5: 43–62.
Smith, D., Odegard, J. and Shea, W. (1984) *Airport Planning*, Belmont, CA: Wadsworth Inc.
Smith, M.J.T. (1989) *Aircraft Noise*, Cambridge: Cambridge University Press.
Smith, S.L.J. (1989) *Tourism Analysis*, Harlow: Longman.
Smith, V.L. (1992) 'Boracay, Philippines: a case study in alternative tourism', in V.L. Smith and W.R. Eadington (eds) *Tourism Alternatives: Potential and Problems in the Development of Tourism*, Philadelphia: University of Pennsylvania Press, 133–57.
Smith, V.L. and Eadington W.R. (eds) (1992) *Tourism Alternatives: Potential and Problems in the Development of Tourism*, Philadelphia: Pennsylvania University Press.
Soames, T. (1997) 'Ground handling liberalisation', *Journal of Air Transport Management*, 3 (2): 83–94.
Sofield, T. and Li, F. (1997) 'Rural tourism in China: development issues in perspective', in D. Getz and S.J. Page (eds) *The Business of Rural Tourism: International Perspectives*, London: International Thomson Business Press, 120–37.
Sommerville, H. (1992) 'The airline industry's perspective', in D. Banister and K.J. Button (eds) *Transport, the Environment and Sustainable Development*, London: E and FN Spon: 161–74.
Speak, C. (1997) 'The new airport: Kai Tak airport', *Geography*, 82 (3): 266–68.
Speakman, C. (1996) 'Britain's changing railways and the tourist package', *INSIGHTS*, A45–A48.
Starkie, D.N. (1976) *Transportation Planning, Policy and Analysis*, Oxford: Pergamon Press.
Stasinopoulos, D. (1992) 'The second aviation package of the European Community', *Journal of Transport Economics and Policy*, 26: 83–7.
Stasinopoulos, D. (1993) 'The third phase of liberalisation in community aviation and the need for supplementary measures', *Journal of Transport Economics and Policy*, 27: 323–8.
Steward, S. (1986) *Air Disasters*, London: Arrow Books.
Stubbs, J. and Jegede, F. (1998) 'The integration of rail and air transport in Britain', *Journal of Transport Geography*, 6 (1): 53–67.
Stubbs, P.C., Tyson, W.J. and Dalvi, M. (1980) *Transport Economics*, London: Allen and Unwin.
Sustrans (1995) *The National Cycle Network: Bidding Document to the Millennium Commission*, Bristol: Sustrans.

Sustrans (1996) *Local Agenda 21 and the National Cycle Network: Routes to Local Sustainability*, Bristol: Sustrans.

Sustrans (1997) *The Tourism Potential of National Cycle Routes*, London: The Tourism Society.

Swarbrooke, J. (1994) 'The future of the past: heritage tourism in the twenty first century', in A. Seaton (ed) *Tourism: State of the Art*, Chichester: Wiley.

Swarbrooke, J. (1997) 'Understanding the tourist – some thoughts on consumer behaviour research in tourism', *INSIGHTS*, A67–A76.

Szyliowicz, J. and Goetz, A. (1995) 'Getting realistic about megaproject planning: the case of the new Denver International Airport', *Policy Sciences*, 28: 347–67.

Taafe, E.J. and Ganthier, H.L. (1973) *Geography of Transportation*, New Jersey: Prentice Hall.

Taneja, N. (1988) *The International Airline Industry: Issues and Challenges*, Massachusetts: Lexington Books.

Taplin, J.H.E. (1993) 'Economic reform and transport policy in China', *Journal of Transport Economics and Policy*, 27 (1): 75–86.

Taplin, J. and Qiu, M. (1997) 'Car trip attraction and route choice in Australia', *Annals of Tourism Research*, 24 (3): 624–37.

Taylor, C. (1983) 'Rail passenger transport in Australia: a critical analysis of the network and services', unpublished PhD thesis, Department of Regional and Town Planning, University of Queensland.

TEST (1991) *The Wrong Side of the Tracks*, London: TEST.

Teye, W.B. (1992) 'Land transportation and tourism in Bermuda', *Tourism Management*, 13 (4): 395–405.

Therivel, R.B. and Barret, B.F.D. (1990) 'Airport development and E.I.A.: Kansai International Airport, Japan', *Land Use Policy*, 7 (1): 80–6.

Thornberry, N. and Hennessey, H. (1992) 'Customer care, much more than a smile: developing a customer service infrastructure', *European Management Journal*, 10 (4): 460–4.

Todd, G. and Mather, S. (1993) *Tourism in the Caribbean*, London: Economist Intelligence Unit.

Tokuhisa, T. (1980) 'Tourism within, from and to Japan', *International Social Science Journal*, 32 (1): 128–50.

Tolley, R. (ed) (1990) *The Greening of Urban Transport: Planning for Walking and Cycling in Western Cities*, London: Belhaven.

Tolley, R. and Turton, B. (eds) (1987) *Short Sea Crossings and the Channel Tunnel*, Keele: Institute of British Geographers Transport Study Group.

Tolley, R. and Turton, B. (eds) (1995) *Transport Systems, Policy and Planning: A Geographical Approach*, Harlow: Longman.

Toms, M. (1994) 'Charging for airports: the new BAA approach', *Journal of Air Transport Management*, 1 (2): 77–82.

Tourism Society (1990) *Tourism and the Environment: A Memorandum to the Department of Employment Task Force*, London: The UK Tourism Society.

Townsend, P.L. and Gebhart, J.E. (1986) *Commit to Quality*, New York: John Wiley.

Tribe, J. (1996) *Corporate Strategy for Tourism*, London: International Thomson Business Press.

Turner, J. (1997) 'The policy process', in B. Axford, G. Browning, R. Huggins, B. Rosamond and J. Turner, *Politics: An Introduction*, London: Routledge, 409–39.

Turton, B. (1991) 'The changing transport pattern', in R.J. Johnston and V. Gardiner (eds) *The Changing Geography of the British Isles*, London: Routledge, 171–97, 2nd edition.

Turton, B. (1992a) 'British Rail passenger policies', *Geography*, 77 (1): 64–7.

Turton, B. (1992b) 'Inter-urban transport', in B. Hoyle and R. Knowles (eds) *Modern Transport Geography*, London: Belhaven, 105–24.

Turton, B. and Mutambirwa, C. (1996) 'Air transport services and the expansion of international tourism in Zimbabwe', *Tourism Management*, 17 (6): 453–62.

Urry, J. (1990) *The Tourist Gaze*, London: Sage.

US Bureau of the Census (1996) *Statistical Abstract of the United States 1995*, Washington: Reference Press Inc.

Usyal, M. and Crompton, V.L. (1985) 'An overview of approaches used to forecast tourism demand', *Journal of Travel Research*, 23 (4): 7–15.

Van Borrendam, A. (1989) 'KLM strives for customer satisfaction', *TQM Magazine*, 1 (2): 105–9.

Van Dierdonck, R. (1992) 'Success strategies in a service economy', *European Marketing Journal*, 10 (3): 365–73.

Veal, A. (1992) *Research Methods in Leisure and Tourism*, Harlow: Longman.

Verchere, I. (1994) *The Air Transport Industry in Crisis*, London: Economist Intelligence Unit.

Viant, A. (1993) 'Enticing the elderly to travel – an exercise in Euro-management', *Tourism Management*, 14 (1): 52–60.

Vickerman, R. (1995) 'The Channel Tunnel: a progress report', *Travel and Tourism Analyst*, 3: 4–20.

Vincent, M. and Burley, S. (1994) 'Delighting the customer', in M. Vincent (ed) *The InterCity Story*, Somerset: Oxford Publishing Company, 107–18.

Wackermann, G. (1997) 'Transport, trade, tourism and the world economic system', *International Social Science Journal*, 151 (1): 23–40.

Wager, J. (1967) 'Outdoor recreation on common land', *Journal of the Town Planning Institute*, 53: 398–403.

Wahab, S. and Pigram, J. (eds) (1997) *Tourism Development and Growth: The Growth of Sustainability*, London: Routledge.

Wales Tourist Board (1992) *Infrastructure Services for Tourism – A Paper for Discussion*, Cardiff: Wales Tourist Board.

Wall, G. (1971) 'Car owners and holiday activities', in P. Lavery (ed) *Recreational Geography*, Newton Abbot: David and Charles.

Wall, G. (1972) 'Socioeconomic variations in pleasure trip patterns: the case of Hull car owners', *Transactions of the Institute of British Geographers*, 57: 45–58.

Wardman, M., Hatfield, R. and Page, M. (1997) 'The UK national cycling strategy. Can improved facilities meet the targets?', *Transport Policy*, 4 (2): 123–33.

Wathern, P. (1990) *Environmental Impact Assessment: Theory and Practice*, London: Unwin Hyman.

Weaver, D. and Elliot, K. (1996) 'Spatial patterns and problems in Namibian tourism', *The Geographical Journal*, 162 (2): 205–17.

Weatherford, L. and Bodily, S. (1992) 'A taxonomy and research overview of perishable asset revenue management: yield management, overbooking and pricing', *Operations Research*, 40: 831–34.

Weiler, B. (1993) 'Guest Editor's Introduction', *Tourism Management*, 14 (2): 83–4.

Wells, A. (1989) *Air Transportation: A Management Perspective*, California: Wadsworth Inc, 2nd edition.

Wen, Z. (1997) 'China's domestic tourism: impetus, development and trends', *Tourism Management*, 18 (8): 565–71.

Wheatcroft, S. (1978) 'Transport, tourism and the service industry', *Chartered Institute of Transport Journal*, 38 (7): 197–206.

Wheatcroft, S. (1994) *Aviation and Tourism Policies*, London: Routledge/World Tourism Organisation.

Wheeler, B. (1991) 'Tourism: troubled times', *Tourism Management*, 12 (2): 91–6.

Wheeler, B. (1992a) 'Is progressive tourism appropriate?', *Tourism Management*, 13 (1): 104–5.

Wheeler, B. (1992b) 'Alternative tourism – a deceptive ploy', in C.P. Cooper and A. Lockwood (eds) *Progress in Tourism, Recreation and Hospitality Management*, Volume 4, London: Belhaven, 140–6.

Wheeler, B. (1994) 'Egotourism, sustainable tourism and the environment – a symbiotic, symbolic or shambolic relationship', in A. Seaton (ed) *Tourism: The State of the Art*, Chichester: Wiley, 647–54.

Whitaker, R. and Levere, J. (1997) 'Web fever', *Airline Business*, February: 26–33.

White, H.P. and Senior, M.L. (1983) *Transport Geography*, London: Longman.

White, P. and Farrington, J. (1998) 'Bus and coach deregulation and privatisation in Great Britain, with particular reference to Scotland', *Journal of Transport Geography*, 6 (2): 135–41.

Whitelegg, J. (1987) 'Rural railways and disinvestment in rural areas', *Regional Studies*, 21 (1): 55–64.

Whitelegg, J. (1993) *Transport for a Sustainable Future: The Case for Europe*, Belhaven: London.

Whitelegg, J. (1998) 'Down the tube', *Guardian*, 29 January, 21.

Wie, B. and Choy, D. (1993) 'Traffic impact analysis of tourism development', *Annals of Tourism Research*, 20 (3): 505–18.

Wilkinson, P.F. (1989) 'Strategies for tourism in island microstates', *Annals of Tourism Research*, 16 (2): 153–77.

Williams, A. (1995) 'Capital and the transnationalisation of tourism', in A. Montanari and A. Williams (eds) *European Tourism: Regions, Spaces and Restructuring*, Chichester: Wiley, 163–85.

Williams, G. (1993) *The Airline Industry and the Impact of Deregulation*, Brookfield, VT: Ashgate.

Williams, P.W. (1987) 'Evaluating environmental impact on physical capacity in tourism', in J.R.B. Ritchie and C.R. Goeldner (eds) *Travel, Tourism and Hospitality Research*, New York: Wiley, 385–97.

Withyman, W. (1985) 'The ins and outs of international travel and tourism data', *International Tourism Quarterly*, Special Report No. 55.

Witt, C. (1995) 'Total quality management', in S. Witt and L. Moutinho (eds) *Tourism, Marketing and Management Handbook: Student Edition*, Hemel Hempstead: Prentice Hall, 229–42.

Witt, C. and Mühlemann, A. (1995) 'Service quality in airlines', *Tourism Economics*, (1): 33–49.

Witt, S.F., Brooke, M.Z. and Buckley, P.J. (1991) *The Management of International Tourism*, London: Routledge.

Witt, S.F. and Martin, C. (1989) 'Demand forecasting in tourism and recreation', in C.P. Cooper (ed) *Progress in Tourism, Recreation and Hospitality Management*, Volume 1, London: Belhaven, 4–32.

Witt, S.F. and Martin, C. (1992) *Modelling and Forecasting Demand in Tourism*, London: Academic Press.

Witt, S.F. and Moutinho, L. (eds) (1989) *Tourism Marketing and Management Handbook*, Hemel Hempstead: Prentice Hall.

Witt, S. and Witt, C. (1995) 'Forecasting tourism demand: a review of empirical research', *International Journal of Forecasting*, 11: 447–75.

Wolmar, C. (1996) *The Great British Railway Disaster*, London: Ian Allen Publishing.

Wood, K. and House, S. (1991) *The Good Tourist*, London: Mandarin.

Wood, K. and House, S. (1992) *The Good Tourist in France*, London: Mandarin.

World Commission on the Environment and Development (1987) *Our Common Future* (Brundtland Commission's Report), Oxford: Oxford University Press.

World Tourism Organisation (1992) *Yearbook of Tourism Statistics*, Madrid: World Tourism Organisation.

World Tourism Organisation (1994) *Global Distribution Systems (GDSs) in the Tourism Industry*, A study prepared for the World Tourism Organisation by O. Vialle, Madrid: World Tourism Organisation.

World Tourism Organisation (1997) *Yearbook of Tourism Statistics*, Madrid: World Tourism Organisation.

World Travel and Tourism Council, World Tourism Organisation and Earth Council (1997) *Agenda 21 for the Travel and Tourism Industry: Towards Environmentally Sustainable Development*, Intercontinental, London.

Yamauchi, H. (1997) 'Air transport policy in Japan: limited competition under regulation', in C. Findlay, C. Sien and K. Singh (eds) *Asia Pacific Air Transport: Challenges and Policy Reforms*, Singapore: Institute of South East Asian Studies, 106–22.

Yardley, L. (1992) 'Motion sickness and perception: a reappraisal of the sensory conflict approach', *British Journal of Psychology*, 82: 449–71.

Zeithmal, A. and Berry, L.L. (1985) 'A conceptual model of service quality and its implications for future research', *Journal of Marketing*, Fall: 41–50.

Zeithmal, V., Parasuraman, A. and Berry, L. (1990) *Delivering Service Quality*, New York: Free Press.

Zhang, A. and Zhang, Y. (1997) 'Concession revenue and optimal airport pricing', *Transportation Research E*, 33 (4): 287–96.

Zhang, W. (1997) 'China's domestic tourism: impetus, development and trends', *Tourism Management*, 18 (8): 565–72.

Zinyama, L. (1989) 'Some recent trends in tourist arrivals in Zimbabwe', *Geography*, 74 (1): 62–5.

Index